VISUAL CULTURES OF **SCIENCE**

INTERFACES: STUDIES IN VISUAL CULTURE

Editors Mark J. Williams and Adrian W. B. Randolph, Dartmouth College

This series, sponsored by Dartmouth College Press, develops and promotes the study of visual culture from a variety of critical and methodological perspectives. Its impetus derives from the increasing importance of visual signs in everyday life, and from the rapid expansion of what are termed "new media." The broad cultural and social dynamics attendant to these developments present new challenges and opportunities across and within the disciplines. These have resulted in a transdisciplinary fascination with all things visual, from "high" to "low," and from esoteric to popular. This series brings together approaches to visual culture—broadly conceived—that assess these dynamics critically and break new ground in understanding their effects and implications.

Luc Pauwels, ed., *Visual Cultures of Science: Rethinking Representational Practices in Knowledge Building and Science Communication*

Lisa Saltzman and Eric Rosenberg, eds., *Trauma and Visuality in Modernity*

For a complete list of books please visit www.upne.com

VISUAL CULTURES OF SCIENCE

Rethinking Representational Practices in Knowledge Building and Science Communication

Edited by Luc Pauwels

Dartmouth College Press
Hanover, New Hampshire

PUBLISHED BY UNIVERSITY PRESS OF NEW ENGLAND
HANOVER AND LONDON

Dartmouth College Press
Published by University Press of New England,
One Court Street, Lebanon, NH 03766
www.upne.com

Printed in the United States of America
5 4 3 2 1

Library of Congress Cataloging-in-Publication Data

Visual cultures of science : rethinking representational practices in knowledge building and science communication / edited by Luc Pauwels.
p. cm.—(Interfaces, studies in visual culture)
Includes bibliographical references and index.
ISBN-13: 978-1-58465-511-4 (cloth : alk. paper)
ISBN-10: 1-58465-511-9 (cloth : alk. paper)
ISBN-13: 978-1-58465-512-1 (pbk. : alk. paper)
ISBN-10: 1-58465-512-7 (pbk. : alk. paper)
1. Visual communication in science.
I. Pauwels, L. (Luc)
II. Series.
Q223.V574 2005
501'.4—dc22 2005023318

CONTENTS

Introduction: The Role of Visual Representation in the Production of Scientific Reality vii

1 A Theoretical Framework for Assessing Visual Representational Practices in Knowledge Building and Science Communications
Luc Pauwels 1

2 The Production of Scientific Images: Vision and Re-Vision in the History, Philosophy, and Sociology of Science
Michael Lynch 26

3 Representing or Mediating: A History and Philosophy of X-ray Images in Medicine
Bernike Pasveer 41

4 The Accursed Part of Scientific Iconography
Francesco Panese 63

5 Images of Science in the Classroom: Wall Charts and Science Education, 1850–1920
Massimiano Bucchi 90

6 Representing Moving Cultures: Expression, Multivocality, and Reflexivity in Anthropological and Sociological Filmmaking
Luc Pauwels 120

7 Arguing with Images: Pauling's Theory of Antibody Formation
Alberto Cambrosio, Daniel Jacobi, and Peter Keating 153

8 Discipline and the Material Form of Images: An Analysis of Scientific Visibility
Michael Lynch 195

9 Edward Tufte and the Promise of a Visual Social Science
John Grady 222

10 Making Science Visible: Visual Literacy in Science Communication
Jean Trumbo 266

Contributors 285
Index 291

INTRODUCTION: THE ROLE OF VISUAL REPRESENTATION IN THE PRODUCTION OF SCIENTIFIC REALITY

Science and (Visual) Representation

The issue of *representation* touches upon the very essence of all scientific activity. What is known and passed on as *science* is the result of a series of representational practices. Visual, verbal, numeric, and other types of representations are used in all sciences and in various types of scientific discourses, by and between specialists (intra-specialist and inter-specialist stages), and with students (the *textbook* stage) and the general public (the popularization stage). Lemke characterizes the science discourse fundamentally as a "multimedia genre" and sees concepts of science as "semiotic hybrids" (1998, 87) consisting of modalities that are "incommensurable" since: "no verbal text can construct the same meaning as a picture, no mathematical graph carries the same meaning as an equation, no verbal description makes the same meaning as an equation, no verbal description makes the same sense as an action performed" (1998, 110). Visual representations are not to be considered mere *add-ons* or ways to popularize a complex reasoning; they are an essential part of scientific discourse.

In his stimulating book *Science: The Very Idea,* Steve Woolgar asserts that the whole idea of representation is often linked with the "essentialist" position with regard to science, that is, with the notion that there are discrete objects "out there" that exist independently of our perception (Woolgar 1988, 30) and that are waiting, so to speak, to be discovered and reflected *truthfully* through various forms of representational devices: graphs, formulas, pictures, descriptions, etc. It is clearly not the objective of this volume to call upon or resuscitate such outmoded and static views of science as an objective body of knowledge that leads to absolute certainty or truth. The often-heard claim that (visual) representations should be "truthful" or "according to nature" is flawed for several reasons. Not only is the notion of reality itself highly contested, but also a degree zero is lacking in visual perception, that is, there is no state where things are perceived in an uncolored and unbiased form. Furthermore, any technique or medium, however sophisticated and advanced, at best can provide some highly mediated *renderings* of that presumed reality. Nevertheless, the recognition that the "Ding an Sich"—to put it in Kantian terms—is not attainable should not lead to its complete denial or be seen as confirmation of the view that all representations are but constructions that refer to basically nothing in reality. Finally, it also should be stressed that visual representational practices are not just about attempts to represent a phenomenon that is observed in a historical and physical reality (thus implying iconic and indexical relationships), but equally may include those more ideogramic activities that seek primarily to express ideas or relations and that have no material referent whatsoever (most representations consist in a combination of mimetic and expressive elements).

Science is not just about truthfully describing or trying to replicate reality, but about making it more understandable and accessible in a myriad ways. Likewise, when seeking to better understand and use visual representations in science, one should not see the process of visualization primarily as an effort to reproduce reality as it presents itself to the scientist, since science is not about copying nature and culture, but about revealing it. We therefore should neither expect nor demand from a representation that it matches reality, that it provides an almost perfect replica of it. Visual representations in science need to serve a number of distinct purposes that go far beyond the act of reproducing nature. Their value is judged by their functionality for resolving a problem, filling gaps in our knowledge, or facilitating knowledge building or transfer. Lynch (1998, 223) astutely remarks that the study of visualization really is about "the production of *scientific* reality."

Recognizing both the perceptual limitations of human observers and the representational limitations of any medium (and of medium languages) vis-à-vis the (in principle) endless number of aspects of an event, object, or concept, we should continue to refine our thinking about different types of media and different types of translations of and relations between a representation and whatever it seeks to represent. This diversity should be linked to the variety of distinct purposes these representational attempts may serve and the extent to which they succeed.

Even if we reduce the broader concern for *representation in science* to the varied and pivotal role of *visual* representation in the social and natural sciences, as this volume aims to do, we are still talking about a vast array of widely divergent practices and products. This virtually limitless scope is well reflected in the wide diversity of work that has been done in this field. Scholars from very diverse disciplinary backgrounds (e.g., sociology of science, medical imaging, philosophy of science, history of science, geographical information systems, geology, biology, physics, visual anthropology, business sciences, information design, mathematics, communication studies, anthropological linguistics) gradually have taken an interest in the complex issue of visualization in science. They have studied a broad range of types, aspects, and uses of visual representations in the sciences, and each of those research traditions have contributed a lot in clarifying the huge potential as well as the many tribulations visual representational practices may implicate. Many aspects of the technical and social development of visual representations and the ways in which they are being employed have been approached from distinct theoretical paradigms and a variety of research methodologies (including among others ethnomethodology, phenomenology, semiotics, social constructivism, conversation analysis, interaction studies, ethnography, semiotic analysis, experiments, surveys, interviews, video recordings, and field notes). Scholars have scrutinized processes of perception and knowledge acquisition and negotiation with representations through detailed studies of practices and discourses of groups of people in different social and professional settings such as those of fieldworkers (Roth and Bowen 1999), laboratory scientists (Knor-Cetina 1981; Latour and Woolgar 1979; Lynch 1985), teachers, engineers, students, publishers, navigators, and lawyers (Goodwin 1994, 1995, 1996; Pea 1994) to name just a few. A variety of products and media of visualization have been studied in great detail such as diagrams, photographs, drawings, and various scanning techniques. In particular, graphs and the social and technical processes that accompany their creation and their use received much attention (Roth, Bowen, and Masciotra 2002; Roth, Bowen, and McGinn 1999; Roth and McGinn 1997).

Scientific visualization—in a broad sense—has been studied in view of its practicability in fostering educational processes and in bridging theory and practice (Gordin and Pea 1995), pointing at both unique opportunities and issues/obstacles. It may be seen as an enculturation device in a community of practitioners or scientists-in-process (Roth and Bowen 2001), but also as an evolving means of expression in science and other disciplines (Pauwels 2000, 2002) and as a prime tool in the development of scientific literacy (Gordin and Pea 1995) among a diversity of audiences.

The many and highly divergent studies of scientific visualization add up to an impressive body of knowledge that occasionally comes together in themed journal issues, edited volumes, and conferences, as well as occupying a niche in the different disciplines and specialized journals. Synthesizing the many contributions could be, as Cambrosio et al. (1993, 663) called it many years ago, a "hopeless task." Yet working toward integrating at least part of this knowledge and translating it into generic and more customized modules for use in many, if not most, academic curricula would be a significant step forward.

Given its complexity and hybridity and the lack of formal training in the visual cultures of science, it is not surprising that many researchers found that students are ill-prepared to perform scientific representational practices thoughtfully and skillfully (Roth, Bowen, and Masciotra 2002) in particular visual representational work, nor to face the challenge of combining the highly divergent representational systems (words, graphs, images, equations) in a meaningful and edifying whole. And probably the same can be said of many scholars, as evidenced by journal articles, text books, and conference presentations.

While there still seems to be an implicit but persistent belief that the rapid spread of visual technologies in almost every sector of society automatically will result in an increased visual literacy or competency, there is at least as much reason to believe that the already vulnerable link between the referent, its visual representation, and the functions it needs to serve will come under even greater pressure. The ever-increasing speed and ease by which visual representations are produced often further conceals the complex issues involved in any type of representational activity.

Chapter by Chapter—a Summary

Given the weight and importance of the above-described issue, this volume aims to present a thorough and conscious investigation of representational practices within the different discourses of scientific data representation, conceptualization, and scholarly and public communication. The emphasis is on the role of the visual dimension of generating, disseminating, and using scientific insight; that is, the book focuses on visual techniques and stylistic features rather than on verbal style, rhetoric, and structure, though these latter aspects also merit further exploration, and in fact cannot be separated from the visual aspect, since the two modes of expression and their complex interplay make up scientific argumentation. This reader seeks to elicit further and more widespread debate about representational issues in scientific research and scholarly communication and to foster a better use of the many representational and expressive properties and opportunities of both old and new media.

The issue of visual representation is explored in this edited volume by an international selection of scholars, coming from and looking at different disciplines in both the social and the natural sciences. All chapters are very critical with regard to the use of visual representations in scientific (and public) discourse. Though the contributors are not necessarily on the same line of thought in every bit of their reasoning, they all position themselves far beyond any naïve notion of science. In addition, they try to move beyond a mere constructivist critique of naïve or essentialist positions by fostering a constructive debate of alternative and future uses of new and old technologies and representational practices.

As *cultures* in general, and thus also the subsets of *visual cultures,* are characterized by change—be it slow or sudden—a historical perspective is essential in trying to understand the impact and future role of visual representation in science. Many articles in this volume therefore deal explicitly with the historical dimension of representational practices, either in a general way or applied to a theme or discipline. While the various chapters in the reader all deal with basic aspects of visualization and often refer to a common set of ideas, they also focus on particular domains, specific disciplines, and uses.

Visualization in science is a complex domain covering many different interrelated aspects that are explored from different perspectives and applied to different disciplines of science in the various articles in this reader. Chapter 1 ("A Theoretical Framework for Assessing Visual Representational Practices in Knowledge Building and Science Communications") develops a general framework for referring to each of these aspects and their complex interrelations. The scheme that is developed allows us to put the particular issues that are discussed in the subsequent chapters into a broader context and draws attention to the cross- and interdisciplinary nature of visualization. Furthermore, it may serve as a theoretical framework for the thoughtful production of visual representations in science or may be used as a tool to assess critically the appropriateness of different aspects of particular representations. This framework may prove useful in examining the complex interdependencies that exist between the nature of the referent, the social, technological, and cultural context of production, the choices with respect to medium and the style of the representation, and the purposes and uses that need to be achieved. In this respect, it also could be read as an outline for a program to develop visual literacy among scientists. The concept of visual literacy is defined as a reflexive attitude (throughout the production process), a specific body of knowledge, and even a certain level of proficiency or skill in assessing and applying specific characteristics (strengths and limitations) of a particular medium or representational technique in support of a scientific undertaking, and taking into account cultural practices (codified uses or expectations) and the actual context of use (including the *cultural repertoire* of the intended audience).

Chapter 2 ("The Production of Scientific Images: Vision and Re-Vision in the History, Philosophy, and Sociology of Science") provides a stimulating introduction to the complex issue of scientific visualization. In this reprint of one of Michael Lynch's many influential articles in the field of the sociology of science, the author rejects the idea that visualization in science may be considered a discrete practice, a well-bounded activity resulting in images and graphic representations with their own particular requirements and properties. Lynch argues convincingly in favor of an approach that situates these visual products in a broader context of scientific activities that includes observation, measurement, description, analysis, and demonstration, and he deals with many of the basic and highly intertwined issues that are linked with this essential activity of scientific discovery and dissemination. Moreover, he adopts a carefully considered position in often very polarized debates (e.g. constructivist, realist, antirealist discourses) in older and more recent studies and theories of visualization. In Lynch's view, visual representations do not simply reflect natural properties, but by the same token they do not merely reflect social and cultural properties or states of mind either. Making progress in any field of inquiry requires that one asks the right kind of questions, and that is exactly what this article does. Its central ideas are further explored and illustrated in the other articles in this reader.

In chapter 3 ("Representing or Mediating: A History and Philosophy of X-ray Images in Medicine"), Bernike Pasveer gives a detailed account of the discovery/invention of X-ray photography as a classic example of visualizations of natural phenomena that are only accessible through specialized techniques that produce their own effects. Pasveer illustrates that this new technology is not a transparent and neutral intermediary that provides straightforward, true-to-nature representations of the inside of living bodies. It is beyond dispute that X-ray photography yields images that are caused by objects in the physical world that are nevertheless hidden from direct view. But how exactly the physical objects and representations were related, what the shadows actually referred to or signified was to be established by a long and cumbersome process of an image in search of clinically relevant content. X-ray images had to be "uploaded" with meaning or encoded on the basis of an existing body of knowledge acquired through other, more traditional, approaches (palpation, percussion, auscultation, and dissection of dead bodies) and many more X-ray images, since the referent itself could not be approached in a more direct way. Given the inaccessibility of the referent (the human body in its living state), it needed to be (re)created, partly by the very instruments and practices that were being used to depict it. Pasveer points at this strange inversion that has occurred whereby the depicted is typified by characteristics of the depiction. The depicted was brought in line with the characteristics of the technology used as least as much as the depiction process was made to comply with the depicted. This reversal is indicative of the great trust that was put in this new representational technique, a faith in *mechanical objectivity* that is put in photographic techniques in general to this day.

In chapter 4 ("The Accursed Part of Scientific Iconography"), Francesco Panese illustrates how, towards the end of the fifteenth and throughout the sixteenth centuries, very different *registers* were used in the visual presentation of the natural world. Images were often mixtures of empirical fact, folk knowledge, magical beliefs, hearsay, and fable that tended to conflate form and meanings. Little by little, the idea emerged that a visual representation should reproduce an object's visible characteristics rather than its essential meanings, and representations that did not comply with this new view were marginalized and maligned by scientists. In his discussion of a number of representations (such as one of a non-existent monstrous creature in scientific and nonscientific publications), Panese clearly demonstrates how the practice of scientific representation is founded on an evolving code that is developed to add verity to an argument and to establish a regime of credibility. This interesting historical account sketches the advent of a particular scientific iconography that gradually established its own rules for depicting the world as separated from other systems of belief and representation, or, in other words, how particular visual cultures of science came into being. Yet it would be a mistake—as the previous chapter and several other chapters clearly demonstrate—to believe that the separation from different cultural spheres that Panese describes so vividly is a *fait accompli*. Visual representations used in scientific discourses always will be influenced by other cultural registers, including art, religion, or popular beliefs. This may be less obvious in the depiction of natural objects than in the representation of more conceptual entities. Today, scientific representation may, for example, still contain metaphorical traits or otherwise borrowed formal characteristics, which need not be a problem as long as they are duly recognized and if they help clarify the issue at hand.

In chapter 5 ("Images of Science in the Classroom: Wall Charts and Science Education, 1850–1920"), Massimiano Bucchi discusses wall charts as a didactic genre that experienced its golden age between about 1870 and 1920. Bucchi offers a captivating view of how a particular medium for visual communication of science originated and blossomed as a result of technical developments (lithographic technology), educational reform (resulting in a dramatic increase in pupil populations), and the changing view of the pivotal role of visualization in education (the need to see and handle an object). In this richly illustrated article, Bucchi argues that the move to a more visual pedagogy in science not only embodied a general idea that visuals are essential devices in any type of knowledge transfer, but also signified a shift with regard to theory. In the field of biology, for instance, it meant challenging the dominant morphological-systematic approach in favor of a visual approach that tried to consider living organisms in their natural surroundings. The wall charts are a visual delight in their own right, but they can also be read as mirrors of a Zeitgeist, the intellectual spirit of a period in history. In those early days, proponents of wall charts clearly saw the benefits of a more visual approach to studying nature and disseminating their knowledge about it, and they fiercely challenged the verbalism of most scientific approaches. While today wall charts largely have been replaced by other means of visual communication in scientific and educational settings, the issues and arguments remain as fresh as ever, as the debate on a more visual science has not yet been decided.

Chapter 6 ("Representing Moving Cultures: Expression, Multivocality, and Reflexivity in Anthropological and Sociological Filmmaking") offers a critical overview of the development and deployment of film in the scientific endeavors of social and cultural scientists. It addresses some of the key issues involved in the use of a rich medium such as film to document, analyze and communicate about society and culture. While the article comments on a distinct tradition and area of scientific inquiry and filmmaking, many of the observations and arguments it contains can be applied to the issue of scientific visualization in general: the tension between mimesis and expression, forms of collaboration among different types of expertise and between scientists and the field they are studying, the need for a more conscious, reflexive approach—in the whole process of data gathering and building of the visual end product, the assumptions and requirements of various purposes and audiences of film in scientific, educational, or broader societal contexts are recurrent themes in this volume. What is discussed in a historical and theoretical perspective is further exemplified and expanded upon through reviews of four films that each represent a somewhat different take on most of the items discussed. While no single scientific approach is advocated, very explicit attention is paid to the far-reaching consequences that the many choices (technical, aesthetic, scientific, ethical) with regard to the constitutive elements of a film may have on its usability for scientific or educational purposes.

In chapter 7 ("Arguing with Images: Pauling's Theory of Antibody Formation"), Cambrosio, Jacobi and Keating conduct an in-depth study of the imagery that is used in immunology. They first provide insight into the broader debate concerning (bio)chemical concepts and processes as they were depicted in their era, and subsequently they focus on a series of images featured in a 1948 article on antibody formation that immunologist Linus Pauling wrote for the journal *Endeavour*, subjecting them to a very detailed semiotic inquiry. The collaboration between Linus Pauling and Roger Hayward, who produced the drawings, is at once an illustration of a fruitful collaboration and an example of how such collaboration is also a source of discontent, in as much as different normative systems are brought into play and confronted. A professional-looking execution by a skilled craftsman gives the visual representation greater credibility and appeal, irrespective of whether it is scientifically sound. Cambrosio et al. clearly consider visual representations to be part of the rhetoric and overall argumentation of a scientific article. Images, drawings, and graphics contribute by their interplay with the text and through the numerous iconic and para-iconic features that they bring into play. Their meticulously executed analysis yields some very important observations and conclusions with regard to the often arbitrary relation between the referent and its depiction. The arbitrariness of certain visual features is often forgotten once a concept has acquired an appealing shape and is thus adopted as "the way it should be depicted." Not unlike texts, visual representations are based on previously existing representations and they build on accepted frames of reference in science that often are no longer challenged or revisited. In contrast with the linear and highly discursive sequence of texts, the receivers of visual representation frequently are left to their own judgement and insight in trying to decode its content and in attributing meaning to each of the iconic and para-iconic features.

In chapter 8 ("Discipline and the Material Form of Images: An Analysis of Scientific Visibility"), Michael Lynch focuses on visualization of natural objects and argues that through the use of complex instruments and careful preparatory procedures in the field or in the laboratory setting, highly artificial objects are created. Discussing two specific cases from the field of biology, Lynch goes through the various modifications that specimens are subjected to in order to provide a representation that is considered fit for scientific purposes and from which scientific knowledge is produced. While these observations challenge naïve realist views on representations as objective reflections, they do not necessarily criticize the use of conventions by laboratory researchers. Still, they warn us that, when we are looking at specific features of visual representations and the way they are organized, we may in fact be looking as much at disciplinary conventions as at reflections of natural objects and relationships. A visually literate scholar (a concept that is explored in other chapters of this book) should be able to develop an eye for the complex merging of graphic properties produced by devices and procedures with the *natural object,* which they ultimately come to embody. This is particularly difficult in fields such as electron microscopy, where "the artificial appearance of a specimen is what enables it to be observed and analyzed in the first place" (Lynch 1985, 38). Lynch describes and typifies the active visual reconstruction of the world as performed by researchers as a process of civilizing or domesticating natural objects in the culture of a discipline (i.e., in its typical ways of doing and thinking). Again, many of the ideas and arguments that Lynch puts forward in this article can be—and have been—applied fruitfully to other disciplines, including the social sciences and other types and purposes of scientific representation.

In chapter 9 ("Edward Tufte and the Promise of a Visual Social Science"), John Grady discusses the seminal work of the statistician and political scientist Edward Tufte in the field of information design. Tufte has been advocating the need for better presentation of information and of scientific fact to the benefit of policymakers and citizens in a democracy through a series of books, each of which is, in itself, a masterpiece of design. Grady's starting point is to question why this fascinating body of work has had such limited influence on scientists and in particular on social scientists. The article is both a presentation of the highlights of Tufte's thinking and design skill and a critical inquiry into the reasons for its neglect, which Grady suggests lie both in peculiarities and idiosyncracies of the work itself and in the culture of the established world of social scientists. Grady's article, then, is a lucid and enticing effort to bridge the gap between the potential that resides in Tufte's work and the approach social scientists take to studying society and reporting on it. For Grady, the master theme of Tufte's books is the idea that aesthetics is inextricably connected with moral, political, and cognitive issues. So Grady goes far beyond reviewing Tufte's work, as he gradually builds a powerful and well articulated case for a more visual science that he hopes will help to convince the mainstream of quantitative (social and natural) science to incorporate visual research methods into their respective disciplines.

The concluding chapter 10 ("Making Science Visible: Visual Literacy in Science Communication") highlights a notion that has been present implicitly in each of the previous chapters: the need to develop a particular kind of visual literacy and media or information literacy as a prerequisite for building more conscious and more effective visual cultures of science. Author Jean Trumbo regrets the absence of a common theory or body of knowledge to address the issue of how to represent scientific data, processes, or discoveries effectively, and suggests in this chapter a basis for building such a program for increasing visual literacy. Her approach is intended to be applicable to the different spheres of communication in science, ranging from communication among specialized scholars to forms of mediated communication with more general publics. Given the ubiquity of scientific representations in more popular media and cultural spheres, Trumbo advocates the training of both public and professional communities in information literacy. This is a very demanding requirement since, as Trumbo well observes, "each form of representation carries its own conventions and potential for interpretation" and often "it is the availability of a technology that dictates how the science is rendered." Trends and many other nonfunctional reasons may determine the choice of medium and the subsequent representational choices, such as those regarding style and execution. Often, the speculative nature of scientific representations (or of some of its aspects: shape, color, etc.) is not fully acknowledged, neither by the general public nor by fellow scientists, and nor for that matter by its creator or initiator. She also acknowledges, as was stressed in chapter 1, the many different purposes a representation in science may serve, as it is not confined to reflecting scientific fact, but may equally take the form of hypotheses that represent subjective knowledge.

Scope and Readership

Together, the articles in this reader—however diverse they may appear judging by the titles and by the subjects, periods, and disciplines covered—complement each other to provide a varied exploration of the multifaceted issue of visual representation in science. Even when dealing with a certain period, theme, field of application, problem, or technological innovation, each of these articles interrogates, exemplifies, and formulates ideas with respect to the broader issue of visual representation in science. Each of the basic issues addressed in these articles—such as the need to provide broader context, the role of aesthetics, forms of collaboration among scientists, educators, and illustrators, the complementary relation between word and image, the diverse nature of the referent, the problem of decoding and encoding, the search for visual meaning, or the adaptability of visual representations to suit different uses or audiences—remain prime issues of scientific visualization until this day.

Several of the chapters provide concrete elements and suggestions with regard to the shaping of the contours of a *visual literacy for scientists*—a program that focuses not only on uncovering misrepresentations and arming academic or broader audiences against purposeful deceit or unsound visual arguments, but also on allowing them to make better use of the expressive capabilities of media and representational techniques on the basis of more explicit and appropriate methodological guidelines.

The title of this volume, *Visual Cultures of Science,* refers to the complex interplay of representational technologies, scientific and cultural traditions, to the culturally molded and historically situated ways of doing and thinking of scholars and their diverse publics. It is aimed at both the advanced student and the seasoned scholar in just about any of the social and natural sciences. It will be of interest in particular to those who are becoming more conscious and critical of the ever-present and growing impact of representational means and practices and who are committed to making better, more transparent and productive use of its expressive capabilities.

References

Cambrosio, A., D. Jacobi, and P. Keating. 1993. Ehrlich's "beautiful pictures" and the controversial beginnings of immunological imagery. *ISIS* 84: 662–699.

Goodwin, C. 1994. Professional vision. *American Anthropologist* 96: 606–633.

———. 1995. Seeing in depth. *Social Studies of Science* 25: 237–274.

———. 1996. Transparent vision. In *Interaction and grammar,* edited by E. Ochs, E. A. Schegloff, and S. A. Thompson. Cambridge: Cambridge University Press.

Gordin, D. N., and R. D. Pea. 1995. Prospects for scientific visualization as an educational technology. *The Journal of the Learning Sciences* 4: 249–279.

Knor-Cetina, K. 1981. *The manufacture of knowledge: An essay on the constructivist and contextual nature of science.* New York: Pergamon.

Latour, B., and S. Woolgar. [1979] 1986. *Laboratory life: The construction of scientific facts.* 2nd ed. Princeton: Princeton University Press.

Lemke, J. L. 1998. Multiplying meaning: Visual and verbal semiotics in scientific text. In *Reading science,* edited by J. R. Martin and R. Veel. London: Routledge.

Lynch, M. 1985. *Art and artefact in laboratory science: A study of shop work and shop talk in a research laboratory*. London: Routledge & Kegan Paul.

Lynch, M., and S. Y. Edgerton. 1988. Aesthetics and digital image processing: Representational craft in contemporary astronomy. In *Picturing power: Visual depiction and social relations*, edited by G. Fyfe and J. Law. London: Routledge.

Lynch, M., and S. Woolgar, eds. 1998. *Representation in scientific practice*. Cambridge: MIT Press.

Ochs, E., P. Gonzales, and S. Jacoby. 1996. "When I come down I'm in the domain state": Grammar and graphic representation in the interpretive activity of physicists. In *Interaction and grammar*, edited by E. Ochs, E. A. Schegloff, and S. A. Thompson. Cambridge: University Press.

Ochs, E., S. Jacoby, and P. Gonzales. 1994. Interpretive journeys: How physicists talk and travel through graphic space. *Configurations* 2: 151–171.

Pauwels, L. 2000. Taking the visual turn in research and scholarly communication: Key issues in developing a more visually literate (social) science. *Visual Sociology: Visual Cultures, Visual Literacies* 15: 7–14.

———. 2002. The video- and multimedia-article as a mode of scholarly communication: Toward scientifically informed expression and aesthetics. *Visual Studies* 17 (2): 150–159.

Pea, R. D. 1994. Seeing what we build together: Distributed multimedia learning environments for transformative communications. *Journal of the Learning Sciences* 3: 285–299.

Roth, W-M., and G. M. Bowen. 1999. Digitizing lizards: The topology of "vision" in ecological fieldwork. *Social Studies of Science* 29: 719–764.

———. 2001. "Creative solutions" and "fibbing results": Enculturation in field ecology. *Social Studies of Science* 31: 533–556.

Roth, W-M., G.M. Bowen, and D. Masciotra. 2002. From thing to sign and "natural object": Toward a genetic phenomenology of graph interpretation. *Science, Technology & Human Values* 27: 327–356.

Roth, W-M., G. M. Bowen, and M. K. McGinn. 1999. Differences in graph-related practices between high school biology textbooks and scientific ecology journals. *Journal of Research in Science Teaching* 36: 977–1019.

Roth, W-M., and M. K. McGinn. 1997. Graphing: Cognitive ability or practice. *Science Education* 81: 91–106.

Woolgar, S. 1988. *Science, the very idea*. Key Ideas Series. London: Tavistock Publications.

VISUAL CULTURES OF **SCIENCE**

1

A Theoretical Framework for Assessing Visual Representational Practices in Knowledge Building and Science Communications

Luc Pauwels

The multifaceted issue of visualization in science basically involves the complex processes through which scientists develop or produce (and communicate with) imagery, schemes, and graphical representations, computer renderings, or the like, using various means (ranging from a simple pencil on paper to advanced computers or optical devices). Therefore, not just the result, but also how it was attained (i.e., the implicit or explicit methodology in the broad sense of the word) and indeed the subsequent uses to which the result is put should all be scrutinized as to their impact on the nature of what is represented visually and the ways in which this representation can be employed. Visual representations in science differ significantly in terms of how they relate to what they purport to represent (i.e., their representational and *ontological* status), the means, processes, and methods by which they are produced, the normative contexts involved, the primary purposes served, and the many ways in which they subsequently are used and combined, to name but some of the more crucial aspects.

Because of the diversity of appearances and applications and the broader contexts in which they need to be placed (e.g., scientific theory and traditions, culture, media, and technology), it is indeed a challenge to make generalizations about the uses and functions of visuals in scientific discourses. This may explain the remarkably few systematic attempts that have been made at devising a common, by necessity rather basic, framework for increasing insight into this complex domain. This chapter seeks to present such a general framework for looking judiciously at visual representations in the context of scientific endeavors. It discusses a number of critical aspects that should be considered when producing, reading, or (re)using visual artifacts. It will not be able to do justice to all aspects of and perspectives on this complex problematic but might at the least help to bring some basic structure into it. The purpose of this model is to stress that the sciences, despite their differences, in fact do have a lot in common—both difficulties and solutions—and that bringing them together may illuminate future practice.

1. The Varied Nature of the Referent

The array of objects or referents of visual representations in science is very broad and of a highly heterogeneous nature. Visual representations in science may *refer* to objects that are believed to have some kind of material or physical existence, but equally may refer to purely mental, conceptual, abstract constructs and/or immaterial *entities*.

Material or physical referents may have visual characteristics that are *directly observable* to the human eye (e.g., various types of human interaction, the external structure of animals, trees, etc.). On the other hand, there are objects and phenomena with aspects that *only become visible with special representational means and devices* (e.g., they can be observed only using special techniques or instruments such as high-speed photography, satellite image transmission, a telescope, a microscope, or an endoscope). These aspects may be too fast (e.g., an explosion, eye movements), too slow (e.g., transformations in a living organism), too big (e.g., stellar configurations), too small (e.g., microscopic organisms), too similar (e.g., color of vegetation), or too far away (e.g., planets) for the human eye to discern, or they may be hidden (e.g., organs of a living body) or inaccessible unless destructive course of action is taken (e.g., by dissection of an organism, creation of a cross section of an object, excavation of remains). Furthermore, physical objects or phenomena may not have visual characteristics as such and still be *translated from a non-visible state* (e.g., sound waves, thermal radiation) *into visual representations* using special devices.

Representational practices in science often do not seek merely to *reproduce* visual or non-visual phenomena but also to provide *visual data representations* (e.g., charts) of aspects of these phenomena based on measurements of some kind (length, weight, thickness, resistance, quantity, temperature, verbal responses, etc.). In the latter cases, *data* are derived from or constructed on the basis of an observed reality and subsequently represented in a visual form that allows one to discern changes or see relationships more clearly. While the resulting representations are based upon empirical observation or interrogation of the field, they are not *reflections* of visual natural phenomena. They are rather visual representations of observations in the physical world that are not necessarily visual in nature. In other words, what is represented are not physical objects or phenomena, but *data* that are constructed by observing aspects of the physical world. The relationships among the data and their representations are much more abstract/arbitrary and conventional, though some aspects may also be *motivated* or iconic (i.e., they may bear some resemblance to the referent). For example, graphical representations of the evolution of the birth rate within a particular population over a certain period of time, temperature fluctuations during one month in summer, or the number of murders per state do not necessarily entertain a visual iconic or indexical relationship with a physical or material referent, as often there is none. Instead, these data representations may have a *mental* referent as far as the source is concerned, since the representations are not so much *depictions* of phenomena in the real world as conceptual translations of aspects of it. Yet, they are based at least in part on quantitative or qualitative aspects of an observed *reality* of some kind and thus are not purely invented or products of the imagination.

The referent of a representation may be even more *immaterial and abstract* in nature: for example, representations that primarily seek to visualize relations among observed phenomena, hypothetical relationships, postulated phenomena (e.g., black holes) or effects, and even purely abstract concepts. The referent of such representations may become an almost purely mental construct that has no *pre-existence* in the physical, historical world whatsoever. Nonetheless, representations of these kinds of referents may play an important role in understanding or influencing that world.

Finally, it should be noted that many representations in science combine several of the abovementioned aspects and thus have *different referents*. Certain aspects of the representation may, for example, refer in an iconic way to an observed visual reality (e.g., it might mimic its shape or color) and at the same time include conceptual structures (such as metaphors) or symbolic elements (arrows, markers, colors, shapes). An edited film will refer iconically to the depicted subject matter (i.e., it will *reflect it* to a certain extent), but at the same time it might allow scientists to express their vision or theory by means of the manner of recording and subsequent editing processes. In fact, as is argued in chapter 6, mimesis without expression is virtually impossible. At the root of every presentation of fact is an implicit or explicit theory, a particular way of looking. In fact, visual representations may not only refer to the material world or to an

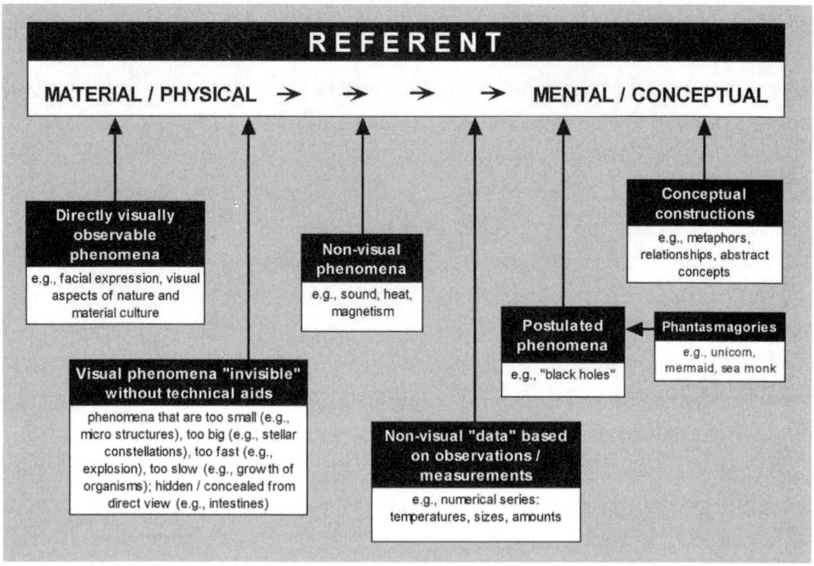

Figure 1.1. The divergent nature of the referent, from material existence to mental construction.

abstract or imaginary world but also may refer to a *possible world,* as is the case when scientists perform simulations to get an idea of what might happen when combining such and such parameters or phenomena.

This concise taxonomic preamble with respect to the nature of the referent already indicates the wide variety of material and immaterial *things* that a visual scientific representation, be it static or dynamic, may *refer* to. But I also have touched implicitly in this brief introduction upon the diversity that exists with regard to types of translation processes and actors in the production cycle, as well as the different purposes and intents of representations and specific contexts of use. The presentation and discussion of the complex interplay of these aspects constitutes an analytical framework for visual representations and related practices in the sciences.

2. Representational Production Processes: Social, Technological, and Cultural Aspects

Inscription, Transcription, Invention, and Fabrication

Every *representational* process involves a translation or conversion of some kind; a process of inscription, transcription, and/or fabrication whereby the initial source (phenomenon, concept) is captured, transformed, or even (re-)created through a chain of decisions that involves several actors (scientists, artists, technicians), technological devices, and normative settings. This complex process of meaning-making has an important impact on what can be known and how, on what is revealed or obscured, and on what is included or excluded.

The ontological relation between a representation and its referent is always somewhat problematic even with so-called *automated processes*. However, all types of translational processes are not problematic to the same degree nor in the same sense. It is therefore not very useful to challenge iconicity in general and thus to ascribe to all types of representation a similar degree of arbitrariness. Also, the use to which a representation is being put constructs its representational status, at least in part.

As I have argued in the previous section, the divergent nature of the referent in science prefigures the crucial importance of the equally divergent processes of producing a visual representation. These processes not only involve technical issues but also encompass important social and cultural aspects. Obviously, technology and each of its products are part and parcel of culture (i.e., they are both a cultural product or *result* and a cultural actor or *force*), both in a broad cultural and a more restricted sub-cultural sense (e.g., scientific disciplinary practices and purposes), and they embody specific norms and values. For example, the specific look and the functionality of a photographic camera (i.e., dominance of the center of the frame for both exposure and focus) to some extent reveals what its inventors and subsequent developers valued the most and what kind of applications they primarily had in mind (e.g., portraits). But apart from the characteristics of the instrumentation, which are to some degree a result of cultural processes as well, a host of other social and cultural influences at the moment of choosing and selecting the objects and samples also have an important impact on how the representation will appear as well as on the purposes it may subsequently serve.

Analyzing the Social and Cultural Setting: Division of Human Labor and Different Normative Contexts

Barbara Rosenblum, in her sociological study of photographic styles, demonstrates quite convincingly how the "look of things," in particular the typical appearance of press, art, and advertising photographs, are to a significant degree a function of social, technological, and cultural factors and constraints that are connected with their creation. Her research into these areas indicated that "photographic styles are directly correlated with systems of production and distribution." Her observations of news photography as a highly conventionalized and comparatively homogeneous "system of images" that is generated within a bureaucratically coordinated system (Rosenblum 1978, 111) provide some interesting parallels with at least some of the ways in which imagery is produced in scientific contexts. After all, the division and standardization of labor, technological constraints, professional ethics, time pressure, as well as economic factors, all play a significant role in their creation, look, and value. Drawing on several fields of expertise and modes of practice, Charles Goodwin (2001, 157) argues credibly that an ethno-methodological perspective provides an essential complement to any study of visual representation that seeks to go beyond textual borders and into the broader contexts of production and uses.

Sociologists of science have studied the complex interactions in a laboratory setting where science is being *produced* (Goodwin 2001; Latour and Woolgar 1979; Lynch 1985b), an approach that yields insight into how an object of inquiry is selected, delineated, and *prepared* to fulfil its role. Lynch has looked at the laboratory setting and the processes by which natural objects are visualized and analyzed. Preparatory procedures that turn the object of investigation into a "docile object" fit to be studied according to the established methods and mores of science, the instrumentation and the laboratory set-up all challenge the idea that scientific visualization provides an unproblematic or uncompromising "window" onto the natural world (Lynch 1985b, 43–44). Similar processes are at work when scientists make observations in the field, as objects likewise are selected and prepared to be subjected to scientific practices or made to participate in data-generating procedures. Not only natural scientists, but also social scientists try to produce "docile objects" through sampling techniques, pre-structured questionnaires, and statistical operations. What cannot be measured, or only very inconclusively, is often overlooked and assumed to be non-existent. The *picture* that is obtained by the established procedures often is presented as a reliable and valid reflection of a broader phenomenon or population. Even in more qualitative types of visual research, such as anthropological filmmaking, the crucial importance of the process of data gathering and processing is recognized, and fierce debates continue to rage about what types of expertise should be combined in the scientist, what are (in)appropriate forms of collaboration with professional filmmakers and/or editors, how the *field* can be involved actively, what purposes a scientific film may fulfil, etc. (Pauwels 1996).

Furthermore, the issues of research funding, academic recognition, peer relations, and societal trends must all be taken into account if one endeavors to reveal and explain the processes that lie at the heart of particular visual representations of facts or ideas. They likewise may influence what is selected and how, and the way in which it is processed.

The Varied Nature of Visual and Non-Visual Transcription

There is a fairly significant though not exclusive or unconditional relation between the nature of the referent and the processes through which a representation is or ought to be produced.

Obviously, *conceptual constructions* that have no material, let alone visual, substance cannot be recorded automatically or according to standardized and repeatable processes (for example, mental images cannot be photographed or scanned electronically), as they are the result of multiple intentional acts that, first and foremost, require a suitable production technique for such highly intentional activity (e.g., pencil and paper or a computer drawing package). The involvement of the originator of the idea is paramount, and a demanding process of translating a mental image into an inter-subjective visible image is required. Aspects or dimensions that cannot in any way be visualized or verbally described are in fact lost to science.

Objects or phenomena that are *visible to the human eye* through direct observation, on the other hand, can be captured by representational devices such as a photographic camera that will produce detailed representations characterized by uniform time and a continuous space. This may result in a kind of *indifference* (some might say *objectivity,* though this may be too burdened a term to use), since all elements and details are treated equally (even though photographers have ways of foregrounding or emphasizing certain aspects at the expense of others, such as through the choice of lens, film, filters, lighting, framing, viewpoint, etc.). However, directly observable phenomena also can be represented through more manual techniques, using simpler media, such as pencils and brushes, which require much more *intentional* series of acts by humans (draftsmen, illustrators) and which produce imagery that do not have a uniform time (in fact quite some time may pass during the creation of the different parts of the representation) and that are not bound by continuous space or a uniform use of scale.

Every representation requires some kind of device or medium. Yet it is useful to make a distinction between mediation processes that are highly automated, or *algorithmic* processes (e.g., photography), and more manually and intentionally performed activities (e.g., hand-drawn or driven representations). However, these are not absolute categories and it is better to think about this useful distinction as two extremes of a continuum. Moreover, current digital technologies have blurred the dichotomy between *machine-generated* and *handmade* imagery and increasingly have allowed for more complex combinations of the two (for instance, digital photographs that can be manipulated at will with the aid of sophisticated software).

The process whereby one works from a *directly visible referent* to a *visual representation of it* would appear to be the most straightforward, but even then a great variety of techniques are available. Moreover, even the more commonly applied techniques have their intricacies, which are easily overlooked. This is true of relatively simple and ubiquitous techniques, including photography of directly observable phenomena, where one often has the advantage of being able to compare the referent (the object or phenomenon with a material existence) and the depiction (a drawing or photograph). However much such devices may differ in terms of the manner in which they *translate* an object or phenomenon into a record of it, it is important to note that both the source or the referent (the natural object or phenomenon) and its representation are *visual* in nature and are respectively captured and constructed by methods or processes that are essentially visual as well. In such instances, there is at least the theoretical option of comparing the source and its representation in order to assess to what degree and in which respects they resemble one another. Thus, a *check of correspondence* can be performed, albeit only to a certain degree.

A much more complex translation process occurs when the referent is visual and physical in nature (though often hidden from direct observation), while the *intermediate steps are not based on reflected visible light* waves. This is the case, for example, when ultrasound scans or X-rays are used: In these instances, it is not light that is reflected by the object that is recorded, but a reaction of other types of *invisible* waves to some characteristic or aspect (such as density) of the structure of the referent. These translations, while equally *indexical* in nature, typically require a more cumbersome process of decoding and calibrating (see, for example, Pasveer 1992) and they do not allow a simple check of *visual correspondence*.

Radiologists, for instance, need to learn how to *read* these images and even then they may differ on how a particular one should be interpreted.

If the translation process is not visual or if the referent is inaccessible or invisible to the unaided human eye, one has to rely on—and thus transfer authority to—the "machine" (Snijder 1989) in order to chart often unknown territory. In such cases, one has to be particularly aware of the possibility that one is looking at *artifacts of the instrumentation,* that is, objects and effects that are generated by the representational processes themselves and that do not refer to anything in the outside world or at least not to the phenomenon that is under scrutiny. In many data-generating processes, it is not always easy to differentiate *noise* from *data.* Artifacts or effects thus may be attributed erroneously to the outside world while in fact they are produced standardly by the instruments or as a result of technical failure. Moreover, an atypical representation also may result from an unexpected and unaccounted for event or coincidence in the physical world.

So, especially if the referent is of an uncertain nature, the problem of artefacts of instrumentation may arise. This may be the case when the existence of the referent is postulated rather than confirmed by fact and the process of representation serves the purpose of providing such evidence, or when complex instruments are being used, or when aspects of reality can only be seen through the instruments, that is to say, as a *representation.* But even with very realistic renderings of directly visible objects (e.g., simple camera images of directly observable phenomena), one should be wary of the possibility of *effects* induced by the instrumentation. Such effects can present themselves to the uninitiated eye as qualities or traits of depicted objects (color, shape, spatiality) while in fact they are merely properties of the instrumentation (e.g., the extremely foreshortened perspective when using telephoto lenses makes objects appear much closer to one another than they are in reality; internal reflections may produce flare and ghosting; etc.).

In a similar way, scientists should be aware of the possibility that important aspects of the referent might not be captured by the instrumentation (e.g., because of an inadequate resolution or insensitivity caused by a limited spectral range) or might mistakenly be weeded out as noise.

Instruments, in addition to capturing or recording data, invariably both reduce (or lose) data and tend to mold (and add) data in a particular way. These two phenomena in themselves should warn against a naïvely realistic view of the merely technical aspect of representation.

Technically sophisticated instruments that produce representations or images in a highly automated and standardized way (such as cameras and scanning devices) are generally thought of as the most suitable for scientific purposes, as they produce coherent, reliable, and repeatable representations with a predetermined level of detail. Moreover, they tend not to rely too much on personal judgement or skills in the process of image generation, unlike manual techniques such as drawing (though the interpretation of such representations may still require a lot of personal judgement and experience!).

However, in some cases more intentional processes and products may be far more convenient. This is true, for instance, if the depiction is *too detailed* for the intended purpose. This may be the case when using a highly automated and *indifferent* process such as a camera recording. Such a recording can be indifferent in the sense that all visible elements in front of the lens receive the same treatment, irrespective of whether they are relevant to the researcher. Thus, the essence of the recording may be obscured by unneeded, distracting, or irrelevant detail that can prevent insight. Furthermore, intentional processes allow a much swifter combination of *different types of signs (iconic, indexical, and symbolic) and levels of signification*. Consequently, they may yield a more functional expressive presentation of fact and vision. A third important consideration is that intentional processes may provide a much needed *synthesis of features* rather than a simple transcript of a particular (snapshot-like) instance of a phenomenon.

For instance, ornithologists who use imagery to determine the species of a particular bird encountered in the field may be better off with well-crafted illustrations of a number of similar-looking species—such as a colored drawing that contrasts a heron (*Ardea cinerea L.*) and a purple heron (*Ardea purpurea L.*). After all, they can derive from such a drawing how the two birds differ *in general*. Color photographs, on the other hand, necessarily show a particular specimen of each type of heron in a particular stage of its development and photographed against a particular background, in particular light conditions, from a particular angle. This photographic *particularity* may be less helpful in determining the species of an individual bird in the wild. On the other hand, purposefully simplified representations and abstractions may instill some misconceptions in people's minds if they are not duly communicated or if they are used for other than the initially intended purpose. For example, medical students

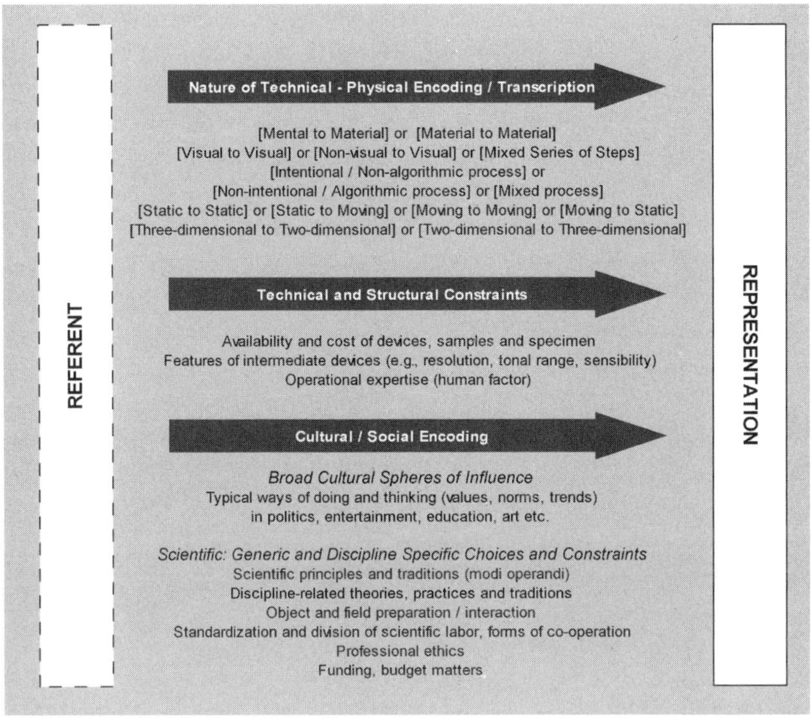

Figure 1.2. Determining aspects of the production/translational processes on the appearance of the visual representation.

may be baffled by the visual difference between stylized and simplified anatomical drawings of heart, lungs, and vascular system in their introductory courses and the *real thing*. Similarly, engineering students may be surprised by the difference between a highly stylized drawing of engine wiring and the three-dimensional reality of a dismantled engine that needs reassembling.

Scientific illustration as a sub-discipline of science is an interesting example of a specialization that has evolved in recognition of the fact that both scientists and artists in general lack the skills to produce renderings of birds, human anatomy, or complex technical artifacts with the required level of detail and generic faculties. Using artists who are very skilled in drawing but largely unaware of the exact purpose of the illustration inadvertently will produce imagery that may thwart that purpose. For example, they may make "corrections" according to their own aesthetic insights, or provide too much unnecessary detail, or fail to highlight elements properly that do not stand out very well in reality because they are buried in other visual stimuli but that are nevertheless central to the purpose. Scientific illustrators, on the other hand, need to be well-versed in both the art of illustration and in specialized fields of science. They are trained to have a thorough and fully integrated knowledge of the subject matter or concepts that they are asked to draw and of the exact scientific and didactic purposes their products need to serve.

3. The Visual Product: The Impact of Medium and Execution

Cultural Impact on Style and Use of Media

Visualization obviously results in a product that can be *seen:* a graphic representation, a photograph, a computer rendering. The term "representation" in this chapter is restricted to what in "social studies of science" is commonly referred to as "inscription" (Latour 1987; Roth and McGinn 1998) as some authors feel that "representation" as a general term is rather ambiguous. However, this should not be the case if one requires that a visual representation has a material substance that is intersubjectively available as a *social object*. Mental images then, have no material and intersubjective character and therefore will not be called representations here. (They may, however, be the referent of a representation.)

The products of a visualization process emanate the characteristics of the (final) medium or successive operations as well as the features of the particular application or instance: the selections and choices of what and how to depict. The end medium or medium of presentation has an important impact on the final appearance of a visual representation. But while each medium has a number of preset characteristics, within each medium there is almost always a great variety in the manner in which a particular referent may be represented (mimetically and expressively). This choice and combination of specific formal options henceforth will be referred to as the "style of execution." The style of execution is only partly determined by the medium. The notion of a wide variety of styles within the same medium is illustrated easily by divergent painterly traditions such as Cubism and Hyperrealism. Similarly, scientists may choose a variety of methods and techniques (ranging from realist to extremely stylized to metaphorical or even phantasmagoric) for depicting a particular subject or idea. These variations in style have to do with genre conventions, cultural schemata, scientific traditions, specific circumstances of the production process, skill, preferences and idiosyncrasies of the maker, as well as the specific purposes the representations need to serve. To complicate matters further, various media and styles may be combined in a particular representation, lending it a highly hybrid character.

Even if the referent is a phenomenon that is accessible through direct observation, this is still not a guarantee for a *faithful* or reliable reproduction, especially if a non-mechanical process such as hand-drawing or painting is involved. This is particularly true if phenomena are drawn from memory after a brief and perhaps exciting encounter (for instance, the early drawings of newly discovered animals). For representations based on first encounters or limited study, it is unlikely that even the scientists know to what extent their representations have a rule-like (general) as opposed to an exception-like quality (deviant). Even if memory is not the major obstacle, perception is always colored by prior knowledge of other phenomena, drawing conventions, cultural representational schemata, matters of skill, and mental processes. The human mind, as Gestalt psychologists revealed, seems very eager to fill in the gaps and to make us see what we want to see. Art historian Gombrich provided a textbook example of this when he commented on Dürer's famous woodcut of a rhinoceros: "He had to rely on second hand evidence which he filled in from his own imagination, coloured, no doubt, by what he had learned of the most famous of exotic beasts, the dragon with its armoured body" (Gombrich 1994, 70–71). But even drawings that are claimed to have been made "from life" ("sur vif"), such as Villard de Honnecourt's *Lion and Porcupine*, may not provide us with depictions that are as faithful as the medium allows, but can be highly ideosyncratic or artistic renderings, which in de Honnecourt's case included a quirky, stylized lion that would better serve heraldic purposes than (naturalistic) representational ones. Gombrich concludes that the claim that it was made "from life" clearly must have had a different meaning at that time (about 1235): "He can have meant only that he had drawn his schema in the presence of a real lion" (Gombrich 1994, 68). In this same classic of art criticism, Gombrich provides another remarkable example of the impact of cultural schemata on the style of a representation when he comments on a strangely oriental-looking illustration of Derwentwater in the English Lakeland by Chiang Yee:

> We see how the relatively rigid vocabulary of the Chinese tradition acts as a selective screen which admits only the features for which schemata exist. The artist will be attracted by motifs which can be rendered in his idiom. As he scans the landscape, the sights which can be matched successfully

> with the schemata he has learned to handle will leap forward as centres of attention. The style, like the medium, creates a mental set which makes the artist look for certain aspects in the scene around him that he can render. Painting is an activity, and the artist will therefore tend to see what he paints rather than to paint what he sees. (Gombrich 1994, 73)

Visual Representational Latitude: Coping with Controlled and Uncontrolled Variation in the Depicted and the Depiction

Though visual media and techniques provide many unique advantages in representing the physical world and in expressing scientific ways of thinking, as soon as a certain level of abstraction or generalization is needed—an essential facet and phase of many scientific undertakings—some distinctive problems may arise.

Verbally, for instance, one can state that a certain bird species may have three to seven spots on its wings. However, when producing a visual representation, one inevitably must draw a definite number of spots. Visuals, unlike verbals, do not offer the option of indicating a range, say "from three to seven." Instead, a choice needs to be made out of the five possibilities when representing in a single drawing a species that exhibits that amount of variation. Moreover, if a photograph is used, one is even forced to show a particular specimen of the species (or a series of photographs of different specimens), of a particular age and sex, and in particular circumstances (habitat, weather, time of day, season). Neither intentional nor more automated (algorithmic) visual images can in themselves express in a simple way the variation (in shape, color, amount) one may expect to encounter in the real world, nor can they fully explain the connections among the particularities of the representation (the variation in the depiction) and what they seek to refer to (the phenomenon and the different forms it can assume in reality).

This multifaceted problem of different types of justified or unjustified variation in scientific representations, combined with both the variation that exists within the species or phenomenon that is depicted and the variation in the depiction of certain phenomena or ideas, could be termed "visual representational latitude." This latitude will be determined partly by the capacities of the medium applied (e.g., intentional versus algorithmic media) in coping with the variation observed within the depicted phenomenon or process, but more importantly by the *manner* in which

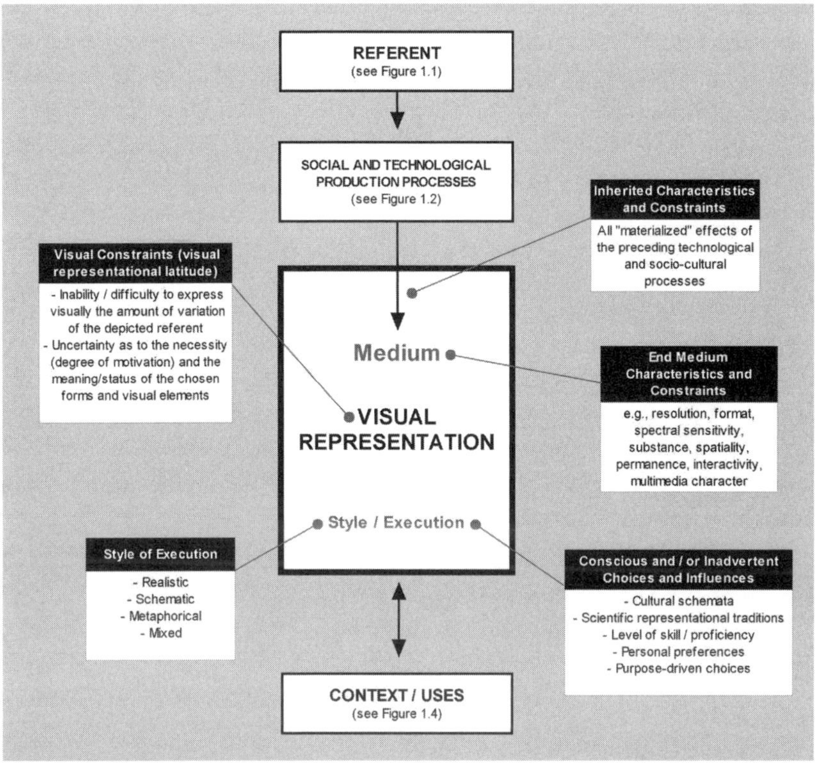

Figure 1.3. Inherited, medium-related, and execution-specific characteristics and constraints of visual representations.

that medium is used, including the stylistic options it offers, the scientifically motivated choices and the various *liberties* that producers allow themselves. The "room for maneuver" or representational margins may or may not be purposeful, functional, and understood.

Visual representational latitude, therefore, is not just a producer's (or sender's) problem, that is, it is not just a matter of deciding how to express variation, of choosing the right level of iconicity or abstraction for a specific purpose. It is also a user's (or receiver's) problem: What kind of variation is to be expected in the real world, and which elements in this particular representation are *motivated* by a perceived reality, and which others are due to specific, intentional, or unintentional choices of the producer, limitations of the medium or larger production context? To what extent is every choice to be interpreted as "necessarily so" or as just "one way of putting it"? If, for instance, a physical phenomenon is depicted as consisting of a core with, say, twenty-three particles revolving around it, one is still uncertain whether this exact number of particles is a unique and thus determining trait of the phenomenon or whether the person who produced the diagram merely meant to indicate that *many* particles are revolving around the core. Similar questions could be raised with respect to the relative distance of the constituting parts of the drawing, their scale, color, and shape, etc.

Verbal comments (e.g., in the form of an extended *legend*) are one way of making sure that users know what they are looking at, what codes are being used, what semiotic variation is being employed, and what representational claims are put in effect by the representation. Another way is to develop further a visual language of scientific representation, which in a sense restricts the ways in which visual elements may be employed but at the same time enables a more visual and less ambiguous form of information transfer and expression.

4. Types and Contexts of Use: Matters of Encoding and Decoding

Representational Constraints

Representations cannot serve adequately just any purpose or intent. Various significant relationships exist between the type of referent, the production process, the medium, and the types of uses and claims that can be attached to them. Visual representations must have the necessary *properties* to fulfil certain functions or uses. These properties, for that matter, refer not only to the characteristics of the medium that is employed but also to the broader contexts of both production and use.

Mitchell distinguishes between two types of representational "constraints" or, put differently, two factors that both the producer and user will have to take into account when trying to apply visuals successfully in a communication and cognition process. First, there is what he calls *"representational commitment,"* by which he means that certain techniques are (more) appropriate for recording certain things and less suited or even totally unsuited for recording others: "different medical-imaging techniques—CT, ultrasound, PET, MRI, and so on—are committed to acquiring different types of data about bony and soft tissue diseases and physiological activities, and so are used for different diagnostic purposes" (Mitchell 1992, 221). Similarly, black-and-white photography may offer the right kind of detail to study naturally occurring phenomena in a social context and thus may be an ideal tool for anthropologists and sociologists, but in some instances this representational choice will be less than adequate. This will be the case, for example, when documenting trends in fashion, home decoration, or the like, where the use of color embodies essential information, or when a detailed account of processes is required, which can only be achieved by means of a continuous record of moving images. A second requirement that Mitchell puts forward is that a visual representation "must have the correct type of *intentional relationship to its subject matter*" (Mitchell 1992, 221, italics added). Some examples may help illustrate the importance of this requirement: The picture of an escaped convict may help police track down that particular individual, but his facial characteristics cannot be used to identify other individuals with criminal tendencies before they can actually commit a crime. A scan of a pathogenic heart may serve as a diagnostic tool to help one particular patient, but that is not to say that it is the most appropriate representation for use in a general biology textbook. A picture of a young blond girl living in Norway could be used as a visual representative of the Norwegian population, and as such she would be *interchangeable* with thousands of other Norwegian girls. However, if one uses a particular picture of a young blond girl on a missing person poster, it is not necessarily the girl who bears the greatest resemblance to the picture who is being traced, but the actual girl who is represented in the photo, even if she might today no longer resemble that image at all. This implies an indexical, not just an iconic relation!

Mitchell's dual requirement that particular representations should be (1) "fit for the particular uses to which they are put" (representational capability) and (2) be "about the right sort of thing" (intentional positioning) may remind us of the long-established scientific requirements of *representativity* and *validity*, and indeed can serve as an important element of their much-needed visual translation. Both terms underscore once again the paramount importance of distinguishing clearly between the varying functional capabilities of different types of visual representations in science. Moreover, they may help overcome the idea that visual media have intrinsic representational qualities irrespective of their use and production context (e.g., the false but persistent view that photography is always better at describing than drawings, or that paintings are always more expressive than photographs, and that the latter are highly "objective" records).

Yet the same medium types of representation may serve a great many purposes and they may entertain widely divergent relations with the depicted matter. Furthermore, a particular visual representation that was made for a specific purpose may be suitable for other purposes, even for some that were not envisioned at the time of production. However, in most cases one needs to know exactly how the images or visual representations came about and what their broader context of production was before one can assess their validity for those other purposes. The use one can make of a representation is determined, to a considerable extent, by its generative process (choice of visual medium and broader production aspects, choices regarding style, selection and preparation of subject, normative systems) vis-à-vis its intended use. So, insofar as this is possible, a predetermined purpose should guide the production process. Some purposes (e.g., the exploration of a naturally occurring phenomenon) may require an indifferent, detailed account of particularistic data in their specific context, while others (e.g., educational aims) may be served better by highly stylized and synthetic representations highlighting only the essence of a more general phenomenon. So the medium and techniques in part will determine the uses that can be made of a representation, but even representations produced with the same medium or technique may have widely divergent intents and representational positions.

Kinds of Intents and Purposes

The intents and purposes of visual representations in scientific discourses are manifold. For one thing, natural phenomena might be visualized for the purpose of *further analysis:* to make a diagnosis, to compare, to describe, to preserve for future study, to verify, to explore new territory, to generate new data. Representations that serve these primary purposes often will be algorithmic in nature and they may have only an *intermediate* function, since they are primary *data.* Visual representations that have no material referent may serve primarily to *facilitate concept development* or to uncover relationships, evolutions (e.g., through charts of all kinds) and, in general, to make the *abstract more concrete* and thus more accessible for further inquiry. Forms of externalized thinking (conceptual graphs) may be useful both on an intra-personal level (for example, to guide researchers in a dialogue with themselves) and an inter-personal/inter-specialist level (to exchange ideas in an early stage, to invite feedback, or to prompt co-operation from peers).

Visual representations not only serve analytical and intermediate purposes, but they are often also used to *summarize or synthesize* empirical findings or a theoretical line of thought. Thus they may provide an overview, display results in their spatial organization or conceptual relations, or clarify the textual or numerical part. More synthesized or purposefully assembled visual representations in science generally serve to facilitate knowledge transfer in a variety of ways and seek to communicate with diverse audiences. They can illustrate, demonstrate, or exemplify features, relations, and processes, or provide mediated experiences in ways that are adapted to the audience (which may vary from highly specialized to lay audiences).

Many visual representations intentionally or inadvertently will embody an implicit or more explicit view on or argument about what is being presented visually, through the many elements and choices that make up the representation. This expressive function of scientific visualizations need not be a problem as long as it is duly acknowledged and, if so required, explained. As intentional forms of communication and through the selection and formal execution of the representations as well as by their thoughtful arrangement in the broader context of an article—a presentation or a multimedia product—visual representations will attempt to exert a certain amount of persuasion. Often, receivers or users of the representation will be invited, seduced or even compelled, in subtle ways, to adopt the views of the sender and to perform the preferred actions (to believe, give approval, appreciate, change opinions, donate money, or support morally). For those reasons, but also for the more acclaimed function of cognitive transfer and education, a visual representation may perform the function of an eye catcher, a means to arouse and maintain attention and interest, or even to entertain the reader/spectator (and thus bring them into the right mood for acceptance). Some aspects of a visual representation in science may even perform no other function than to appeal to the aesthetic feelings of the receivers or just be an expression of the personal aesthetic preferences of the maker. These latter functions, though not readily associated with a scientific discourse, are not necessarily detrimental to the mission of a scientific undertaking, as long as these traits do not interfere with the more fundamental functions of data or cognitive transfer, and on condition that transparency is provided.

While one can never be complete in the listing of possible functions and intents of a scientific visual representation, this brief discussion of functions demonstrates that the idea that scientific visualizations and representations are solely meant to generate and present *objective* data or to facilitate pure cognition should be abandoned. It should be clear that most functions and intents that are found in human communication in general also will be found in scientific representation, though some functions and

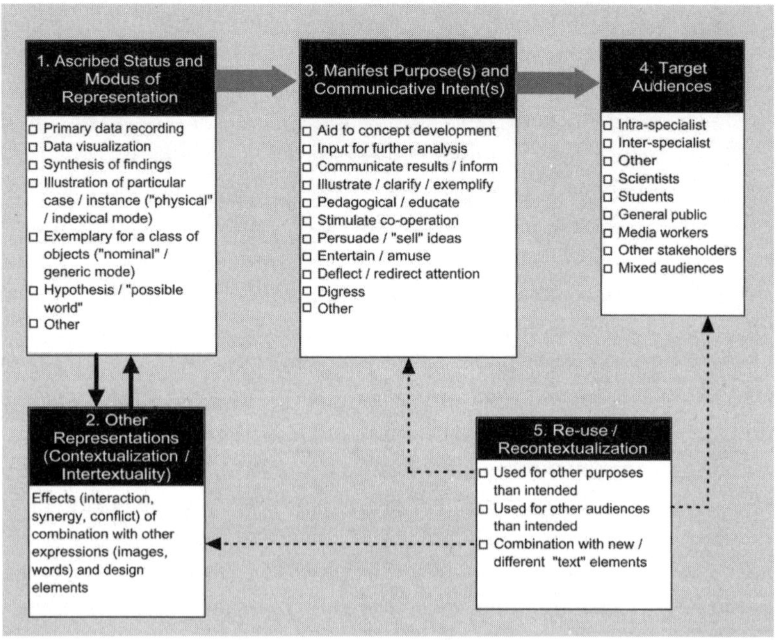

Figure 1.4. Representational status, context, and use: connections between representations, purposes, and audiences.

Chart flow: Visual representations with a particular ontological status (1), in interaction with other representational and presentational elements (2) may be used for different purposes and intents (3) with specific target audiences (4) in mind. Subsequent uses (5) may involve new goals (3) and new *textual* combinations (2) for different audiences (4) and may challenge the ontological status and the representational *fit* of the original presentation (1).

intents obviously will serve a more central role while others will not feature prominently or may be intended to perform an auxiliary function. Moreover, it should be clear that any visual representation used as part of a scientific discourse will serve and combine different functions at the same time, whether intentionally or unintentionally. These purposes may be scientific in a narrow sense, but they may also have to do with intents that lie outside the realm of the acknowledged scientific purposes, such as to serve vested interests of persons and institutions (see, for example, Lynch and Edgerton 1988). Finally, it should be stressed that the different functions that are embodied by aspects of the visual representation may be read or *decoded* in many different ways by different receivers (based on their intents, experience, formal background, etc.) in different contexts and over time. This need not always pose a problem. As scientific work routinely is characterized by many different skills, viewpoints, and goals spread over as many actors, the vagueness and multi-interpretability and multi-usability of the same representations may at times even play an important role in mitigating possible tensions among these heterogeneous contexts of actors, viewpoints, and functions (specialized and non-specialized audiences: scientists, designers, sponsors, management, etc.) and promote co-operation and mutual understanding. Star and Griesemer (1989) have coined the concept "boundary object" to denote such interfaces between multiple social worlds that serve to facilitate the exchange of skills, knowledge and materials among different social actors (Roth and McGinn 1998, 42). For instance, a map of an area, which is always a highly reductive representation, may be a "boundary object" to the degree that it is accepted and used by multiple parties to serve their needs. It then becomes a communication device across social and disciplinary boundaries.

5. Conclusion: Developing Visual Scientific Literacy

The principle underlying this chapter is that representations and representational practices may be extremely helpful in developing, clarifying, or transmitting scientific knowledge. However, when not produced and used with extreme care and competence, they may create at least as much confusion and misunderstanding. If one considers scientific representations and the ways in which they can foster or thwart our understanding, it is clear that a mere *object approach,* which would devote all attention to the *representation* as a free-standing product of scientific labor, is inadequate. What is needed is a *process approach:* Each visual representation should be linked with its context of production. Moreover, it cannot be understood sensibly outside a particular and dynamic context of use, re-use, and reception. However, given the great many types of referents, representational techniques, purposes, and uses, it seems fair to assume that the vast consequences of this requirement are hardly grasped by the growing number of people who produce and use visual representations on a daily basis.

Scientists should develop a sensitivity for the wide variety of visual representational practices and products and the many ways in which they can be deployed in scientific discourse. Furthermore, a real set of skills is needed in order to be able to *assess* the usability of given representations based on a thorough knowledge of their generic processes, and to be able to *produce* visual representations with the required representational and expressive properties in relation to their purpose(s). Visual representations invariably have a strong communicative function, certainly with regard to the originator (e.g., to guide his or her thinking, or to serve as data for further analysis), but often also toward a variety of specialized and non-specialized audiences. Unconsciously applied and/or unmotivated use of aesthetics and unexplained use of certain conventions are a potential hazard, while well thought out and reflexive use of aesthetics, formal choices, and well-explicated representational codes and conventions may create hitherto not fully exploited opportunities to further scientific knowledge building and communication. Modern technology offers many complex ways of generating images, but few users have a clear understanding of all the steps involved. To counter this emerging "black box syndrome," it is clear that scientists need to keep track of new media technologies to the extent that they offer new ways of looking and (not) knowing.

This complex set of requirements involving specific knowledge attitudes and skills may be understood as a specific kind of visual literacy for scientists. Visual literacy for scientists therefore can be defined as a reflexive attitude (throughout the production process), a specific body of knowledge, and even a certain level of proficiency or skill in assessing and applying specific characteristics (strengths and limitations) of a particular medium, and awareness of cultural practices (codified uses, expectations) and the actual context of use (including the *cultural repertoire* of the intended audience). In other words, a visually literate scholar should be aware of the impact of the social, cultural, and technological aspects involved in the production and handling of representations, as well as the different normative systems that may be at work and how they exert a determining influence on the eventual appearance and the usefulness of representations.

Visual scientific literacy shouldn't just imply establishing a clear division of labor (every person keeping to his or her trade) and then linking together those various types of expertise, as in fact they need to be merged rather than developed and applied according to a separate logic for each specialized aspect. The different normative systems (e.g., scientific, technical, creative, cultural) that are employed consciously or unconsciously need to be combined skillfully with a view to the ultimate purpose of the representation. While expertise obviously cannot be accumulated endlessly in one and the same person, a serious effort should be made at providing a unifying framework whereby each contributor develops a knowledge about and sensitivity for the bigger whole. What they should not do is lock themselves up in their own area of expertise, as hardly any choice that is made along the way is without epistemological consequence.

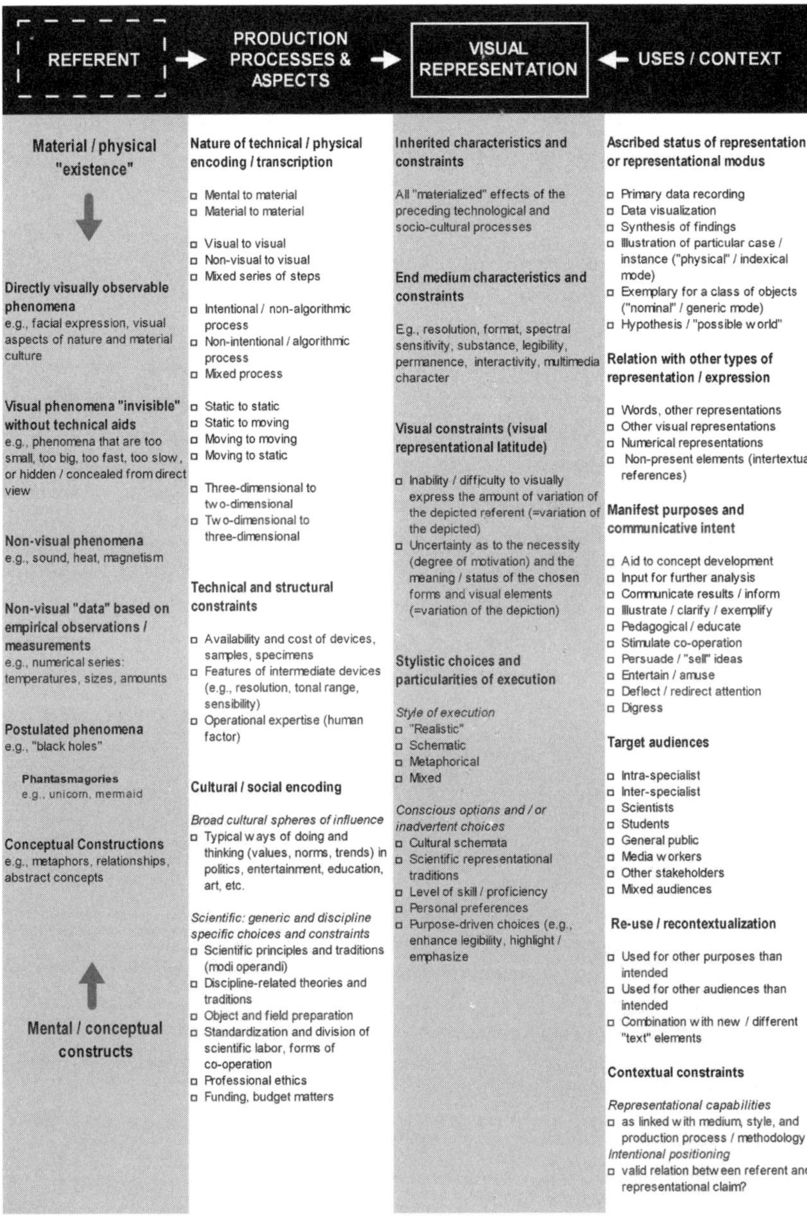

Figure 1.5. A conceptual framework for analyzing and producing visual representations in science.

The aspects and issues that have been discussed in this chapter may serve as a theoretical framework for the thoughtful production of visual representations in science or they may be used as a tool to assess critically the appropriateness of different aspects of particular representations. Such a framework may prove useful in examining the complex interdependencies that exist among the nature of the referent, the social, technological, and cultural context of production, the choices with respect to medium and style of representation, and the purposes and uses that need to be achieved. Figure 5 is an attempt to summarize and in a very limited way visualize the elements and arguments of this framework as gradually developed from figures 1 through 4. While these schematic representations may help to map the complex issue of visualization and visual communication in science, to some extent they will remain inadequate for visualizing the complex interrelations among its constituent parts. Nor will they ever be complete and fully self-explanatory in any of their dimensions. Interestingly, while these figures may serve to clarify and promote insight, they inevitably will also obscure and hinder insight to some degree, but this in itself actually illustrates one of the arguments put forward in this chapter. Visual representations always will be used to enlighten and broaden our understanding, while at the same time, they will obscure it. It is the task of visually literate scholars to make sure that the enlightening aspects gradually gain the upper hand.

References

Gombrich, E. H. [1960] 1994. *Art and illusion: A study in the psychology of pictorial representation.* London: Phaedon.

Goodwin, C. 2001. Practices of seeing visual analysis: An ethnomethodological approach. In *Handbook of visual analysis,* by T. van Leeuwen and C. Jewitt. London/Thousand Oaks/New Delhi: Sage Publications.

Latour, B. 1987. *Science in action. How to follow scientists and engineers through society.* Cambridge: Harvard University Press.

Latour, B., and S. Woolgar. 1979. *Laboratory life: The social construction of scientific facts.* London: Sage.

Lynch, M. 1985a. *Art and artifact in laboratory science: A study of shop work and shop talk in a research laboratory.* London: Routledge & Kegan Paul.

———. 1985b. Discipline and the material form of images: An analysis of scientific visibility. *Social Studies of Science* 15: 37–66.

Lynch, M., and S.Y. Edgerton. 1988. Aesthetics and digital image processing: Representational craft in contemporary astronomy. In *Picturing power: Visual depiction and social relations,* by G. Fyfe, and J. Law. London: Routledge.

Mitchell, W. J. 1992. *The reconfigured eye*. Cambridge: MIT Press.

Pasveer, B. 1992. Shadows of knowledge. Making a representing practice in medicine: X-ray pictures and pulmonary tuberculosis, 1895–1930. Ph.D. diss., University of Amsterdam.

Pauwels, L. 1996. *De verbeelde samenleving: Camera, kennisverwerving en communicatie*. Leuven/Apeldoorn: Garant.

Rosenblum, B. 1978. *Photographers at work, a sociology of photographic styles*. New York/London: Holmes & Meier Publishers, Inc.

Roth, W-M., and M. K. McGinn. 1998. Inscriptions: Toward a theory of presenting as social practice. *Review of Educational Research* 68: 35–59.

Snijder, J. 1989. Benjamin on the reproducibility and aura: A reading of "The Work of Art in the Age of its Technical Reproducibility." In *Benjamin. Philosophy, aesthetics, history*, edited by Gary Smith. Chicago: University of Chicago Press.

Star, S. L., and J. R. Griesemer. 1989. Institutional ecology, "translations" and boundary objects: Amateurs and professionals in Berkeley's museum of vertebrate zoology, 1907–39. *Social Studies of Science* 19: 387–420.

2

The Production of Scientific Images

Vision and Re-Vision in the History, Philosophy, and Sociology of Science

Michael Lynch

Until the late 1970s, historians and philosophers of science paid scant attention to visual images. This is no longer the case, as there are now many excellent studies of pictures, diagrams, graphs, and other scientific images. The increased attention to visual images is part of a more general interest in the detailed contents of scientific texts, discourses, and practices. For many social historians and sociologists of science, visual images are less significant, in themselves, than are the practices that compose and establish the evidential significance of such images. These practices include setting up and preparing observational fields and materials, building and operating observational equipment, processing and reprocessing images, designing and selecting exhibits for presentations and publications, and discussing what such exhibits reveal or demonstrate. The themes of tacit knowledge and scientific craft have been familiar to social scientists for decades, but the more recent studies are distinguished by the specificity with which they approach historical cases, and by their attempts to respecify the methodical basis of discovery, observation, experiment, and representation (Garfinkel 1991).

In this chapter, I will address the question of what studies of visualization might be about. I aim to clarify what the topic of visualization can and

This chapter initially appeared as an article in a special issue on Studies in Visual Analysis, edited by Mike Ball, in *Communication & Cognition* 31(2/3), 1998: 213–228. I thank Mike Ball for his help with earlier drafts of the paper.

should include, to explain why it is an interesting topic, and to suggest why it is important for understanding scientific practices. My initial reason for raising the question of what studies of visualization might be about is that I believe that it presents interesting difficulties that are worth exploring. The problem is not that I am unable to come up with an appropriate word for the topic. For my purposes, "visualization" serves as well as any term to gloss the various practices associated with making objects observable and intelligible. The initial problem is to delimit what the topic includes. In part—but only in part—visualization is about maps, diagrams, graphs, photographs, and other pictorial documents produced as evidence in laboratory research and published as illustrations in scientific texts. Studies of visualization in science are not just about pictures or images, because the documents in question are deeply integrated with a nexus of activities that include observation, measurement, description, analysis, and demonstration. Rather than being a discrete, well-bounded aspect of science, visualization is intertwined with observational and experimental practices, literary representations, and methods for disseminating scientific results. It might be said that an attention to the practices and documents through which researchers visualize phenomena is a way to gain perspective on the whole field of scientific practice instead of singling out a particular aspect or phase of it.

1. Conceptual Preparation

It may seem odd to begin a discussion of visualization with an examination of language. After all, one of the notable features of visual images is that they are not words. However, when we write about visualization it would be foolish to ignore the different connotations and implications of words associated with practices of depicting, imaging, illustrating, demonstrating, and so forth.[1] When considered pragmatically, visualization includes practices for disclosing and publicly exhibiting research objects. A number of other terms have been used to describe such practices. "Discovery," "observation," "representation," and "reference" are some of the most familiar terms, but each of them carries epistemological baggage that might better be unpacked and redistributed. Other terms, including "materializing," "imaging," "illustrating," "depicting," "indexing," and "demonstrating" are used less frequently in general epistemological discussions, but when taken together they alert us to diverse practices that do not fit easily under a single conception of observation or representation. Some of these practices involve pictorial, graphic, and non-verbal images and materials, but it would be a mistake to consider them to be limited to visual objects and inscriptions that are seen with the eyes. Auditory soundings and graphic recordings of non-sensory phenomena such as X-rays are also registered by instruments. Typically, such data are exhibited in pictorial or graphic form on photographic plates, chart recordings, or digital displays, and so in that sense they result in visual imagery, but no conception of a distinct domain of visual form or visual cognition can take into account the various practices of making phenomena pictorially observable and graphically accountable. A whole complex of instruments, practical operations, graphic formats, and communicative media come into play when laboratory scientists visualize and display phenomena.

A lesson to be taken from recent sociological and social-historical studies of science is that visualization should not be equated with sense perception and cognition. The philosophies of science developed by Feyerabend (1975), Hanson (1967), Kuhn (1970), and Polanyi (1967) strongly influenced recent sociology and history of science, but a major difference can be found between their accounts of visualization and some of the later ethnographic and social historical studies. Hanson's discussion of theory-laden observation and Kuhn's (1977) account of acquired similarity relations analyze observational activities in terms of individual perception and cognition. Polanyi (1967) treats observational instruments as extensions of individual sense organs through which the skilled scientist learns to see or feel what the instrument discloses or contacts. Feyerabend (1975) interprets visual representations as evidence of historically and culturally relative cognitive and perceptual tendencies. Although such treatments are instructive, they convey a limited, and in some respects a misleading, understanding of visualization.[2] The various activities that I have glossed under the term "visualization"—materializing, imaging, illustrating, depicting, referencing, indexing, and demonstrating—cannot be reduced to perceptual processes. It may be the case that models of perception make use of technological imagery (e.g., the optics of a camera, and more recently digital image processing hardware and software), but the relationship does not necessarily work the other way around. The collective practices, complex technologies, and pictorial artefacts used in particle physics, X-ray astronomy, or molecular biology include material artifacts and communicative practices that would be misunderstood if they were reduced to perceptual or cognitive mechanisms. Graphs, chart recordings, and digital images are public communicative devices. They have a material organization of their own, and are collectively discussed, reproduced, and worked over. The public accountability of what is shown or disclosed by such displays cannot be traced back to a private idea or a mental representation. Borrowing from Marx, we might say that in modern science the production of displays is a labor process that sets up, and creates materials for, individual perceptions and judgements. In order to understand how such displays are produced and reckoned with, it is necessary to examine complex assemblages of verbal, numerical, geometrical, textual, material, instrumental, and pictorial phenomena.

2. Some Cases and Themes

Historical and ethnographic studies of visualization in science date back to the 1970s. Martin Rudwick's (1976) pictorial essay on the "visual language" that developed in early nineteenth-century geology, and Samuel Edgerton's (1976) book on the interconnections between painterly techniques and scientific discoveries in Renaissance Europe set the stage for many later studies. Rudwick and Edgerton were not the first historians to take an interest in scientific images, but their studies went into far greater depth than had earlier discussions in showing how visual techniques and iconographic devices were featured in scientific research and communication. For example, Rudwick described how early geologists selectively oriented to cliff-faces, mineshafts, and other exposures of strata. The arrangement and ordering of strata in an exposure provided a kind of protolithographic account that was upgraded systematically with diagrams, maps, and geological sections. By tracing the strata, imaginatively extending them into unseen depths, and developing indices and codes, geologists were able to exploit and enhance the legibility of the earth.

Starting in the late 1970s, a series of ethnographies of scientific laboratories examined the day-to-day work of composing, interpreting, and arguing about literary inscriptions and visual renderings of research objects (Knorr-Cetina 1981; Latour and Woolgar 1979; Lynch 1985a). These studies emphasized the significance of such inscriptions and renderings as intersubjective sites of agreement and disagreement in laboratory research, and vehicles for disseminating results in a scientific community. The detailed attention to day-to-day laboratory practices enabled a vivid appreciation of how research objects (e.g., mammalian tissues, plant proteins, hormones) were transformed over time and gradually shaped into graphic and pictorial data that could be analyzed, quantified, copied, printed, and circulated. These descriptions of the temporally extended and collective work of producing objective displays contrasted with the established view of observation and representation as individual, and largely instantaneous, perceptual acts. Instead of being a confrontation between a world and a prepared mind, the research *act* began to resemble a form of factory production in which material inputs were transformed into readable data to be disseminated widely in a community. In addition, close descriptions of the production and dissemination of pictures and graphs exposed the rhetorics and technologies of "virtual witnessing": the pictorial and verbal means through which readers are given a sense of having experienced the events described in a report (Shapin 1984; Shapin and Schaffer 1985).

These case studies, along with many others by historians of science, art historians, literary scholars, philosophers, and social scientists, addressed a variety of different themes, but one lesson that can be drawn from them is that visualization is not a single kind of practice or process. Visual images can play a number of different roles in an activity or text. It is sometimes, but not always, pertinent to think of visualizing as a matter of disclosing, exposing, uncovering, or revealing phenomena that were hidden from view. Optical instruments like microscopes, telescopes, and specialized cameras enable images to be made of things that are too small, too far away, too slow moving, or too fast developing to be seen or noticed with naked eyes, but while such images may provide prototypes for one widely recognized mode of scientific visualization, they do not exhaust the field. Some figures, like the line drawings of duck-rabbits and reversible cubes in perceptual psychology texts, act as templates for elucidating perceptual effects.[3] When a reader views the image appropriately, it elicits optical effects. A reversible cube is not just a line drawing of a three-dimensional figure, nor is it an icon that signifies "reversal"; it is a textual artefact with which the viewer interacts to produce a visible effect. Other figures, like the conceptual diagrams in social theory and cognitive science texts, visualize what the text already says without the use of such figures, and it would be inappropriate to imagine that the sketches resemble material objects or that the arrows and geometric figures in a conceptual diagram stand for concrete vectors and shapes. Such graphic figures seem to function more as translations of verbal arguments than as illustrations or evidentiary supplements, and in a few cases they are ornaments that add little to what a text otherwise says (Lynch 1991). Moreover, pictures, graphs, and models that represent distinct natural phenomena like planetary orbits, cellular membranes, geological strata, antibodies, or molecular structures invariably include features that do not depict what a hypothetical human would see if she was far enough away, small enough, or otherwise in a position to see the phenomena (Cambrosio, Jacoby, and Keating 1993; Francoeur 1997; Gilbert and Mulkay 1984).

A useful analogy can be made between the pragmatic uses of linguistic expressions and the various uses of graphs, models, and pictures in scientific work. Just as Wittgenstein (1958) disrupted philosophy's obsession with referential meaning by making a topic of the pragmatic uses of language in different language games, studies of visualization can disrupt traditional obsessions with representation by investigating an array of pictorial and graphic *games.* It may seem confusing at first to consider such diversity, but one reason for doing so is to clear up confusions that arise from forcing heterogeneous orders into a single conceptual framework. When such confusions give rise to heated debates that have no clear prospect of resolution—as seems to be the case with the realist-constructivist debate—a greater degree of clarity may be precious indeed. So, let us begin with the theme of construction.

Construction, Craft, Art, and Science

There is no question that visualizing research objects involves intensive constructive work. It also can be appropriate to speak of artistry and craft. Some scientific practices are fairly direct adaptations of pre-existing artistic skills and industrial crafts (Edgerton 1991; Sibum 1995). In larger projects, the craft can take a less romantic, more industrial form. So, for example, in contemporary diagnostic firms and forensic organizations in which DNA profiling techniques are employed, different grades of technician are hired to perform routine tasks, some of which are designed to require limited, if any, understanding of molecular biological principles (Jordan and Lynch 1998). Depending upon the circumstances, the production of observational data can be likened to a highly skilled craft, a cottage industry, or a factory operation with a complex division of labor between managers, staff scientists, technicians, and production workers.

The analogy between laboratory practice and artistic or industrial forms of constructive activity becomes contentious when it is paired with arguments that question or deny the referential reality of any or all scientific representations. Examples of such arguments can be found throughout the constructivist literature, but they are especially prominent in the *classic* ethnographies of scientific laboratories by Latour and Woolgar (1979) and Knorr-Cetina (1981). For example, after noting that scientific representations materially differ from the objects they depict, Latour and Woolgar (1979, 237) assert that whatever scientists can know of their objects is a consequence of the way they represent them. The only way they can see or know the object is through the medium of the representation. Although Latour, Woolgar, Knorr-Cetina, and other ethnographers of science no longer trade in such simplistic metaphysical inversions,[4] such statements are now firmly entrenched in *postmodern* critiques of scientific realism. Just as the distinction between fictional and non-fictional writing collapses in postmodern literary theory, so does the distinction between representative and non-representative artefacts collapse in postmodern conceptions of science (Ward 1996). Studies of visualization are often cited in support of the alleged collapse, but they also can support contrary arguments. I will mention two related arguments briefly.

The first argument has to do with the relationship between constructivist arguments and the vernacular idioms of laboratory science. In disputes among scientists—both the kind that occur among colleagues at the bench, as well as the more formal exchanges at conferences, in letters, and in articles—an emphasis on the construction of one or another aspect of data often is connected with substantive uncertainty, doubt, and challenge. The term "artefact" often is used in reference to an inadvertent product of laboratory procedures that is separate from phenomena of interest.[5] Arguments about what is or is not an artefact are commonplace in laboratories (Lynch 1985a). When reading the relevant research literature or listening in on meetings among co-practitioners, it does not take long to find questions and debates about possible artefacts that may confound experimental results. A practical concern with artefacts is more than a matter of argument; it is part of the work itself. So, for example, when astronomers at image processing centers work with digital data, they attempt to separate source radiation from noisy backgrounds, and more generally they attempt to distinguish immediate conditions of observation from the stable properties of the objects observed (Lynch and Edgerton 1988). The adequacy of such work can be challenged at many points, and often is challenged. The practices and arguments are technical and circumscribed; they are not about *reality* in general. But, when social scientific researchers challenge the reality of scientific facts—both particular facts and facts in general—and argue that all facts are artefacts, it may seem as though they are raising a familiar kind of challenge to the technical adequacy of research results (Lynch 1993, 78ff; Sharrock and Anderson 1991). But since social constructivists are rarely in a position to challenge research results on technical grounds, it can seem unclear what kind of challenge they are raising.

The second argument has to do with a distinction between constructions that enable laboratories to visualize phenomena and the artefacts that inhibit or confound visualization. For example, in electron microscopy, a sequence of "renderings" is accomplished to prepare anatomical specimens (Lynch 1985b). Laboratory animals are bred for uniformity and systematically modified during experiments; their tissues are extracted, stained, counter-stained, plastinated, and then sectioned, in order to prepare them for viewing; constituents in the electron micrographic field are positioned, outlined, coded, and counted. All of these practices turn the specimen into a fragmented, abstracted repository of data. However, the formats, codes, labels, and other inscriptions that facilitate the production of data are distinguished from the splodges, grains, scratches, plastic deformations, blurred membranes, and other artefacts that confound interpretations and ruin the appearance of data. The distinction is not absolute, and making it is often problematic in particular cases, but it is relevant to the way laboratory researchers produce and analyze their research materials. Even if one were to assert that practitioners' efforts to recognize and eliminate artefacts reflect an "ideology of representation" (Woolgar 1988), or that they represent an incessant effort to "purify" the primordial heterogeneous field (Latour 1993), such assertions would be far too general, absolute, and abstract to come to terms with the material production of data. The preparation of particular data displays is answerable to relative considerations: What kind of object is being studied? What is the state of the art? What are the limits of the instrumentation? What counts as a precise enough measure under the circumstances? What are the consequences of error, and what alternatives are available for investigating the object? Even if one were to argue that these questions presuppose the possibility of representational *truth*—a possibility that can only be presupposed and never proved—a metaphysical rejection of that possibility does not, and cannot, come to terms with the specificity of the representational activities in question. Consider an analogy with debates about the analysis of magical practices. In a well-known essay, Peter Winch (1970) objected to descriptions of "primitive" magical practices that likened them to scientific experiments. Such "experiments" supposedly lacked the essential element of falsifiability. The persistence of magical beliefs was "explained" by saying that such beliefs were reinforced by spurious confirmation and perpetuated in the face of contradictory results. Winch argued that magical rituals and scientific experiments were different forms of life, and that it is inappropriate to use the rules of the one to understand the other. To treat a magical ritual as though it is a primitive experiment would be like insisting that a score should be kept in order to properly understand a game in which keeping score is specifically absent, and even avoided.[6] Winch's arguments were controversial, and some philosophers dismissed them as relativistic, but they were taken on board by social constructivists. What constructivists are reluctant to acknowledge, however, is that Winch's arguments undermine their own conflation of metaphysics with empirical investigation. It is one thing to reject the ultimate possibility of representational *truth,* but quite another to reject a particular research finding by saying that what appears to be a discovery is actually an artefact of the instrumentation. The possibility of a (relatively) adequate account is a constitutive feature of the latter form of argument. A realist, anti-realist, or irrealist metaphysic cannot engage, let alone evaluate, the argument's adequacy. So, instead of objecting to general constructivist arguments for being relativistic, one might wonder if constructivist arguments are less relativistic than the activities they purport to explain.

Enframing

Descriptive or field sciences are often distinguished from laboratory sciences. In standard methodological terms, field studies are naturalistic, descriptive, and non-mathematical, whereas laboratory projects employ experiments, fancy equipment, and mathematical models. Several studies of visualization have shown that, while the distinction between laboratory and field remains relevant, the pragmatic and semiotic connections between lab and field may be more interesting than the distinction itself. When taken together with ethnographic studies of sciences in which the *field* is comprised of laboratory specimens and/or instrumental data, such studies suggest a radically different picture of observation and representation than is given by the familiar philosophical conceptions of minds confronting worlds and words referring to objects. The picture we now have is of material chains and relays involving many hands and many devices that link the laboratory to the field. In addition to being an alternative picture of scientific practice, it may offer a way out of the impasse between realist and constructivist versions of representation.

To summarize a complicated topic quickly, I will introduce a mundane conception of the Heideggerian theme of "enframing" (Heidegger 1977). Devoted readers of the Heideggerian corpus are likely to find my use of the theme banal, but in any case what I include under "enframing" are the various jobs associated with placing a natural object (or an entire field of objects) into a material/literary matrix that facilitates the systematic work of collecting, recording, comparing, and other forms of mathematical ordering. Such ordering often extends from the beginning to the end of a naturalistic project. The links between the naturalistic field and the abstractions of laboratory science include the following:

- *Ready-mades.* In many projects, the natural field is found to be already ordered, as a kind of tableau vivante or "ready-made" graphic ensemble. Examples include geological exposures with particularly well-aligned strata (Rudwick 1976), and a stratified region of the mammalian brain (Lynch 1985b) that enabled systematic reading and manipulation of the tabular order.

- *Cultivations and enhancements.* Natural fields rarely are allowed to remain fallow for long. Instead, they are emplotted with grids that transform the ground surface or the archaeological pit into a map or graph (Lynch 1985b; Goodwin 1994). Tags are put on plants and radio transmitters on animals to keep track of their identities and movements. Boundaries of fields and objects are clarified and enhanced to facilitate their recognition and categorization. Such practices are not limited to science, as they can be found when police investigate a crime scene and collect samples that are transported to a forensic lab. It often is said that observation is theory laden (Hanson 1967), but this is too crude a formulation to describe the various ways in which objects become identity laden, category bound, and graphically embedded. Methods of enframing are not closely tied to particular theories or disciplines, as similar procedures for gridding, marking, counting, and enhancing boundaries can be found in numerous fields, many of which do not pretend to be sciences.

- *Abstraction and restoration.* In some cases, visualization involves a progressive abstraction or simplification of research materials. Specimens in the field progressively become reduced to extracts, indices, graphic points, and other literary products. These, in turn, facilitate calculation, comparison, standardization, circulation, and theoretical analysis. As Latour (1995, 181) observes, what the field specimen loses in specificity and materiality it gains in legibility and (relative) universality. Although many cases do exemplify progressions of this sort, abstraction is not the only game in town. Yoxen (1987), for example, describes how ultrasound imaging involves a progressive naturalization of the electronic image, so that the form of a foetus emerges with greater clarity from a noisy field, and Lynch and Edgerton (1988) describe how astronomers progressively naturalize electronic data taken from telescope detectors. With procedures and programs similar to those used in digital photo restoration, the astronomers progressively would *restore* images that, in the case of X-ray images, were only *seen* with a detector. Raw data taken from a detector was purified progressively, as flaws in the detecting instrument were compensated for, and extraneous sources of background light were subtracted from the data frame. But the product was not an abstraction from an original object. Instead, it was held to be a more natural depiction of *the physics* of a source object that had appeared only in a confused way in the raw data.

Chains and Paper Trails

In many field studies and also in criminal investigations, evidence is extracted from the field (or crime scene) and brought to a lab for analysis. Evidence often undergoes a series of transformations, starting at the site itself and continuing through a series of renderings, many of which occur within a lab. In criminal forensics, the terms "continuity" and "chain of custody" are used to denote the methods for certifying that samples analyzed in the lab are the same as those collected from a crime scene or taken from a suspect. A declared match between DNA profiles of suspect and crime scene samples has no evidentiary value if a break in the chain of custody can be demonstrated. A vast array of contingencies can be cited in support of such breaks, and an equally vast array of precautions and procedures is designed to certify in particular cases that no such breaks occurred. Continuity describes a material identity: the blood collected at a crime scene is the *same* material from which the DNA analyzed in the lab was extracted. This material identity is certified by an array of tags, bar codes, and forms, which constitute a paper trail, in which signatories attest to proper work, orderly transfers, and adequate controls from the initial point of collection through the entire chain of custody. The legal auspices of forensic investigation make perspicuous a feature of chains in many other field and laboratory sciences: material identities are entrusted to bureaucratic procedures, records, and audits.[7] As Latour (1995, 190) puts it, perhaps too concretely: "If interrupted at any point, [the chains] cease to transport truth—cease, that is, to construct, and to conduct it. Reference is a quality of this chain in its entirety, and is no longer *adequatio rei et intellectus*. Truth circulates here like electricity through a wire, so long as it is not interrupted." This image is slightly misleading, in so far as it implies that *truth* travels from one end of the chain to the other, like electricity in a wire. Instead, what one finds is a series of relays in which different materials (blood in a vial, an extracted DNA pellet, an electrophoretic gel, an auto-radiograph, etc.) are transferred from hand to hand, place to place, machine to machine, and machine to paper. The links in the chain are overlapped by numerical indices, bar codes, signatures, and other elements of the paper trail. Referential truth is not an essence that is transported concretely from beginning to end; rather, it is a contingent, certified assessment of a sequentially ordered and routinely organized production.

3. Conclusion

In this chapter, I emphasize what is distinctive about recent social and historical studies of visualization in science: they are about practices that make things visible and accountable. This emphasis differs from an interest in a discrete domain of visual representation, because it includes the practices of rendering things and fields into representative subjects. The emphasis also differs from interpretations of visual representations as surfaces on which hidden ideological, metaphysical, cognitive, and cultural orders are projected. It is always possible—indeed, it is quite easy—to treat pictorial and graphic renderings as representations that reflect antecedent motives, interests, identities, agencies, structures, processes, or mechanisms. But, while much can be said for the idea that graphs, photographs, and so forth do not simply reflect natural properties, much also can be said for the idea that they do not simply reflect properties of mind, society, or culture. The analogy with a mirror, and related conceptions of images as projective surfaces the details of which can be traced back to discrete cultural standpoints or cognitive processes, is no more or less problematic than the idea that such details can be traced back to natural attributes and processes. What is most distinctive about visualization is its intermediary status. Visual images are intersubjective: they stand between persons, and between persons and things; they are both material and symbolic; they integrate things with projects; they incorporate verbal references into their frames and supply scenic contexts for interpreting them.

Visualization is not simply a mediation between an independent world and a human perceiver. An image does not stand between the thing itself and a knowing subject (Latour 1995, 181). The selection of specimen sites, the re-ordering of fields into grids, the operations of tagging, encoding, and marking—these activities reach into the world, arrange it in a pose, and cultivate its accountability. These artifices do not necessarily negate the world's natural accountability, although in some cases they are blamed for distortions and illusions. Instead, they are the very means through which natural order is revealed, ordered, compared, encoded, and quantified. Consequently, in answer to the question of what studies of visualization are about, we can say that they are about the production of scientific reality.

Notes

1. The same applies to the different verbs associated with visual perception (Coulter and Parsons 1991).
2. Ian Hacking's (1983) emphasis on the extent to which representation is a matter of actively engaging and intervening in a material complex, as opposed to understanding a visual spectacle or discursive formulation, has greatly influenced more recent discussions. For a study that systematically develops an interventionist view of scientific activity, see Rheinberger (1992).
3. Perceptual effects also can be elicited when a text is read with a prism. Bjelic and Lynch (1992) experiment with a series of cards originally devised by Goethe to refute Newton's theory of color. The cards are black-and-white figures that must be viewed through a prism, and the series of cards demonstrates an alternative way to generate the color green by mixing yellow and blue fringes. The color fringes are seen along black/white edges when the cards are viewed appropriately with a prism.
4. See Amann and Knorr-Cetina (1990); Woolgar (1990); and Latour (1995).
5. Artefacts are not necessarily unnatural in origin. A radiation source that occludes or obscures another source can be held responsible for cosmetic defects, misidentifications, miscalculations, and so forth. The intrusive effect is reciprocal to the phenomenon of interest.
6. A similar impasse arises in anthropological analyses of the economy of gift exchange (Bourdieu 1977).
7. These remarks on chains of custody are based on an ongoing study in collaboration with Ruth McNally, funded by the Economic and Social Research Council: *Science in a Legal Context: DNA Profiling, Forensic Practice and the Courts*. For an original philosophical treatment of *trails* as irreducible conditions of reference, see Cussins (1992).

References

Amann, Klaus, and Karin Knorr-Cetina. 1990. The fixation of (visual) evidence. In *Representation in scientific practice,* edited by Michael Lynch and Steve Woolgar. Cambridge: MIT Press.

Bjelic, Dusan, and Michael Lynch. 1992. The work of a (scientific) demonstration: Respecifying Newton's and Goethe's theories of prismatic color. In *Text in context: Contributions to ethnomethodology,* edited by Graham Watson and Robert Seiler. London: Sage.

Bourdieu, Pierre. 1977. *Outline of a theory of practice.* Translated by Richard Nice. Cambridge: Cambridge University Press.

Cambrosio, Alberto, Daniel Jacoby, and Peter Keating. 1993. Ehrlich's "beautiful pictures" and the controversial beginnings of immunological imagery. *Isis* 84: 662–699.

Coulter, Jeff, and E. D. Parsons. 1991. The praxiology of perception: Visual orientations and practical action. *Inquiry* 33: 251–272.

Cussins, Adrian. 1992. Content, embodiment and objectivity: The theory of cognitive trails. *Mind* 101: 651–688.

Edgerton, Samuel Y. 1976. *The Renaissance rediscovery of linear perspective.* New York: Harper & Row.

———. 1991. *The heritage of Giotto's geometry: Art and science on the eve of the scientific revolution.* Ithaca, N.Y.: Cornell University Press.

Feyerabend, Paul. 1975. *Against method: Outline of an anarchistic theory of knowledge.* London: Verso.

Francoeur, Eric. 1997. The forgotten tool: The design and use of molecular models. *Social Studies of Science* 27: 7–40.

Garfinkel, Harold. 1991. Respecification: Evidence for locally produced, naturally accountable phenomena of order, logic, reason, meaning, method, etc. in and as of the essential haecceity of immortal ordinary society (I)—an announcement of studies. In *Ethnomethodology and the human sciences,* edited by Graham Button. Cambridge: Cambridge University Press.

Gilbert, G. Nigel, and Michael Mulkay. 1984. *Opening Pandora's box: An analysis of scientists' discourse.* Cambridge: Cambridge University Press.

Goodwin, Charles. 1994. Professional vision. *American Anthropologist* 96: 606–633.

Hacking, Ian. 1983. *Representing and intervening: Introductory topics in the philosophy of natural science.* Cambridge: Cambridge University Press.

Hanson, N. R. 1967. An anatomy of discovery. *The Journal of Philosophy* 64: 321–352.

Heidegger, Martin. 1977. *The question concerning technology and other essays.* New York & London: Harper & Row.

Jordan, Kathleen, and Michael Lynch. 1998. The dissemination, standardization, and routinization of a molecular biological technique. *Social Studies of Science* 28: 773–800.

Knorr-Cetina, Karin. 1981. *The manufacture of knowledge: An essay on the constructivist and contextual nature of science.* Oxford: Pergamon Press.

Kuhn, Thomas S. 1970. *The structure of scientific revolutions.* Revised ed. Chicago: University of Chicago Press.

———. 1977. Second thoughts about paradigms. In *The structure of scientific theories,* edited by F. Suppe. Urbana: University of Illinois Press.

Latour, Bruno. 1986. Visualization and cognition: Thinking with eyes and hands. *Knowledge and Society: Studies in the Sociology of Culture Past and Present* 6: 1–40.

———. 1993. *We have never been modern.* Translated by Catherine Porter. London: Harvester Wheatsheaf.

———. 1995. The "pédofil" of Boa Vista: A photo-philosophical montage. *Common Knowledge* 4: 144–187.

Latour, Bruno, and Steve Woolgar. 1979. *Laboratory life: The social construction of scientific facts.* London: Sage.

Lynch, Michael. 1985a. *Art and artifact in laboratory science.* London: Routledge and Kegan Paul.

———. 1985b. Discipline and the material form of images: An analysis of scientific visibility. *Social Studies of Science* 15: 37–66.

———. 1991. Pictures of nothing: Visual construals in social theory. *Sociological Theory* 9: 1–22.

———. 1993. *Scientific practice and ordinary action: Ethnomethodology and social studies of science.* New York: Cambridge University Press.

Lynch, Michael, and Samuel Y. Edgerton. 1988. Aesthetics and digital image processing: Representational craft in contemporary astronomy. In *Picturing power: Visual depictions and social relations,* edited by Gordon Fyfe and John Law. London & New York: Routledge.

Polanyi, Michael. 1967. *The tacit dimension.* London: Routledge and Kegan Paul.

Rheinberger, Hans-Jorg. 1992. Experiment, difference, and writing, part I: Tracing protein synthesis. *Studies in History and Philosophy of Science* 23: 305–331.

Rudwick, Martin. 1976. The emergence of a visual language for geological science, 1760–1840. *History of Science* 14: 149–195.

Shapin, Steven. 1984. Pump and circumstance: Robert Boyle's literary technology. *Social Studies of Science* 14: 481–520.

Shapin, Steven, and Simon Schaffer. 1985. *Leviathan and the air pump.* Princeton: Princeton University Press.

Sharrock, Wes, and Bob Anderson. 1991. Epistemology: Professional scepticism. In *Ethnomethodology and the human sciences,* edited by Graham Button. Cambridge: Cambridge University Press.

Sibum, Otto. 1995. Reworking the mechanical value of heat: Instruments of precision and gestures of accuracy in early Victorian England. *Studies in the History and Philosophy of Science* 26: 73–106.

Ward, Steven C. 1996. *Reconfiguring truth: Postmodernism science studies, and the search for a new model of knowledge.* Lanham, Md.: Rowman & Littlefield.

Winch, Peter. 1970. Understanding a primitive society. In *Rationality,* edited by Bryan Wilson. Oxford: Basil Blackwell.

Wittgenstein, Ludwig. 1958. *Philosophical investigations.* Translated by G. E. M. Anscornbe. Oxford: Basil Blackwell.

Woolgar, Steve. 1988. *Science: The very idea.* London: Tavistock.

———. 1990. Time and documents in researcher interaction: Some ways of making out what is happening in experimental science. In *Representation in scientific practice,* edited by Michael Lynch and Steve Woolgar. Cambridge: MIT Press.

Yoxen, Edward. 1987. Seeing with sound: A study of the development of medical images. In *The social construction of technological systems: New directions in the sociology and history of technology,* edited by Wiebe Bijker, Thomas Hughes, and Trevor Pinch. Cambridge: MIT Press.

3

Representing or Mediating

A History and Philosophy of X-ray Images in Medicine

Bernike Pasveer

> A good radiograph in some respects may be said to resemble a painting by Turner. Without intuition or previous study the one is almost as incomprehensible as the other, but as we gaze the wealth of detail rises before our vision until finally we are able to interpret the meaning of streaks and shadows that to the untrained eye are meaningless.—J. F. Halls Dally (1903, 1806)

When Wilhem Konrad Röntgen invented the phenomenon he named X-rays at the end of 1895, and published about it in the early days of 1896, the response was enthusiastic.[1] "All the world seems to have gone off on two crazes—bicycles and X-rays. With the latter I have myself been badly bitten," Dr. Silvanus Thompson, president of the newborn British Roentgen Society, wrote in 1897 (Thompson 1897, 29). There were few people who doubted that the new rays provided spectacular ways of visualizing inner structures of objects—bodies among them. There were even fewer people who dared doubt that the hazy grey photographs of transradiated objects really depicted those objects. It was, they thought, simply a matter of carefully adjusting the technology's key ingredients: the technology itself, the object to be transradiated, and the spectator, before a "wealth of detail rises before our vision" (Dally 1903, 1806).

Contemporaries had good reasons to believe that the new images somehow represented a reality *out there*. One of these good reasons was that the new technology employed a photographic technique. Quite quickly after its invention, photography had been endowed firmly with the characteristic of being a truthful record of something that had really happened, that existed in front of the lens at the time the picture was taken. The X-ray image of Mrs. Bertha Röntgen's hand in Röntgen's first publication on the rays surely encouraged a principal reading of the images similar to the reading of other photographs: as the visual evidence of the presence, the existence, out there, of the phenomenon depicted.

Another explanation for contemporaries' trust in the "new photography's" ability to depict reality may be found in the epistemological and methodological paradigms of the day. Daston and Galison (1992) have described the constitution of what they call "mechanical objectivity" from the late eighteenth century onward. "[T]his ideal of objectivity attempts to eliminate the mediating presence of the observer: some versions of this ideal reign in the judgments that select the phenomena, while others disparage the senses that register the phenomena, and still others ward off the theories and hypotheses that distort the phenomena" (Daston and Galison 1992, 83–84; see also Maas 2001).

Mechanical objectivity consisted, grosso modo, of rendering redundant the scientist's senses in the production of scientific knowledge, through two movements that reinforced one another. The first was the invention of technologies that *substituted* a machine for the human sensory apparatus: "[machines] produced not only more observations, but better observations" (Daston and Galison 1992, 84). The second movement consisted of the invention of technologies that would render visible things that the human eye simply could not see; technologies, that is, that did not substitute but rather *were* a new sensory apparatus themselves, and that so produced new objects of inquiry. X-ray technology was one of these latter apparatus.[2] It was believed to produce objective renderings of existing, given phenomena invisible to the eye; it promised to shed a light on these phenomena as they appeared on screen in ways not seen before; and it did so without the senses of the scientist acting as mediators in the dispositive.

And there was more that acted in favor of the X-rays. The new photography opened up the *living* human body as an object about which new knowledge might be produced. It has often been pointed out that medical doctors had always wanted to *see* inside the living body (see, for example, Reiser 1978; Rosenberg 1987). Before the nineteenth century, the insides of the living body were hardly a source of medical knowledge at all; it came from dead bodies and atlases made on the basis of dissections and prepared body parts only. But in the course of the nineteenth century, various kinds of instruments were being devised that allowed medical doctors to "inspect the architecture of the internal organs during life with the ease and clarity possible after death" (Reiser 1978, 45). X-ray images promised to be the best of these technologies of inspection.

Thus the new photography entered the stage endowed with excellent credits. However, these credits alone cannot explain how the images came to represent anything in particular, how they became signifiers of specific parts and diseases of the human body's insides. But when looking for an answer to that question, not much will be found. From historians of radiology (and of other medical visualization technologies), we learn that "there were meanings implicit in the images" that X-ray workers had to "learn to decode" (Brecher and Brecher 1969, 110); or that "as the machines improved, physicians gradually pushed back the veil in front of the internal organs" (Kevles 1997, 3).[3] It is as if the images are *encoded by nature;* hence historians concentrate on what happens next: the formation of the profession of radiologists, the diffusion of the technology into the medical world, etc. An epistemology is taken for granted in which what is the case *precedes* the work of science (which ultimately is considered to be an act of unveiling). It is assumed that images or other forms of knowledge claims *represent* what was always already the case.

Technologies of visualization appear in such accounts as transparent, neutral intermediaries between nature and scientists. Although it is unmistakable, even for historians, that an X-ray image is *not* a simple, true-to-nature representation of a body's insides, but only a particular perspective on that body, it seems impossible to go beyond this world "out there" waiting to be discovered. Accounts from within this epistemology thus always render invisible what is required for an object to be depicted at all, and what is required before an image refers to anything in particular. They ignore the work of mediation in which an image comes to have a referent, and instead pretend that the scientific image delivers a clear and unmediated access to the referent *out there.*

These ignored questions are central to this chapter. I show what must transpire in order for X-ray images to have a referent—for "il faut bien comprendre qu'une image nue n'a pas de referent" (Latour 1993, 164). I argue that the objects to which (medical) images refer are always mediated by the very instruments and methods used to depict them. I wish to describe what these processes of mediation consist of—and consisted of in the early decades of the "new photography." I will recount how X-ray images of the human lungs, in particular, were encoded, and so acquired a referent, an object to signify. I argue that decoding the images in terms of a patient's body could only be done because of the specifics that went into the process of encoding in the first place—processes in which a body is brought in line with the potentialities of the technology and vice versa.

There was no body "out there" waiting to be unveiled with the new technology—there never was or will be. This body had to be crafted carefully out of historically specific other bodies, in order to become a referent for the images. Only through the subtle work of rendering an object that matches an imaging technology, and vice versa, could and can images be read in terms of that object: as referring to it, as signifying it.

So my account is historical in that it empiricizes much of what has been left untold in many other accounts of (medical) imaging.[4] The account is also conceptual in that it proposes another view of how medical images refer to bodies or diseases. What follows is an analysis of three different strategies of encoding. In the conclusions, I wrap up my argument, and distill some of the conceptual points and consequences from it.

1. Encoding X-ray Images

How was the content of the new X-ray images constructed? How did they come to refer to the human lungs *in vivo?* We will see that modes of likeness were being crafted between the shadows as they appeared on the screen and various other, stable, well-known renderings of the lungs as they were used in medical practice. No lungs *out there,* that is, but lungs *in here*. For the images, three *working objects* were set up, which came to *fill in* the content of the photographs, and which allowed for a reading of the images in terms of a patient's inner body and its ailments. The three procedures are mutually constitutive and at work simultaneously.

Anatomy Dead and Alive

In the course of the nineteenth century, pathological anatomy gradually had become a stable way of presenting and defining the body: the skeleton, the place of the organs, its normal and its pathological states. Anatomy ordered the normal and the pathological, it taught students of medicine, and increasingly it came to order relations between patient and doctor, directing both in their activities when a complaint was uttered.[5]

Before that, in the seventeenth and eighteenth centuries and earlier, a physician determined what was wrong with his patient in other ways. He would listen to a patient's complaints, he would observe the outward appearance of the body and its fluids, and in exceptional cases he might examine the body physically. Generally speaking, the patient's story, told to the physician at the patient's home or through the mail, was by far the most important source for a diagnostic judgment (Duden 1991). A Galenic body prevailed, consisting of different fluids the relative amounts of which would constitute that body's healthy and diseased states.

Due to the specific ways in which medical doctors increasingly sought to use the dissected body, relations among the patient's narrative, the physician's skills of listening, smelling, and seeing, and the disease diagnosed, changed from the late eighteenth century onwards (Sawday 1995; Stafford 1991; Wilson 1987). Dissecting the dead was not new, of course, nor was the interest in the boundaries of the normal and the pathological. But where pre-nineteenth-century physicians might have dissected bodies or used an anatomy atlas to find evidence of disease, they generally had refrained from *dissecting the living,* that is, from relating the dead to the living body. In pre-nineteenth-century medical discourse and work, the location of a disease seldom was included in its aetiology. In the course of the nineteenth century, this looseness, this disconnectedness between anatomy and physiology, between structure and function, changed significantly. Doctors systematically began to relate findings of autopsies with disorders found in living patients, on the basis of symptoms told and signs found at examination. The names of Bichat and Morgagni are bound closely to the onset of this *anatomization* of the living human body. Reiser notes that it was based on two premises: "[f]irst, disease often leaves telltale footprints on the tissues of the body; second, study of these footprints is the best way for physicians to verify judgments made on the living, and the key to achieving clinical excellence" (Reiser 1978, 17).[6]

In 1761, the Viennese physician Auenbrugger was one of the first clinicians to describe a method of physical examination he called "percussion." "He percussed by striking the chest 'by the fingers brought close together and stretched out straight'" (Adam 1902, 500) in order to find altered body structures in a living patient. Auenbrugger's treatise on percussion did not attract much attention in his own day. But the activity of looking for pathological changes inside the living patient through physical examination persisted and became more popular in the nineteenth century, as did the employment of various material technologies that were devised in order to enhance the physician's senses of sound and sight. One of the first such instruments was the stethoscope, invented in 1816 by Laennec. He used the instrument to "examine the sounds emanating from the chests of healthy and of ill patients, most of which he could relate to anatomical faults he found by dissecting their bodies after death" (Reiser 1978, 25). For Laennec, anatomical lesions were "the least variable and most positive of the phenomena of local disease" (quoted in Reiser 1978, 25), and this foundational role of pathological anatomy for clinical medicine was to strengthen greatly during the nineteenth century.

Anatomy and pathology, and the increasing methodical, material, and causal links with the anatomical sites of a patient's complaints found during physical examination, changed the human body in medicine. A different model of disease partly replaced the old one in which the patient's own story, her social surroundings, and the physician's passive observations had been decisive. Active intervention on the body, performed more and more inside hospitals, and virtually impossible without the physical presence of the patient, came to reconstitute the normal and the pathological, as well as the patient's experiences, and the relations between his body's inside and outside.

X-ray workers were entrenched within this renewed relation of dead and living anatomy and pathology, and they tried to render the content of the new photography in accord with it. The anatomical body was used to *map out* both the normal and the pathological X-rayed body, and to define its boundaries. The anatomical body was used as the primary frame of reference according to which early X-ray workers sought to see and *read* the shadows.

Mobilizing the anatomical body as an encoder of the images was not simply a matter of looking carefully though, or of simply *decoding* one body image in terms of another. When, in May 1896, the first handbook on the rays was published, its frontispiece showed a chest skiagram of "'The Human Heart, *in situ*' but upside down. The author of the book proudly called it 'the greatest triumph up to date in this direction'" (quoted in Posner 1971, 234). The event shows the looseness of the connections between shadows and flesh: *that* they were related was beyond dispute; *how* the one would re-present the other, was to be established.

In all publications about new X-ray organs and body parts, we can see the explicit effort to translate the shadows in terms of an anatomical body, and to *read out loud* before the reader the correct translation between image and body. When the first X-ray picture of the heart itself was published in 1896 in the *Archives of Clinical Skiagraphy*,[7] the editor of the brand new journal, Sidney Rowland, wrote: "In Plate XVII, a reproduction of the first photograph of the heart attempted will be seen. There the outline of the organ in its contained sac is perfectly distinct. Below it one can see a curve representing the diaphragm, surrounding it there are the ribs, and in the neck there is an indication of the large blood vessels" (Rowland 1896, 31). In just the same way, J. Ed. Stumpff, medical superintendent at the Amsterdam University Hospital, described precisely how he had made his lung pictures, and what, if produced properly, a normal lung picture would reveal: "the spinal column in the centre, . . . clavicula and scapula, . . . more obscured the shoulder-belt, . . . the bending of the ribs, . . . of the organs of the thorax one sees the heart and the upper limit of the liver as dark shadows opposite to the relatively clear lungs" (Stumpff 1899, 960–961). Many others too dwelled extensively upon the translation of the shadows in terms of the known and stable anatomical body in order to construct an X-ray image of that body. The anatomical body was used as the *given* to which the images would refer. It was the shadow's working object, and it took little to make it that. But, as I have indicated, this *given* and stable anatomical body was in no way a body *out there:* It was a highly rendered body itself.

X-ray workers did not employ this stable anatomical version of the body only to encode the images. In particular, when they were unable to translate a shadow into an anatomical structure, the anatomical body was modified explicitly in order to produce likenesses: to become as much *like* the image as to be able to be its referent. Two British X-ray workers, Walsham and Beale, in 1901 demonstrated some X-ray pictures of healthy chests, in which, they stated, "the heart, diaphragm, and mediastinal tissues were clearly seen, but in addition certain faint streaks near the left side of the heart, and both sides of the mediastinal shadow. These were shown to be due to the pericardium and visceral pleura" (Walsham and Beale 1901, 150). They had reached that conclusion "by removing the different structures seriatim from the thorax of the cadaver, and finding by successive skiagraphs that they still persisted unchanged when everything except the pericardium had been removed" (Walsham and Beale 1901, 150). In 1902, Walsham wrote of the same image that "in order to test the question" of the vertical streaks that appeared on X-ray pictures of the chest, "four skiagrams were taken of the body of a boy who had died of heart disease the first with the body unopened and placed on the plate in the same position as in the living subject; the second with the chest opened and the heart alone removed, the body lying on its back; the third with the lungs removed, but the pericardium left in position, and some blood allowed to lie in the thoracic cavity; and in the last skiagram the thorax was

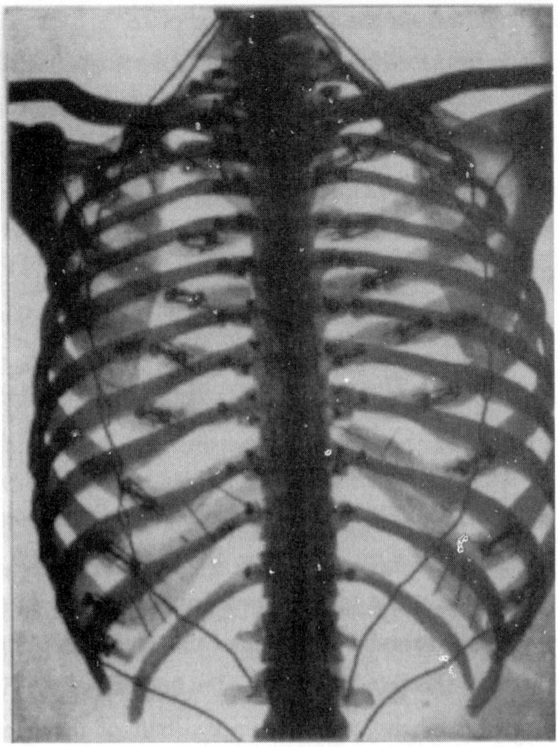

Figure 3.1. X-ray photograph of a skeletonized cadaver. (*Nederlandsch Tijdschrift voor Geneeskunde*, II, 11 November 1899, 972)

carefully cleaned out, the pericardium alone remaining with the upper mediastinal contents undisturbed . . . The recognition of these streaks as normal is of considerable importance in the interpretation of skiagrams of incipient disease" (Walsham 1902, 93).

And Stumpff would set up the thorax of a cadaver, put a lead thread in the right bronchus, X-ray the thorax, and so be able to read another X-ray image of what he supposed to be bronchial shadows: the shadow of the lead thread "corresponded with the above mentioned shadows" (Stumpff 1899, 961).

In America, still another method of creating a working anatomical object for the X-ray images of the body existed. It was called "retrospectroscopy" (Brecher and Brecher 1969, 109). The X-ray worker would X-ray a living patient's body, and make a diagnosis on the basis of the picture. When this same patient would be operated on, or die and the body be dissected, the physician would attend the operation or obduction. He would then make another X-ray of a part of the dissected body, and so have comparable shadows and anatomical lesions, and establish whether his radiological diagnosis was consistent with the findings at the operation or the dissection table. Thus the process of dissection, of taking apart the body, enabled the X-ray worker to construct that body's X-ray anatomy.[8]

This way, shadows and the anatomical body swapped places, and of course this swapping itself indicates powerful trust in the new photography. Still, the anatomical body, both in its *given* state and in situations expressly produced to perform new referents for unknown shadows, was employed to encode the images. Passive acquaintance with the well-known anatomical body, as well as an active interference into dead bodies to make them a proper match for the new images, worked to create likenesses between shadows and anatomical structures.

The body—be it the stable anatomical body or the anatomical body in transition—moved into the position of a working object for the X-ray images to acquire a content, something to refer to, to signify. This body was not a natural body outside any medical frame whatsoever. It was always already a mediated body, a body of knowledge, a body *within*. It acquired specific radiological characteristics as it moved closer to the shadows, but in the end, the images would not refer to a body outside of this web of interrelated translations.[9] On the contrary, they would have this translated body as its referent.

Sound and Sight

Still, the anatomical body could be a working body only to a limited extent. It resisted being brought to life, just as the living body—breathing lungs, beating heart, food being digested—could not always find a counterpart in the still bodies of the dead. This was not problematic when the new technology was to fulfill a role in teaching anatomy to students. "Any one who is watching the progress of anatomy . . . must have noted a very definitive change—almost a revolution—in the outlook of our younger anatomists. It is not difficult to define the nature of the changes which are taking place in our dissecting-rooms. The older school studied and described the *dead* human body; the aim of the newer school is to study and teach the anatomy of the *living* body . . . The discovery of Roentgen rays was the factor which ultimately killed the dead, unimaginative study of the human body," someone wrote enthusiastically in 1915 (Clinical anatomy 1915, 782).[10] The anatomical body was problematic, however, if the rays were to be useful in the clinic for the diagnosis of functional diseases like pulmonary tuberculosis. "Owing to the disappearance of air from the lung at death, there are relatively few conditions which can be confirmed this way," a roentgenologist wrote in 1913 (Lawson 1913, 57).

Technologically speaking, images could be produced of organs in motion, if the fluorescent screen was used together with the use of the photographic plate as a carrier of still images: it produced moving shadows. In this way, the rays promised to visualize the moving insides of the body. "[F]urther advances might be made in parts of the body—such as the heart and the lungs—which it is impossible to maintain in a position of absolute rest; the screen can give accurate information where the skiagraph can only give the net result of the variations in opacity which occur during the time the exposure lasted" (Skiagraphic diagnosis 1897, 99). But just as with the still images, there were no self-evident ways of reading these moving images: The insides of the living body up to then had not been mediated visually. And again, it would be one thing to infer from the positioning of patient and X-ray machine *that* shadows would refer to the area of lungs and heart; it was quite another to say *what* the moving shadows signified in any clinically relevant manner.

Just as X-ray workers made the dead body into a working object to which the new photography could refer, so they worked with the living body and the new machinery. Again, the process went two ways almost simultaneously. On the one hand came a concise reading of the moving images in terms of the results of other, stable methods of physical examination. On the other hand came an equally concise work of mobilizing these diagnostic methods and their results in terms of the shadows. We will move a bit closer to the tubercular lungs to find out about this mutual matching of old and new.[11]

At the time, pulmonary tuberculosis commonly was diagnosed in several ways. A patient's history was taken, her sputum could be examined bacteriologically for the presence of bacilli, and the physician possessed four ways of physically examining someone he thought could be tubercular. Through *inspection* (looking at the build, the form, length, width, and curve of the thorax, the position of collarbones and ribs, etc.), the physician "would be surprised by the number of things seen. Very small disorders of the lungs will report themselves to the examining practitioner in this way" (Nolen 1908, 610). *Palpation*—carefully touching and sensing the thorax—if performed seriously and symmetrically, would give sharp indications of disorders. Inspection and palpation both allowed one to "observe signs which are *indirectly* dependent upon abnormalities of the lungs" (Nolen 1908, 612; emphasis added). *Percussion* (striking the chest by the fingers brought close together and stretched out

METHOD.

Radioscopy		Inspection.
Radiography		Palpation.
Stereoscopic { Radioscopy / Radiography	=	Percussion.
		Auscultation.
Localisation		Mensuration.

RESULT.

Roentgen rays.		*Percussion.*
Brightness	=	Hyper-resonance.
Transradiancy	=	Normal resonance.
Faint shadow	=	Impaired resonance.
Dense shadow	=	Dulness.
Opacity	=	Absolute dulness (tanquam femoris).

Figure 3.2. Tables devised to show the correspondence between X-ray technologies and findings and methods of physical examination and findings. (Dally, *The Lancet,* 27 June 1903, 1800)

straight) and *auscultation* (listening to sounds enhanced by the stethoscope) would result in sounds that were "*directly* caused by the physiological, respectively pathological condition of the lungs" (Nolen 1908, 612, emphasis added).

Although the sounds provided clues to the geography of the lungs, they were not images. In order to be able to show convincingly how the X-ray images of the thorax were related to the sounds of percussion and auscultation, a distinct comparison between the old and the new was sought. Both methods were held to render macroscopical information of the lung's (dis)functioning, as well as a clue to the place where the lungs were infiltrated by lesions caused by the disease. In this sense, the anatomical encoding of the new photography also worked for its potential clinical uses. This was as far as it got. But there would come a time, people thought, when X-rays alone might be the basis for the diagnosis of pulmonary tuberculosis. For the present, however, work needed to be done to bring the pictures "into line with the older methods" (Dally 1903, 1800). Dally, a young X-ray worker, provides us with a telling example of this work, in an article published in *The Lancet.* He devised "the accompanying tables" shown in figure 3.2.

In the first table, five different X-ray methods are listed in the left-hand column, and Dally makes them correspond to five different methods of physical examination. Thus he makes radiology equal to physical examination. In the second table, he compares the density of shadows obtained when making an X-ray image of the lungs with the sounds heard upon percussion. Given the fact that lungs are full of air during inspiration, and that the rays' trajectory would be obstructed by anything but air, a *normal* lung picture would show bones and the contours of the fleshy heart as white shadows, and the lung area as black. An X-ray picture with no shadows apart from the *normal* shading of ribs and internal organs would correspond to the sound of *hyperresonance* upon percussion, a dense shadow to a dull sound, and so on.

Dally subsequently describes the making of a good X-ray picture of the thorax, and then describes an X-ray image of a normal thorax. He discusses shadows with known anatomical referents, and shadows held to be normal without yet having acquired a solid basis in anatomy or physiology. Finally, he describes shadows that are supposed to be pathological. Although Dally argues for the diagnostic superiority of X-rays, the older methods are the explicit framework against which the content of the shadows is determined.

Through comparing sight with sound in the graphic from *The Lancet* reproduced in figure 3.3, Dally sought to define a possible means of translating sounds and shadows into a common image—and subsequently to argue for the superiority of X-rays.

We see, again, a double move being carried out. In order to read the shadows on the image, well-known and firmly established instruments of physical examination of the lungs and their results were being mobilized. The results of these other methods were the obligatory passage point through which the shadows could be interpreted, and their content was constructed out of these results. But whereas at first shadows were read in terms of sounds, next came the effort to *visualize the sounds:* sounds and shadows were, quite literally, made to be like one another inside the convention of the drawn thorax. In figure 3.3, sounds are illustrated in the left row, shadows in the right. Here we see a gradual change in a fixed frame of reference: it is being un-fixed, moved, and changed into a new working object for the images. Results obtained by physical examination acquire analogues of X-ray qualities, instead of (or rather, besides) the other way around.

Halls Dally was not alone in encoding X-ray pictures with the visual analogues of the result of other, well-established diagnostic methods, nor was he alone in trying thus to swap the relative diagnostic value of old and new. Many others actively tried to fabricate resemblances between shadows and sounds. It was only on the basis of the likeness crafted between sound and sight that proposals concerning the differences between the diagnostic potential of sight and sound could appear. In the early years, there had been men who had stated bluntly that "examination by radioscopy and radiography supersedes all other methods" (Discussion 1903, 318). But before images and sounds had been made comparable by Dally and others, no such superiority could be proven. The much contested nature of the new machinery in medicine made the increasing number of physicians who

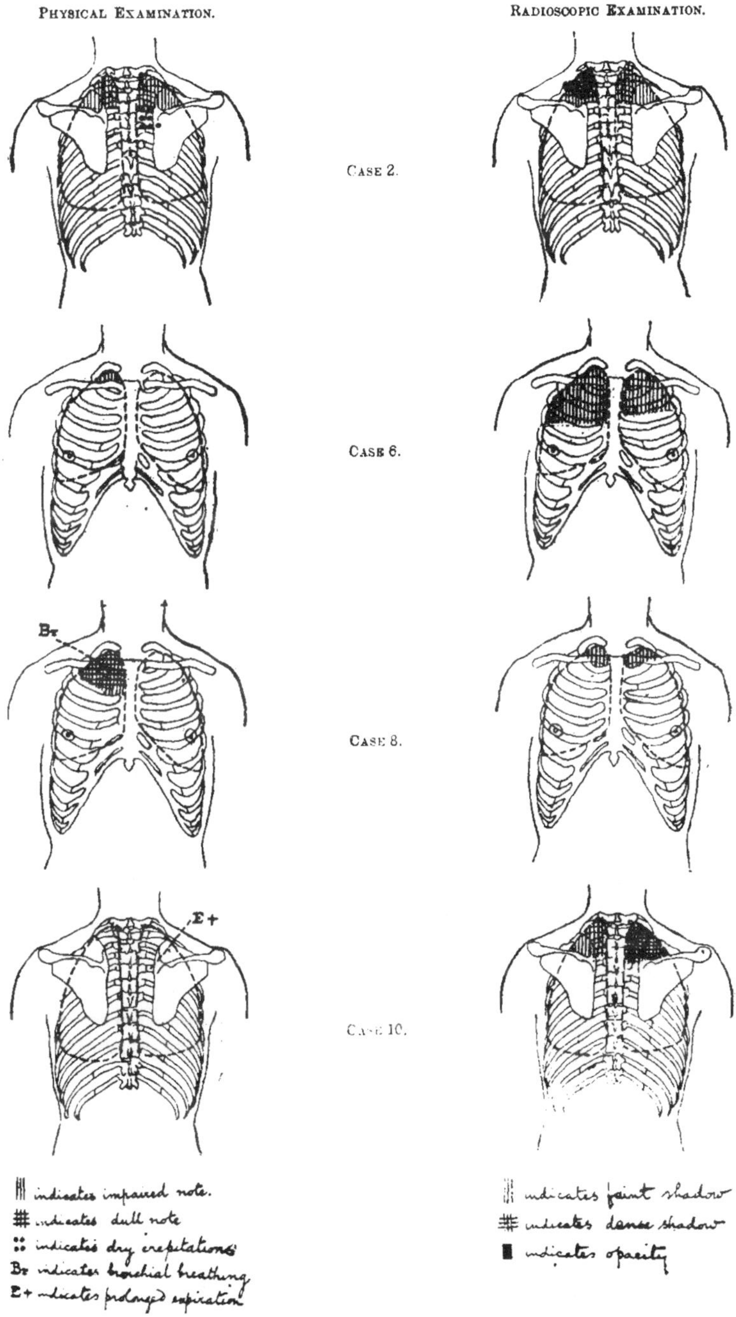

Figure 3.3. A graphic comparing renderings of the results of physical examination to X-ray images. (Dally, *The Lancet,* 27 June 1903, 1805)

liked the new technology careful in their evaluation of the relative merits of the new and the old, and compelled them to search for ways to combine the old and the new instead of bluntly replacing the one with the other. "[T]hat an increase of our diagnostic abilities can . . . be expected, is not difficult to foresee," Stumpff wrote in 1899 (955). And Nolen declared in 1908: "it must furthermore be noted that the radiological examination is an extraordinary objective method. The abnormalities are delineated on the photographic plate and can be demonstrated to everyone" (615).

To the carefully proposed superiority of the images in detecting more and earlier signs of pulmonary tuberculosis were added its other superiorities. X-ray pictures showed the anatomical or pathological conditions of the lungs—just as percussion and auscultation did—but they were reproducible, public, comparable, visual, and due to their mechanical objectivity, less subject to the individual physician's subjectivity. The photographs "yielded objective results that can be controlled by anyone, that can be saved, and used as a comparison with later photographs" (Stumpff 1899, 957–958). As Hugh Walsham had asserted, "[s]imple inspection, palpation, percussion and auscultation open to the examiner a range of signs which the X rays cannot equal . . . [But] such is the supremacy of the eye, and such the importance of things seen, that even after a careful physical examination the skiascopic image comes like a revelation" (Walsham 1900, 88). And even though there would be "numerous cases where the rays will discover nothing which was not appreciable by the more usual methods of physical examination . . . it is satisfactory to have the condition of the internal organs confirmed by sight" (Walsham and Orton 1907, 206).

Again, it was not a given and unmediated human body that was used as the body in terms of which the new images were being *encoded* and could be *decoded*. Rather, auditive renderings of body parts were mobilized to encode the images. This resulted in a swapping of priorities—the X-ray image became the means to *re-code* a physical examination.

Shadows of Shadows

The swapping of priorities between the new technology and the old methods of physical examination was more radical in the clinic than it became in the dissection room. Live anatomy was not replaced by X-ray anatomy.[12]

But the new images did make a claim to clinical relevance, and in pragmatic terms their value would be limited severely if they only could show the same, or even less, than the other methods.

The claim that the rays would make a difference for the study of anatomy and for the clinical diagnosis of several diseases, including lung diseases, came from several directions at the same time. It came, as we have seen, from the technology's ability to provide medical doctors with a more objective view of the inner body, undisturbed by their own subjectivity, undisturbed by the patient's equally subjective complaints. It also came from more specific characteristics of the technology: X-ray photographs could be examined in the absence of the patient, they could be carried around and studied by anyone interested and competent. But it was not a rhetoric of technological progress and objectivity alone that made the new photography strong and promising. It was, as we saw, through the work of encoding the new images in terms of old and well-established renderings of the body, and of transforming the old in terms of the new, that X-ray images acquired specific potential—sometimes even the potential to be the best diagnostic tool available.

To demonstrate the claim that X-ray images could allow more accurate and earlier diagnosis of disease than other methods, a different referent than the renderings of physical examination was needed. And again, we see X-ray workers carefully encoding the images in this way too. The working object now became the X-ray photograph itself. Radiology was, in the words of Reiser, "the transcendent instrument of visualization in medicine" (1978, 58), and it was within this third rendering practice of comparing photographs with photographs that it became so. It is here that one may speak of the establishment of a *roentgenology* of specific organs and diseases, as the various body parts that were X-rayed acquired radiological characteristics. To get there, "[d]ozens, if not hundreds, of chests must be examined with the most careful attention to detail, before one can become expert in grasping all the facts made plain in a carefully carried out X-ray examination. In viewing the X-rays as an instrument of diagnosis, it may be compared with the ophthalmoscope and microscope, both of which instruments require the possession of considerable knowledge and much skill if they are to be used with effect" (Hall-Edwards 1904, 143).

Apart from examining "dozens if not hundreds of chests," an accurate *reading* of an isolated X-ray meant having to compare shadows with shadows. This was done in three ways. One was to compare the left lung with the right—here the double organ could provide a frame of comparison. A second was to compare a patient's chest X-rays over a period of time and so potentially find changes in the lung shadows that could indicate the progression of a pathological process. A third method was to lay side by side the shadow images of different lung patients.

An early example of encoding the images through comparing right with left came from France, and was published in the *British Medical Journal* in 1896. A Professor Bouchard had examined some patients who suffered from pleurisy (the diagnosis had been made with the use of other methods) with the Röntgen rays. "In three male patients with effusion on the right side," the correspondent says, "he observed that the cavity filled with the effusion presented a dark sombre tint that *contrasted with the clear aspect presented by the healthy side*. Moreover when the cavity is not filled with pleural effusion, the apex of the cavity remains clear in aspect, and the sombre tint marks the upper level of the effusion. In a fourth case in which the effusion was no longer present, the pleural cavity was retracted and cast a shadow" (Special correspondence 1896, 1852; emphasis added). This was a kind of in-between rendering of the pictures: the clinical diagnosis was used as a known and stable background against which to assess the pictures, but so was the *healthy side* as seen on the photograph.

Increasingly, roentgenologists tried to read the shadows-as-such, alone, disconnected from or before they were informed of a clinical picture. And they did so by quite methodically describing the procedures through which they arrived at the conclusion that specific abnormalities would have specific, exclusive X-ray qualities. The method was a combination of the three rendering practices described so far: anatomy, physical examination, and the comparison of series of X-ray images among themselves. In his 1899 article on the radioscopy and radiography of the lungs, Stumpff enables us to follow such an attempt at defining the exclusively radiological qualities of the normal and the pathological lung. The different extent to which several media can be transradiated, he states, "is, for the examination of the lungs, to which I wish to confine myself, of greatest importance. For the air-containing lung will fairly easily allow X-rays to pass through it, so that one, when X-raying the chest of a patient, will see the bright intercostal spaces sharply delineated against the relatively dark rib-shadows; however, this brightness decreases as soon as the lungs contain less air. Irrespective of the causes of such increased opacity, be it the formation of tumours, infiltration, retraction, everything will resist the X-rays more than air, and will therefore betray itself through a darkened area on the screen" (Stumpff 1899, 955). Stumpff concludes that "apart from the shadows already mentioned [of the spinal column, the shoulder belt, the ribs, the organs of the thorax] the image of the healthy lung is evenly

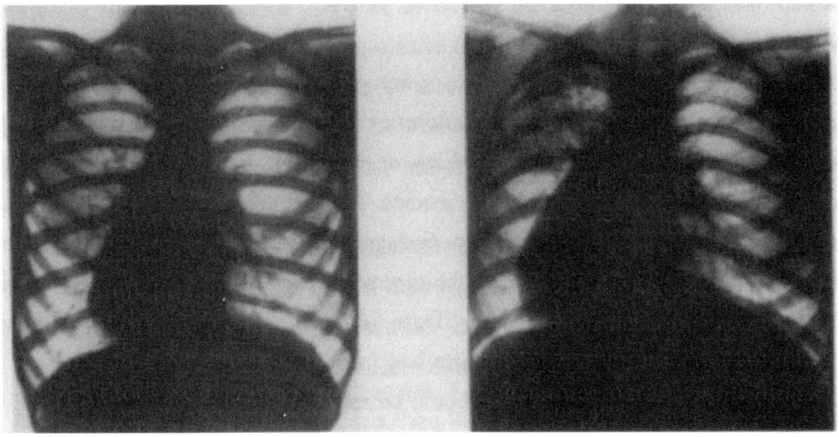

Figure 3.4. X-ray photograph of a normal thorax (left), and of a patient with tuberculosis of the left upper lobe. (Josselin de Jong et al., *Leerboek der Tuberculosebestrijding,* 's Gravenhage, 1926, 102)

clear, *so that we may eventually regard appearing dark spots as pathological*" (955, emphasis added). He goes on to discuss the radiological image of a diseased lung for which "the diagnosis could be entirely made with percussion and auscultation, so that the screening was not absolutely necessary, although through it we obtained a clear *insight into the extent* of the process" (955, emphasis added). And he concludes with a photograph of a patient of whom "with our normal methods of examination *nothing was found* in the lungs . . . *Radiographically the focus in the right upper lobe is clearly demonstrable, whereas continued observation could also establish recovery*" (962, emphasis added).

Stumpff was not alone in proceeding this way. Again and again we see first an image drawn of the normal chest, "for the sake of comparison" (Finney and Watson 1902, 633). "To interpret morbid shadows it was necessary to study those thrown by the normal thoracic tissues" (Walsham and Beale 1901, 151). At lectures given in the early years, X-ray pictures often were shown to demonstrate the appearances of the normal and the pathological, and to compare them explicitly (see figure 3.4). "These cardiac lines were seen more or less *pronounced by all skiagraphs of the healthy chest, and must, therefore, be considered normal*" (Discussion 1903, 318, emphasis added).

In this way, a radiography of the healthy and the diseased body was being designed, with visible boundaries of normal, and, as a consequence, visible indications of the abnormal. In this visual process of encoding the signs of disease with radiological qualities, *swapping* was not the issue—although it helped the swapping described above. That is, the creation of a radiology of specific body parts and processes did enforce the rays' trustworthiness and their diagnostic superiority in an increasing number of cases. What is at issue in this third practice of encoding is a series of simultaneous events. What is crafted here is a solid place for the technology, for images as being superior to sounds, and a stable frame of reference that was to be used increasingly as a prime signifier of pulmonary tuberculosis: its onset, its location, its increase and decrease.

2. Decoding Images: Conclusions

"A Roentgen-ray picture is a silhouette only. The difficulty is to tell whether a shadow is normal or not . . . Thorough anatomical knowledge, the greatest care, and a vast amount of experience as to the different modes of delineation in various projection plates, are needed to form a correct idea of the significance of skiagraphs" (Beck 1901, 52).

We have seen that it is possible to substantiate the claim that the new photography was not already encoded by *nature,* but rather that this encoding consisted of a historically traceable process in which other bodies in use were being mobilized and rendered docile in terms of the new images. The process described here, of aligning bodies and images/imaging technologies, has resulted to some extent in a match that is stable and fixed. The work has been done and need not be done all over again every time a patient presents herself with a specific complaint. The profession, the technology, and our bodies, once connected, are loaded with the traces of this encoding work. X-ray images are not *naked* anymore: they come with this history and they bring it along. They can be *decoded* with the use of that history in terms of the body of a particular patient. But there is always an open space, part of the image to be encoded by the specific body of the particular patient with her particular complaint. It would require another article to describe this final work of encoding, of *dressing-up* the image, but it is important to note here that this final work of making the image signify *my* lungs does not come naturally either.

In order to make an X-ray image of, say, my lungs, my body must be modified so as to acquire the characteristics that make it correspond to the X-ray technology. I must undress, I must remove silver and gold, I must hold my breath for a while and keep still. Only then can the image that results signify my lungs. Only through the encoding work done before combined with this last and always particular phase of matching body and image/technology, can an X-ray image be read in terms of my body.

So decoding an image is very simple—if only we take into account that decoding an image always comes second, not first, and that *second* involves a new body, not an old. Put otherwise: First comes the elaborate, skillful work of encoding medical images in terms of historically specific mediations. Second comes a new body, a body that has the potential of carrying with it phenomena that can be rendered visible through specific technological mediations. In order to refer back to a body, that body too is being loaded with these historical and contemporary characteristics. An image can only signify when it has been transformed by what it refers to, and when what it refers to has been transformed by the imaging technology.

Notes

1. Of course, this way of presenting what happened is controversial: Röntgen is supposed to have *discovered* the rays. But as the word "discover" implies taking away a *cover* that hides a given phenomenon, I prefer the term "invention": Röntgen invented both the cover and a method to dis-cover. Inventing is to be understood not as making up, but as bringing into existence.
2. Daston and Galison do not distinguish these two kinds of technologies—they only consider the first. Snyder, however, points to this second kind of machines: "displacements registered by mechanical monitors and traced by clockwork-driven inscribers fall outside the scope of human sensibility. Consequently, they do not permit even the possibility of human intervention" (Snyder 1998, 380). Although Snyder takes Marey's chronophotography of movement as his prime example, the argument fits technologies like the X-ray quite well. Contrary to Snyder, though, I do not think that the matter is an either/or case: Both kinds of technologies are productive of mechanical objectivity, and of the new epistemology it contained; the core issue is about the appearance of technologies as mediators/translators of nature to science.
3. See Pasveer (2000) for a critical review of Kevles (1997).
4. I gratefully use the work and terminology of, among others, Beaulieu (2000), Daston and Galison (1992), Latour (1993 and 1998), and Lynch (1990). The story of X-ray images of the lungs is based on data gathered some time ago for my dissertation (Pasveer 1992). In part, this chapter reruns an earlier and more primitive version of the argument (Pasveer 1989).
5. Although we now take this anatomical body as a natural, given body, this body did not always exist, nor is its existence a universal phenomenon—not even the term "body" is. Other peoples have different bodies, and what the body is, is never separated from the ways the body is made to be. Sawday recounts the story in which a Melanesian,

asked what the West had contributed to the culture of the islands, answered: "What you have brought us is the body" (Sawday 1995, 11).

6. For some fine *anatomies of anatomy,* for the merging of the surgeon's localized notion of disease and allied nosologies and diagnostics, and the broad, usually global doctrines on health and illness, diagnosis and therapy common in medical circles under the flag of specific, localized and anatomy-based pathology, see Maulitz 1987; Reiser 1978; and Warner 1986.
7. What we now call radiography was called many names in the early days, including Röntography, Shadowgraphy, Skiagraphy, Ixography, Pyknoscopy, Electrography, Scotography, Kathography, Fluorography, Actinography, Diagraphy, and Radiography.
8. The distinction is Wilson's: "dissection, considered as the taking apart of the body, and anatomy, considered as its reconstitution" (Wilson 1987, 63).
9. For a precise anthropology of signifier and signified, see Latour 1993.
10. The quotation suggests that the technology would not be one means among others to represent the dead body, but that it would be anatomy's way of expression. In a similar vein, Marey had considered his chronophotography of movement to be the language of physiology itself. My point is that presenting the representation and its technology as *the real thing* ignores the enormous amount of epistemologically relevant work of isolating the phenomenon of motion from the body through a huge transformation of the body, and of adapting the technology so as to trap the phenomenon. This work does not forbid an image to speak the essence, the core, of the object depicted, but it makes the essence of the phenomenon appear as an achievement, a result of the work that produces it (see Braun 1992; Douard 1995).
11. As occurs with the invention of any new medium or technology, in the early days of radiology prophecies of a revolution in medicine were abundant. Yet, as with any new medium, the only way the new technology could become important and significant was through a careful crafting of continuity between the old and the new. See Manovich 2001 and Bolter and Grusin 1999 for further conceptualizations of relations between old and new media.
12. The Visible Human Project (VHP) has engendered anew a lively discussion over the relative values of dissecting real cadavers versus dissecting the dissected and digitally rebuilt male and female that, up to now, make up the Visible Human Family. See Cartwright 1998; Csordas 1997; Dijck, J. van 2001; and Waldby 1999 for fine analyses of these and other issues concerning the VHP.

References

Adam, J. 1902. Medical diagnosis and modern discoveries. *British Medical Journal* I 1 (March): 499–502.

Beaulieu, A. 2000. The space inside the skull: Digital representations, brain mapping and cognitive neuroscience in the decade of the brain. Ph.D. diss., University of Amsterdam.

Beck, C. 1901. Errors, caused by the false interpretation of the Roentgen rays and their medico-legal aspects. *Archives of the Roentgen Ray* 5 (2): 52.

Bolter, J. D., and R. Grusin. 1999. *Remediation: Understanding new media.* Cambridge: MIT Press.

Braun, M. 1992. *Picturing time: The work of Etienne-Jules Marey*. Chicago: University of Chicago Press.

Brecher, R., and E. Brecher. 1969. *The rays: A history of radiology in the United States and Canada*. Baltimore: Williams & Wilkins.

Clinical anatomy of the gastro-intestinal tract, the. 1915. Review of T. Wingate Todd. *British Medical Journal* I 26 (June): 1085.

Cartwright, L. 1998. A cultural anatomy of the visible human project. In *The visible woman: Imaging technologies, gender, and science*, edited by P. Treichler, L. Cartwright, and C. Penley. New York: New York University Press.

Csordas, Th. 1997. Computerized cadavers. Shades of being and representation in virtual reality. Paper presented at the conference Biotechnology, Culture and Body, 23–26 April, University of Wisconsin, Milwaukee.

Dally, J. F. H. 1903. On the use of the Roentgen rays in the diagnosis of pulmonary tuberculosis. *The Lancet* (June): 1800–1806.

Daston, L., and P. Galison. 1992. The image of objectivity. *Representations* 40: 81–128.

Dijck, J. van. 2001. Virtuele dissectie en digitale kadavers. In *Het transparante lichaam*. Amsterdam: Amsterdam University Press.

Discussion on the use of the Roentgen rays in the diagnosis of pulmonary tuberculosis. 1903. *British Medical Journal* II 3 (August): 313–335.

Douard, J. W. 1995. Etienne-Jules Marey's visual rhetoric and the graphic decomposition of the body. *Studies in the History and Philosophy of Science* 26 (2): 175–204.

Duden, B. 1987. *Geschichte unter der haut*. Stuttgart: Klett-Cotta.

———. 1991. *The women beneath the skin: A doctor's patients in eighteenth-century Germany*. Cambridge: Harvard University Press.

Finney, J. M., and E. J. M. Watson. 1902. A report of cases illustrating the aid of the Roentgen rays in the diagnosis of intrathoracic tumor. *British Medical Journal* I 15 (March): 633–626.

Hall-Edwards, J. 1904. Editorial. *Archives of the Roentgen Ray* 8 (9): 143.

Kevles, B. 1997. *Naked to the bone: Medical imaging in the twentieth century*. Reading, Mass.: Addison-Wesley.

Latour, B. 1993. *La clef de Berlin*. Paris: La Découverte.

———. 1998. How to be iconophilic in art, science, and religion? In *Picturing science, producing art*, edited by C. A. Jones and P. Galison. London: Routledge.

Lawson, D. 1913. X-rays in the diagnosis of lung disease. *Practitioner* 1: 53–58.

Lynch, M. 1990. The externalized retina: Selection and mathematization in the visual documentation of objects in the life sciences. In *Representation in scientific practice*, edited by M. Lynch and S. Woolgar. Cambridge: MIT Press.

Maas, H. B. J. B. 2001. Mechanical reasoning: Jevons and the making of modern economics. Ph.D. diss., University of Amsterdam.

Manovich, L. 2001. *The language of new media*. Cambridge: MIT Press.

Maulitz, R. C. 1987. *Morbid appearances: The anatomy of pathology in the early nineteenth century*. Cambridge: Cambridge University Press.

Nolen, W. 1908. De diagnose der beginnende long-tuberculose. *Nederlandsch Tijdschrift voor Geneeskunde* II 22 (August): 607–619.

Pasveer, B. 1989. Knowledge of shadows: The introduction of X-ray images in medicine. *Sociology of Health and Illness*. 11 (4): 360–381.

———. 1992. Shadows of knowledge: Making a representing practice in medicine: X-ray pictures and pulmonary tuberculosis, 1895–1930. Ph.D. diss., University of Amsterdam.

———. 2000. Affected bodies: On medical imaging technologies and the bodies imagined. Review-essay of *Naked to the bone*, by B. H. Kevles (1997). *Studies in History and Philosophy of Biological and Biomedical Sciences* 31: 659–668.
Posner, E. 1971. The early years of chest radiology in Britain. *Thorax* 26: 233–239.
Reiser, S. J. 1978. *Medicine and the reign of technology*. Cambridge: Cambridge University Press.
Rosenberg, Ch. 1987. *The care of strangers: The rise of America's hospital system*. New York: Basic Books.
Rowland, Sidney. 1896. Editorial. *Archives of Clinical Skiagraphy* 1 (1): 3–31.
Sawday, J. 1995. *The body emblazoned: Dissection and the human body in Renaissance culture*. London: Routledge.
Skiagraphic diagnosis applied to soft parts. 1897. *British Medical Journal* I 9 (January): 99.
Snyder, J. 1998. Visualization and visibility. In *Picturing science, producing art*, edited by C.A. Jones and P. Galison. New York: Routledge.
Special correspondence from Paris. 1896. *British Medical Journal* II 26 (December): 1852.
Stafford, B. 1991. *Body criticism: Imaging the unseen in enlightenment art and medicine*. Cambridge: MIT Press.
Stumpff, J. ed. 1899. Radioscopie en radiographie van de longen. *Nederlandsch Tijdschrift voor Geneeskunde* II 11 (November): 954–963.
Thompson, S. P. 1897. Presidential address to the Roentgen Society. *Archives of Clinical Skiagraphy* 2 (2): 23–30.
Waldby, C. 1999. IatroGenesis: The visible human project and the reproduction of life. *Australian Feminist Studies* 14 (29): 77–90.
Walsham, H. 1900. The Roentgen rays in diseases of the chest. *Archives of the Roentgen Ray* 4 (4): 85–89.
———. 1902. The interpretation of radiographs of the thorax. *Archives of the Roentgen Ray* 6 (4): 93.
Walsham, H., and C. Beale. 1901. Skiagraphy in the diagnosis of chest disease. *British Medical Journal* I 19 (January): 150–152.
Walsham, H., and G. H. Orton. 1907. *The Roentgen rays in the diagnosis of diseases of the chest*. London: H. K. Lewis.
Warner, J. H. 1986. *The therapeutic perspective: Medical practice, knowledge, and identity in America 1820–1885*. Cambridge: Harvard University Press.
Wilson, L. 1987. William Harvey's *Prelectiones:* The performance of the body in the Renaissance theater of anatomy. *Representations* 17: 62–95.

4

The Accursed Part of Scientific Iconography

Francesco Panese (translated by Digby Thomas)

> In the selection of our reports and experiments, we consider that we have been more cautious for mankind than any of our predecessors. For we admit nothing but as an eyewitness, or at least upon approved and rigorously examined testimony; so that nothing is magnified into the miraculous, but our reports are pure and unadulterated by fables and absurdity.—Francis Bacon (1854b, 341)

The emergence of modern scientific practice during the seventeenth century was accompanied by the establishment of a peculiar manner of reproducing and broadcasting knowledge. Experiments carried out in the laboratory were translated into words and images. This allowed them to be communicated to others who we might term "virtual witnesses." These then were capable of reproducing the experiments mentally—generally without actually undertaking them themselves—as though they had nature before their eyes, freed from the contingencies of any particular perspective.

For Francis Bacon (1561–1626), the first task of science was the constitution of a grand "archive of nature" in which all the paragraphs of nature's laws—which he termed its "forms"—would be assembled and organized. In a movement analogous to Bacon's propositions, an important part of the scientific enterprise consisted then in the attempt to translate these forms into images. From that time forward, the practice of science

This paper was previously published in French as "La part maudite de l'iconographie scientifique," *Espaces Temps* 62–63 (1996): 76–89. It is republished here by permission.

has been inseparable from the activity of thinking by and with images. Thus, one way of considering the invention of modern science is as a transformation of our way of looking.

1. A Change in Our Way of Looking

The *Historia plantarum universalis* by Jean Bauhin (1541–1612) was published after his death in 1650 by the Swiss physician D. Chabrey (Bauhin 1650). It is possible to see this work as something truly groundbreaking. Its appearance coincides with the first attempts to establish a universal botany that would include the plants of the New World that were unknown to the classical authors—even though those classical authors remained omnipresent in commentaries. The book's frontispiece illustrated this tension perfectly (figure 4.1). Bauhin and his son-in-law Cherler (who had helped him compile the book) raise a curtain that opens upon the scene of a well-ordered garden answering to the precepts of not only the classical and modern authorities—from the Greek Theophrastus to Conrad Gesner—but also, and in the final analysis, to those of the Almighty himself, whose angels hymn his universal glory. A faithful illustration of the text that follows it, the image presents the visual and ideological synthesis of a system of knowledge in which things seen and things said are linked to one another and give rise to descriptions that seem but orisons to God's glorious Creation.[1]

The way that modern science looks at nature seems to be born of the dissolution of this link. This can be observed in another ideal image, the one that serves to open the *Anatomie animalium* by Gerhard Blaes (Blaes 1681; see figure 4.2). Here we no longer find any trace of the ancients at all, nor of time, nor history. In an open space, the lantern of Reason, armed with a magnifying glass and a scalpel, unveils a bountiful Nature around whom presses a docile crowd of animals. The entrails of these are displayed for us by two innocent cherubs. The right-hand angel rolls them out like a portraitist across a canvas, where they acquire a new visibility. He holds his scalpel—or his lancet—as we would a pen to fix the form of an organ on paper.[2] This allegory of the world in writing deploys an idealized knowledge in which description *immediately* adheres to the real without recourse to any other register.

The difference between the two frontispieces is replicated in the bodies of the texts. Bauhin's *Historia* begins with the apple *(De malis),* veritable first fruit of the Bible, around which are aggregated commentaries and topological outlines. The rudimentary xylographs fail to convey the particularities of the

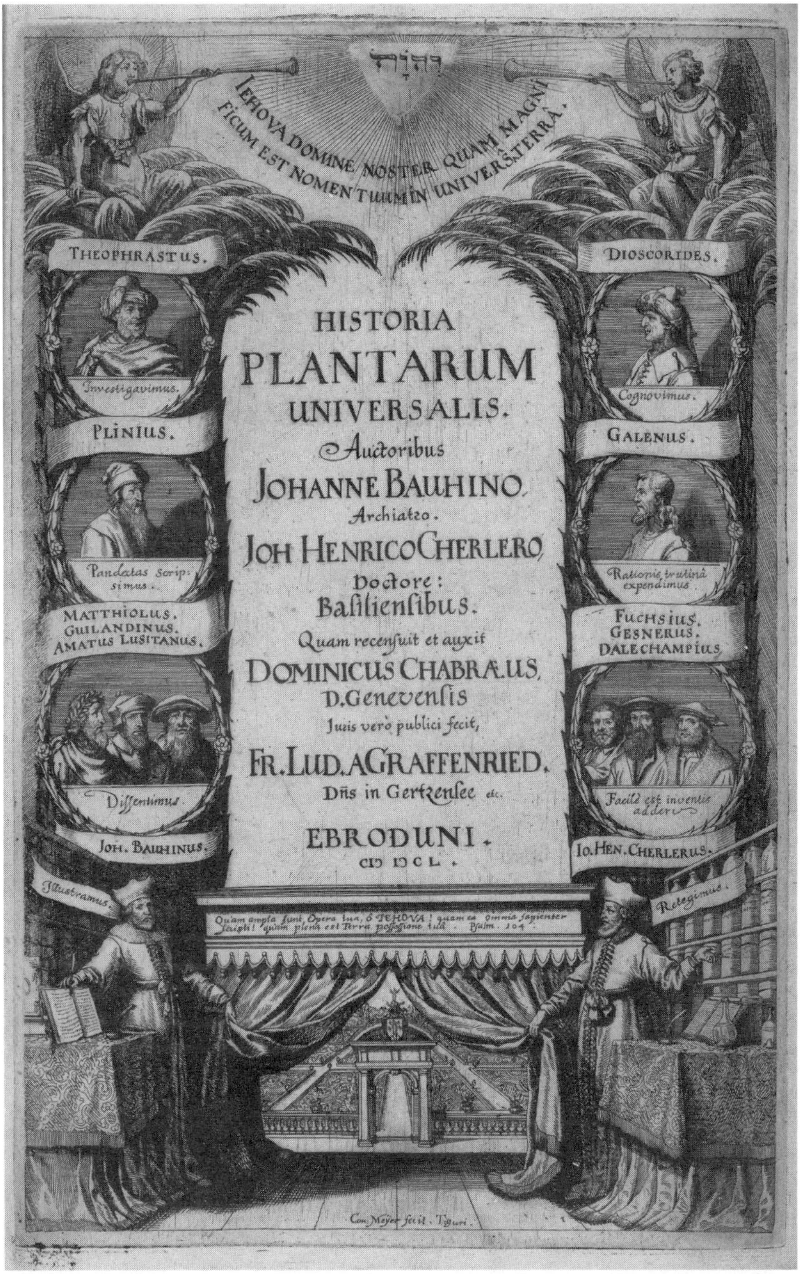

Figure 4.1. Frontispiece to Jean Bauhin, *Historia plantarum universalis nova et absolutissima cum consensu et dissensu circa eas* (Ebroduni, 1650, 3 vol., in folio). (Courtesy, Library of the University of Lausanne)

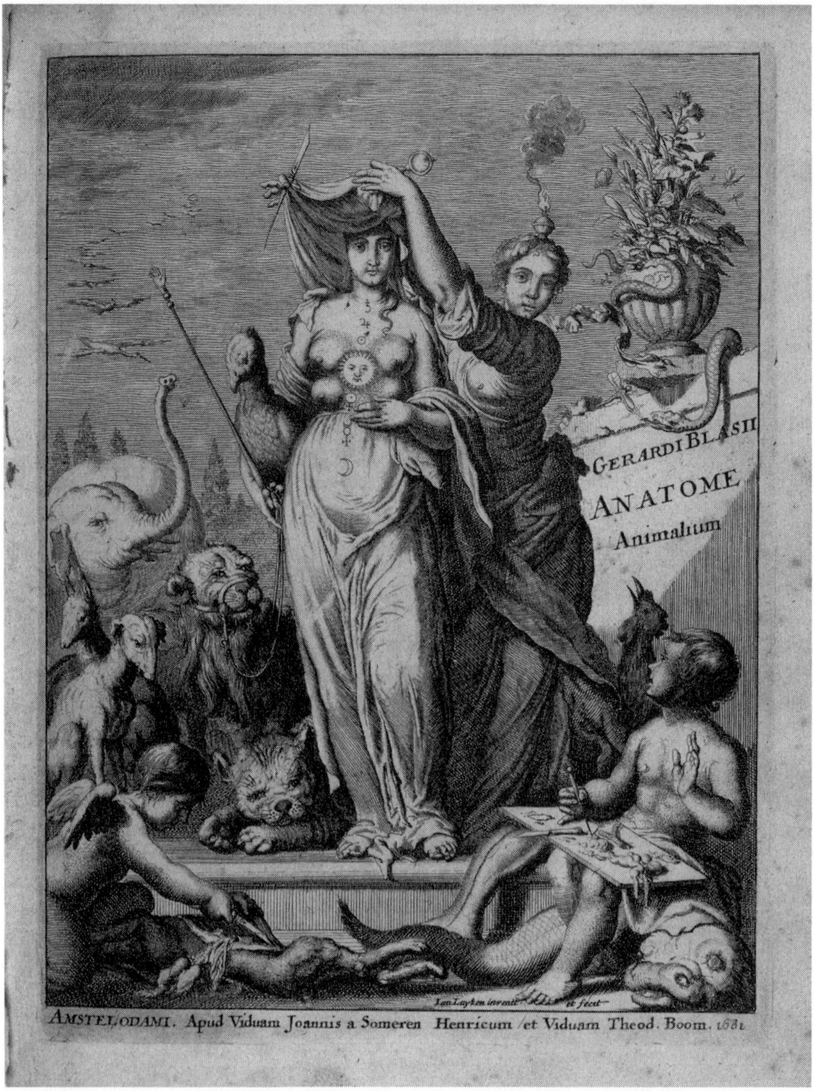

Figure 4.2. Frontispiece to Gerhard Blaes, *Anatome animalium terrestrium variorum, aquatilium, serpentum,insectorum, ovorumque structuram naturalem, ex veterum, recentiorum, propriisque observationibus proponens* (Amstelodami, 1681, 1 vol., 4^{0}). (Courtesy, Library of the University of Lausanne)

different varieties evoked in the text. The *Anatome,* on the other hand, proposes wordy descriptions that adhere to the visible without recourse to the commentaries of preceding observers or commentators. Its descriptions are organized in classes, within which species are listed in alphabetical order.[3] The image, here, is inseparable from the text. Every textual nuance is reflected in the softly etched engravings, which give both general and detail views. The better the species is known, the more numerous are its illustrations. Thus, following a path first mapped out by Francis Bacon in the seventeenth century, an ideal of knowledge in which text and images become transparent vectors from seeing to knowledge replaced an unattained ideal of completeness and fidelity to Holy Scripture along with the commentaries produced by the long tradition of the naturalists.

The nudity of Nature constitutes the emblem of this new vision. As shown by Shapin and Schaffer, far from offering itself *immediately,* nature is the fruit of an important work of construction (Shapin and Schaffer 1985). These two authors, in effect, based their celebrated study on the controversy between Hobbes and Boyle regarding the *natura aeris*. Their dispute emerged from the English tradition of experimentalism that depended on the invention and implementation of three "technologies"—material, linguistic, and social—serving an ideal of the production, transcription, communication, and ratification of scientific facts.[4] These actions, taken as a whole, took place within a social space characterized by a system of norms, conventions, and procedures that so guaranteed the conditions of confidence prevailing among the gentlemen-philosophers who were their protagonists that they were able to share an intersubjective agreement about the facts of their observations and experiences (for more on this point see Shapin 1994 and Hall 1991).

This mode of construction and communication of scientific facts gradually imposed itself in the first scientific journals, which emerged almost simultaneously in France and England during the 1660s. The journals constitute, in this way, the birthplace of a new iconography and, more generally, of a new writing about nature that, in turn, gradually imposed themselves at the expense of the other modes of description. As a result, these latter rapidly disappeared from legitimate scientific discourse. In this chapter, we intend to evoke some of the characteristics that provoked the marginalization of that accursed part of iconography: the harnessing of images to the word; the mixing of the different registers of knowledge that preside over their construction and their reading; and finally, the marginal space occupied by the mathematical translation of observations. All of these elements attributed to the resulting incompatibility of two manners of seeing.

2. Eyes for Talking

At the end of the fifteenth century, images that appeared in manuscripts and then in the first printed books treating the natural world illustrated texts that were a mixture of empirical fact, practical magic and alchemy, secrets and *folk* knowledge—texts that, in other words, engaged with a wide range of different registers.[5] Throughout the sixteenth century, these registers multiplied further to include hearsay, fable, and innumerable remarks about everything and anything bearing an even remote similarity to the object described. This new knowledge surprises us with the desire for a kind of totality that is revealed in the lengthy additions of things said and things seen or imagined, which already had been considerably augmented through the rediscovery and translation of classical authorities at the beginning of the fifteenth century and through the increased distribution of these texts that the invention of printing allowed. As was the case with the other intellectuals of their age, for naturalists, things were the catalysts of crystallization, while certainty was the child of accumulation.

The mixing of registers that characterized the works of this epoch can be found in its iconography: Things are created in an interaction of perspective and language capable of subjugating the image to its own syntax. Emblematic examples include the monkfish that—faithful to the story that describes it as such (figure 4.3)—became the fruit of the superimposition of a fish and a monk. Another example is the first images of the chameleon *(Khameleion)* that made visible its etymology, which was "lion which crawls on the ground."[6]

This mixing sometimes organized itself in more complex ways. The long tradition of the mandrake would be an example of this (figure 4.4). Here, the image not only presents a caricature resembling a root with a human body, but pays visual witness to the supposedly supernatural powers infused in its flesh and that, as a result, are invisible. Thus, when Abraham Bosse draws the mandragora with the form of a female body whose torso is made of flowers, "[h]is representation was locked into a system of beliefs in which form and meaning must be regarded as inseparable—a system in which accidental configurations of mandragora roots could indeed act as effective talismans, with or without the assistance of a human sculptor" (Kemp 1990, 129).[7]

We can associate these images to those mental figures used in the "art of memory" that allow us to travel through the re-visualization of spaces memorized by means of an ordering of information according to a mnemotechnical scheme (Rossi 1983; Yates 1999). They also and equally

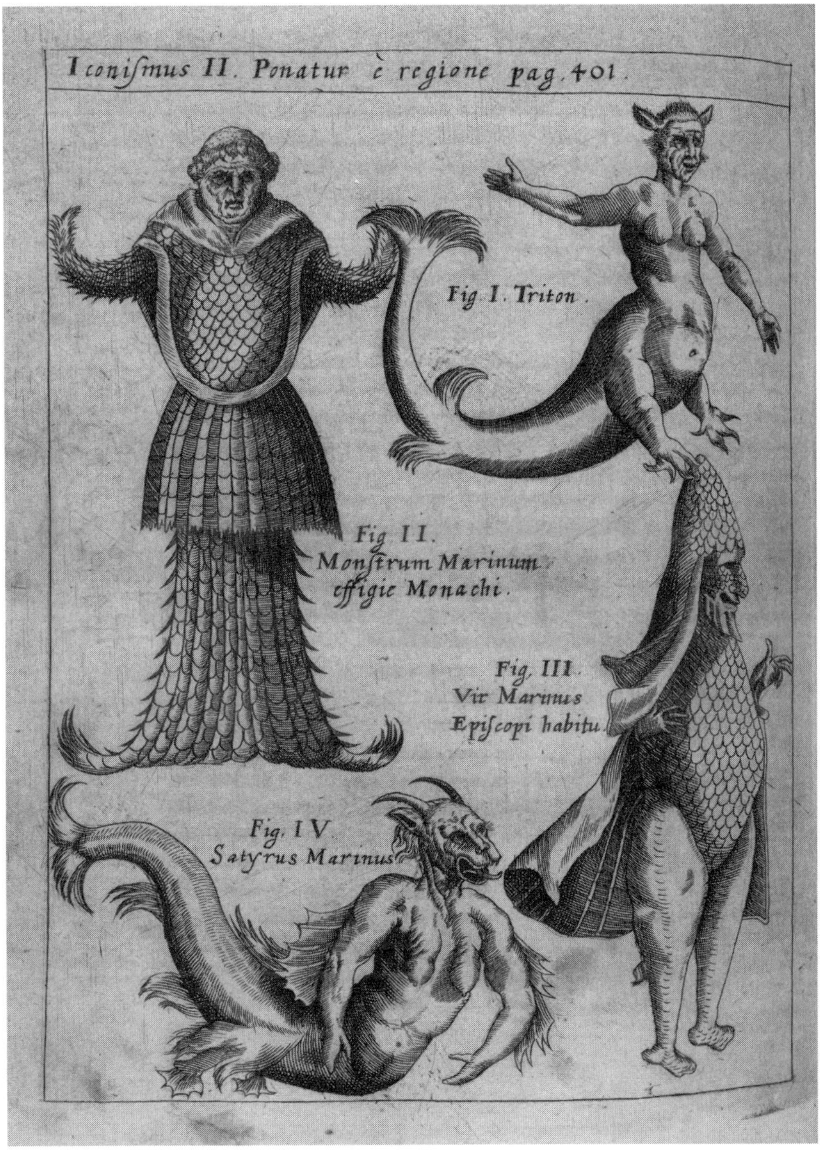

Figure 4.3. Monstrum Marinum effigie Monachi (Caspar Schott, *Physica curiosa, sive mirabilia naturae et artis libris XII comprehensa: quibus pleraque quae de angelis, daemonibus, hominibus, spectris, etc. rara, arcana, curiosaque circumferuntur, ad veritatis trutinam expenduntur,* Herbipoli, 1662, 2 vol., 4⁰). (Courtesy, Library of the University of Lausanne)

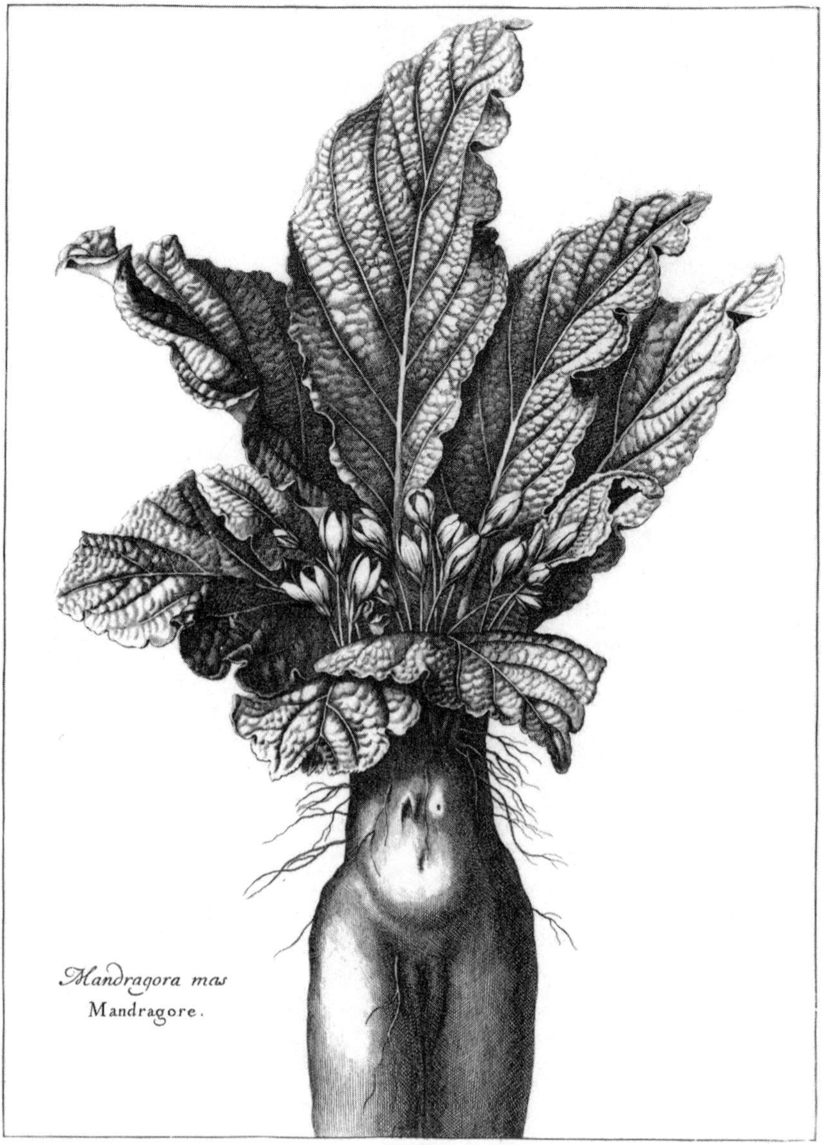

Figure 4.4. Mandragora (Abraham Bosse, in Nicolas Robert, *Recueil de plantes gravées par order du Roi,* Imprimerie Royale: Paris, 1701). (Courtesy, Library of the University of Lausanne)

may be associated with the manner of description employed in the work of Giovanni Battista Della Porta, whose *Phytognomonica* was published in Naples in 1588 and in which the figures of plants and roots reveal their particular attributes on the basis of visual signatures (Della Porta 1650; see also Foucault 1973; Panese 2003; Tomasi 1984, 42). Through a subtle interplay of iconographic elements, these images reveal the long chains of resemblance that bind the wide diversity of registers. Subservient to the power of speech, such images are incompatible with any scheme of systematic classification. To put it another way, they are incompatible with any system of hierarchy and derivation that might by applied to the character of the things themselves (see Daston and Park 1998; Defert 1987, 47–67). According to Martin Jay, the possibility of classifying beings in this way is linked to the "de-narrativization" of the way we look at things. Thus, the perspective is no longer one concerned with a thing considered as an essence of meaning that the image must visually reproduce. Rather, it is viewed as an observable object, subject to an order to which it belongs by the delegation of certain specific, visible characteristics that are sufficient for the drawing up of a table of classification (Blumenberg 1981; Foucault 1973, chap. 2; Jay 1993, 1988, 3–28).

The first classifications proposed by naturalists in the sixteenth century participated in the same movement that resulted in the ever more central role devolved to visual description in their work. As a signal of this transition, the naturalists adjoined captions to the new images such as *ad vivum expressus* (Gesner) or *au naturel* (Belon).[8] We should note that captions of this sort would soon evolve to become a label of distinction and *quality* in comparison with the images of the older tradition. Indeed, so much was this the case that we find such captions accompanying images that seem very far from having been recorded according to nature, but upon which their authors wished by this expedient to confer the conviction of authority.[9] The exemplary case of this transformation was the publication of Gesner's *Icons* (1553), in which he replaced alphabetical ordering—which he had employed in the complete edition of his *Historiae Animalium* (Gesner 1551–1558) from which *Icons* was extracted—with categories based on similarity. Gesner's categories no longer are based on the idea of the essence of the thing but on its visible surface. He said, "In the books of images, I have strived to adhere to a natural order, and to group together animals of the same class or species" (Gesner, quoted in Braun 1990).

Strictly speaking, this transformation did not constitute a reversal of the text-image duality. It seems more likely that it was the fruit of a more and more strict compartmentalization of the registers and modes of description within which movement the visual imposed itself as the chief criteria of classification. Gesner's attitude thus announced a new manner of seeing: The similarities—still intuitive at that time, but that nevertheless allowed *practical* groupings to be created—rapidly would be domesticated by the establishment of criteria founded on observation, and then would be entered into an abstract, organizational—and ideally automatic—paradigm (Daston and Galison 1992).

The new connective relationship that was evolving here between the images of various species and the place they occupy in a system of classification profoundly reconfigured the way we visually describe nature. With regard to the study of plants, this new relationship was especially responsible for generating a multiplication of illustrations of detail that could serve for their classification. By so doing, this process reinforced the importance of the image in description. On the other hand, by founding the identifying criteria of plants on a fixed palette of characteristics corresponding to a preconceived "theory of species," the visual descriptions of systematic botanists, following Carl von Linné, re-engaged with textual description (Knight 1985). Here, it was not so much a question of describing plants *in real life* as of determining systematically the genres to which they, as species, belong. The key criteria of such determination thus referred increasingly to the structure of the plant in question (number of petals, form of the corolla, number of stamens, etc.) and could, when taken to their logical conclusion, do without images altogether and depend exclusively on classification tables.[10]

The transformation of perspective—this new way of seeing—whose advent we have observed in the generation of naturalists to whom we have referred and whose development we have tracked through their use of images, by no means constitutes a radical rupture with the past. Nor can it, strictly speaking, be reduced merely to the establishment of intellectual technologies. The transformation in question participates rather in an historical movement of configuration that is marked by numerous other transformations. With regard to sixteenth-century naturalists in general and to Gesner in particular, Hoeniger drew up what appears to be at first glance a rather disparate list. In it, we note the impact of hermetic writings edited by Florentine neo-Platonists, of the Protestant Reformation, of the discovery of America and other new worlds, of the advent of the printing press and the diffusion of information on a grand scale and, finally, of the formation in Northern Europe of a pragmatic and realist merchant class (Hoeniger 1985, 130–148). If all these contextual elements are indeed crystallized in the work of the naturalists, we also find plentiful clues indicating an evolution in their readership that was not perhaps without influence on the content of their works. In general—and without it being possible here to distinguish between editorial strategy and intellectual frame—all this took place as if the evolution in the ways in which things were described was linked to the emergence of a vaster, non-specialized public who discovered in the plurality of registers drawn together in the work of scientists a vehicle for their desire for knowledge about the curiosities and wonders of nature.

3. From Aldrovandi's Stag to Leibniz's Roe

As Foucault notes, the method of description that we evoked earlier is the same as that which Buffon traces back to Ulysses Aldrovandi. More generally, moreover, it is the method used in the compilations and criticized by natural history at the very time that it was inaugurating a space of pure visibility in which were deployed characteristics of living beings ordered by a new way of looking at things:

> Until Aldrovandi, History was the inextricable and perfectly unified fabric of what is to be seen in things and of all the signs that are to be discovered in those same things or are applied to them. [Whereas] natural history is nothing but the naming of the visible.[11] (Foucault 1973, 141, 144)

For the Bolognese scholar, the description of an animal is always its refraction through a plural range of registers, registers that one may find in the succession of chapters devoted to that animal: proverbs, symbols, its uses in food and medicine, etymologies, names, classifications, and, finally, the *forma descriptio* generally accompanied by an image (figure 4.5).

Aldrovandi often borrows his images from other naturalists. This practice, seemingly innocent, is yet clear evidence of his belonging to an epoch of understanding that is drawing to a close. Ashworth explains that there is nothing exceptional in this but rather that it reflects the normal life of images, which can, over long periods of time, become prototypes that other authors integrate into their works in the same way that they would integrate a quotation by Pliny or a contemporary author. In most cases, the shelf life of these images depends on the authority accorded to the texts accompanying them. The texts of Gesner, Belon, or Aldrovandi, for example, therefore marked or influenced the naturalist's perspective for a long time (see Ashworth 1985).[12]

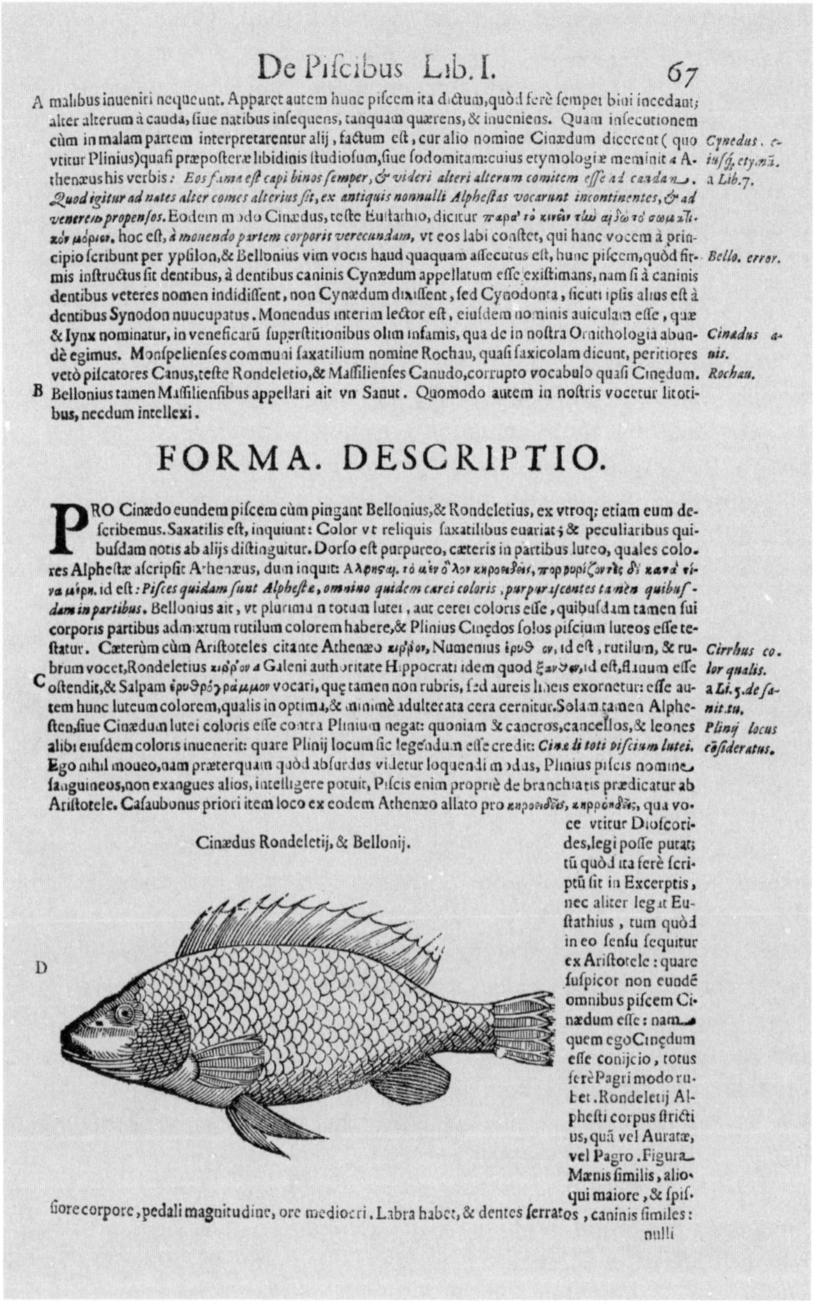

De Piſcibus Lib. I. 67

A malibus inueniri nequeunt. Apparet autem hunc piſcem ita dictum, quòd ferè ſemper bini incedant; alter alterum à cauda, ſiue natibus inſequens, tanquam quærens, & inueniens. Quam inſecutionem cùm in malam partem interpretarentur alij, factum eſt, cur alio nomine Cinædum dicerent (quo vtitur Plinius) quaſi præpoſteræ libidinis ſtudioſum, ſiue ſodomitam: cuius etymologiæ meminit a Athenæus his verbis: *Eos fama eſt capi binos ſemper, & videri alteri alterum comitem eſſe ad caudam. Quod igitur ad nates alter comes alterius ſit, ex antiquis nonnulli Alpheſtas vocarunt incontinentes, & ad venerem propenſos.* Eodem modo Cinædus, teſte Euſtathio, dicitur παρὰ τὸ κινεῖν τὴν αἰδῶ τὸ σωματικὸν μόριον, hoc eſt, *à mouendo partem corporis verecundam,* vt eos labi conſtet, qui hanc vocem à principio ſcribunt per ypſilon, & Bellonius vim vocis haud quaquam aſſecutus eſt, hunc piſcem, quòd firmis inſtructus ſit dentibus, à dentibus caninis Cynædum appellatum eſſe exiſtimans, nam ſi à caninis dentibus veteres nomen indidiſſent, non Cynædum dixiſſent, ſed Cynodonta, ſicuti ipſis alius eſt à dentibus Synodon nuucupatus. Monendus interim lector eſt, eiuſdem nominis auiculam eſſe, quæ & Iynx nominatur, in veneficarũ ſuperſtitionibus olim infamis, qua de in noſtra Ornithologia abundè egimus. Monſpelienſes communi ſaxatilium nomine Rochau, quaſi ſaxicolam dicunt, peritiores verò piſcatores Canus, teſte Rondeletio, & Maſſilienſes Canudo, corrupto vocabulo quaſi Cinędum.
B Bellonius tamen Maſſilienſibus appellari ait vn Sanut. Quomodo autem in noſtris vocetur litotibus, necdum intellexi.

Cynedus. eiuſq; etymũ.
a Lib.7.
Bello. error.
Cinædus auis.
Rochau.

FORMA. DESCRIPTIO.

PRO Cinædo eundem piſcem cùm pingant Bellonius, & Rondeletius, ex vtroq; etiam eum deſcribemus. Saxatilis eſt, inquiunt: Color vt reliquis ſaxatilibus euariat; & peculiaribus quibuſdam notis ab alijs diſtinguitur. Dorſo eſt purpureo, cæteris in partibus luteo, quales colores Alpheſtæ aſcripſit Athenæus, dum inquit: Ἀλφησταί. τὸ μὲν ὅλον κηροειδεῖς, πορφυρίζοντες δὲ κατά τινα μέρη. id eſt: *Piſces quidam ſunt Alpheſtæ, omnino quidem cerei coloris, purpuraſcentes tamen quibuſdam in partibus.* Bellonius ait, vt plurimu n totum lutei, aut cerei coloris eſſe, quibuſdam tamen ſui corporis partibus admixtum rutilum colorem habere, & Plinius Cinędos ſolos piſcium luteos eſſe teſtatur. Cæterùm cùm Ariſtoteles citante Athenæo κιῤῥὸν, Numenius ἐρυθρὸν, id eſt, rutilum, & rubrum vocet, Rondeletius κιῤῥὸν a Galeni authoritate Hippocrati idem quod ξανθὸν, id eſt, flauum eſſe
C oſtendit, & Salpam ἐρυθρόγραμμον vocari, quę tamen non rubris, ſed aureis lineis exornetur: eſſe autem hunc luteum colorem, qualis in optima, & minimè adulterata cera cernitur. Solam tamen Alpheſten, ſiue Cinædum lutei coloris eſſe contra Plinium negat: quoniam & cancros, cancellos, & leones alibi eiuſdem coloris inuenerit: quare Plinij locum ſic legendum eſſe credit: *Cinædi toti piſcium lutei.* Ego nihil moueo, nam præterquam quòd abſurdus videtur loquendi modus, Plinius piſcis nomine ſanguineos, non exangues alios, intelligere potuit, Piſcis enim propriè de branchiatis prædicatur ab Ariſtotele. Caſaubonus priori item loco ex eodem Athenæo allato pro κηροειδεῖς, κιῤῥοειδεῖς, qua voce vtitur Dioſcorides, legi poſſe putat; tũ quòd ita ferè ſcriptũ ſit in Excerptis, nec aliter legat Euſtathius, tum quòd in eo ſenſu ſequitur ex Ariſtotele: quare ſuſpicor non eundẽ omnibus piſcem Cinædum eſſe: nam quem ego Cinędum eſſe conijcio, totus ferè Pagri modo rubet. Rondeletij Alpheſti corpus ſtrictius, quã vel Auratæ, vel Pagro. Figura Mænis ſimilis, alioqui maiore, & ſpiſſiore corpore, pedali magnitudine, ore mediocri. Labra habet, & dentes ſerratos, caninis ſimiles: nulli

Cirrhus color qualis.
a Li.5. de ſanit. tu.
Plinij locus cõſideratus.

Cinædus Rondeletij, & Bellonij.

D

Figure 4.5. A page from Ulisse Aldrovandi., *De piscibus libri V, et de cetis lib. unus* (Bononiae: apud N. Thebaldinum, 1638, in folio). (Courtesy, Library of the University of Lausanne)

His confidence in the descriptions of other authorities—erudite or profane—and the diversity of the registers that one finds in Aldrovandi's book at the level of the organization of its chapters are expressed by some of the images it contains in a particularly emblematic way in the form of visible analogies. This is the case, for example, with the images of a shark or an alarming sea serpent, both of which display their cruelty etched on their faces in accordance with their reputation. Similarly, a fish carries on its flank a fleet of sailing ships in accordance with a verbal description. And it is the case with Aldrovandi's strange leafy-antlered stag that we will now discuss in greater detail (figure 4.6).

This image seems to be the iconic recombination of a reported description that has been translated perhaps a thousand times. It demonstrates the perfect adherence of the engraver's tool to the reporter's words. It is surely illusory to think that the picture's public at that time really believed that what they were seeing was a faithful representation of a real stag. What Aldrovandi proposes here to his numerous readership is rather the expression of the subjective experience of a particular observer who is both collector and amateur. Those whom he invites into his great collection of curiosities are chosen both for questions of prestige and so as to participate in a shared and intimate experience of the world. In other words, if a collection can be open to individuals other than the collector himself, those others are not present in order to exercise their reason in the sense that the Enlightenment would give to the term, but in order to be moved. The important thing is not the order of the world as such, but the experience of belonging to a world that is diverse, secret, and marvellous. Such a space, populated with a miscellany of objects, is very close to those images in which we saw such a great diversity of registers mingled together. Those images, just like the objects that fill the shelves and drawers of curiosity collections, can be thought of as "object-semaphores" (Pomian 1990), whose value is proportional to the possible connections that they enable with ancient texts, history, the personal experiences of amateur scientists, and, in a more general way, with the diverse miracle of Creation as it is reconstituted in the unity of private space:

> The heterogeneous matrix of these collections constitutes—by the eclecticism and the "extravagance," as they used to say, of the objects assembled therein—an attempt to systematize and recompose knowledge through a microcosmic reconstruction of the macrocosm. (Tomasi and Tongiorgi 1984, 129)

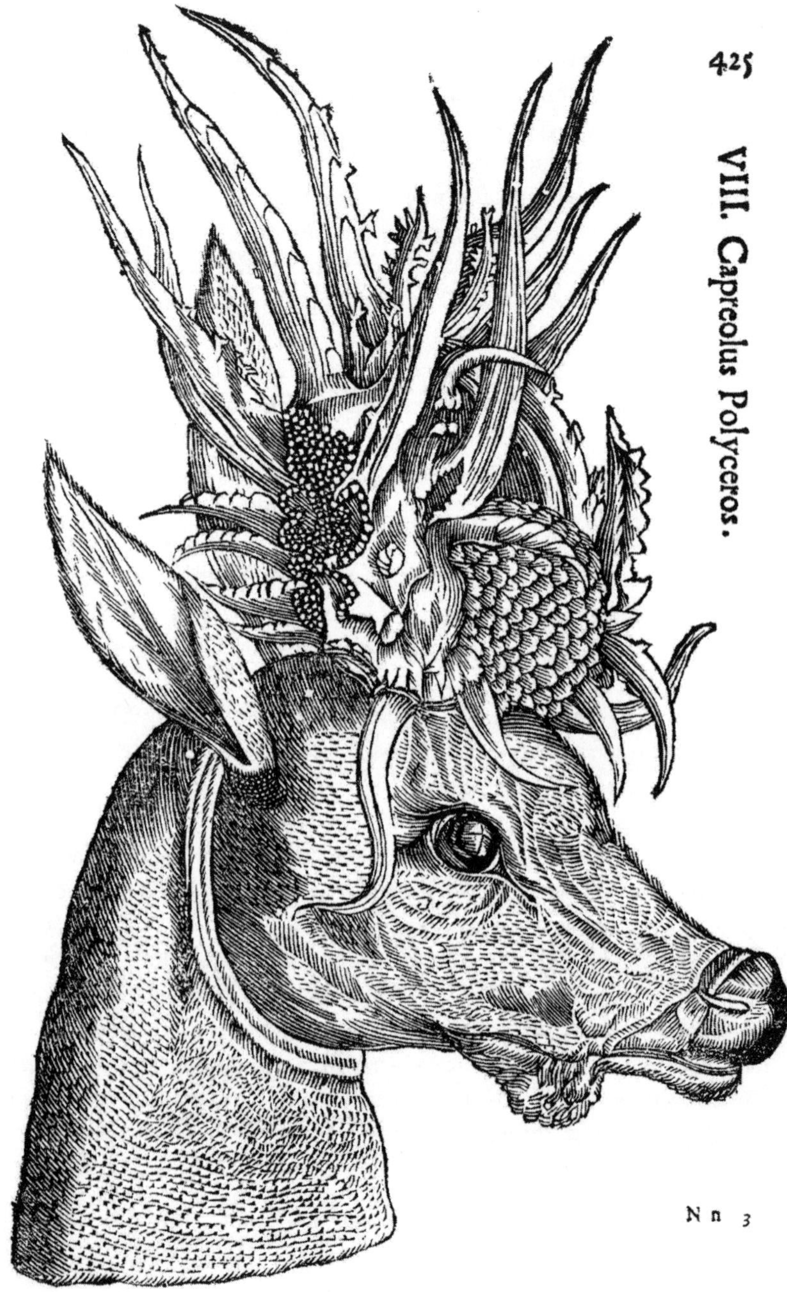

Figure 4.6. Capreolus Polycoros, in Ulisse Aldrovandi, *Monstrorum historiae cum paralipomensis historiae omnium animalium* (Bononiae, typis Nicolai Tebaldini, 1642, 2t. in 1 vol., in folio). (Courtesy, Library of the University of Lausanne)

Modern science—or at least that version of it left to us by English experimentalism in the seventeenth century—can be thought of as the boiling down of the dense network of significations that allowed all things to hold together until there remained only those threads that appertained to the register of natural phenomena. This transformation had a clearly identifiable extension in scientific iconography. Another image that demonstrates the phenomenon is one similar to Aldrovandi's stag; it affords us a deeper understanding of the transformation in question (figure 4.7). It is the image of a roe deer that was published in the *Journal des Sçavants* on 12 July 1677 (Leibniz 1677). The letter that gave rise to its publication is significant because it bears the signature of Leibniz, who resided at that time in Hanover, and was serving as librarian to Johann-Friedrich, Duke of Brunswick-Lunebourg.

Leibniz's letter reports that the Duke of Hanover, "prompted by the monstrous hare that he had seen in [the] eleventh Journal [of Sçavants] in the year 1677, had had brought into his possession the portrait of a roe deer whose hair was done in a strange manner, along with a narration." In this "narration," we learn that the young roe had been fettered, "because he charged passers-by. It was then that they saw this strange head dressing that seemed to be growing around his head." Leibniz believed that it was the deer's loss of liberty and its resulting sadness that generated its deformity by analogy with "the stories [which] teach us that great sadness or anxiety can cause the colour of a prisoner's hair to change overnight, so that a youth comes to resemble an old man." To these explanations, the *Journal* added a hypothesis as to "the physical cause of this excrescence, which could be attributed to the fact that its aqueous humour could no longer be dissipated once the animal was attached as it usually is by the heat that such animals generate by means of their charges, leaps and runs. This great humidity mixed with the volatile juice and salt which constitute its horns thus drew down the matter of the horns by its weight and rendered it glutinous and of a cooler temperature."

Numerous registers mingle in this text and accompanying image. The deer is spawned out of the courtiers' curiosity about all things unusual, exceptional, and rare, and most especially about objects whose strangeness is visible or demonstrable. A web of analogies, correlations, and similarities with humankind is woven around the roe—liberty and sadness, intellectual notions enabled by the revelation of the supposed causes of the deer's strangeness. The culture of curiosity takes precedence over the dominion of the law "because a notion of nature that is supposed to be subject to

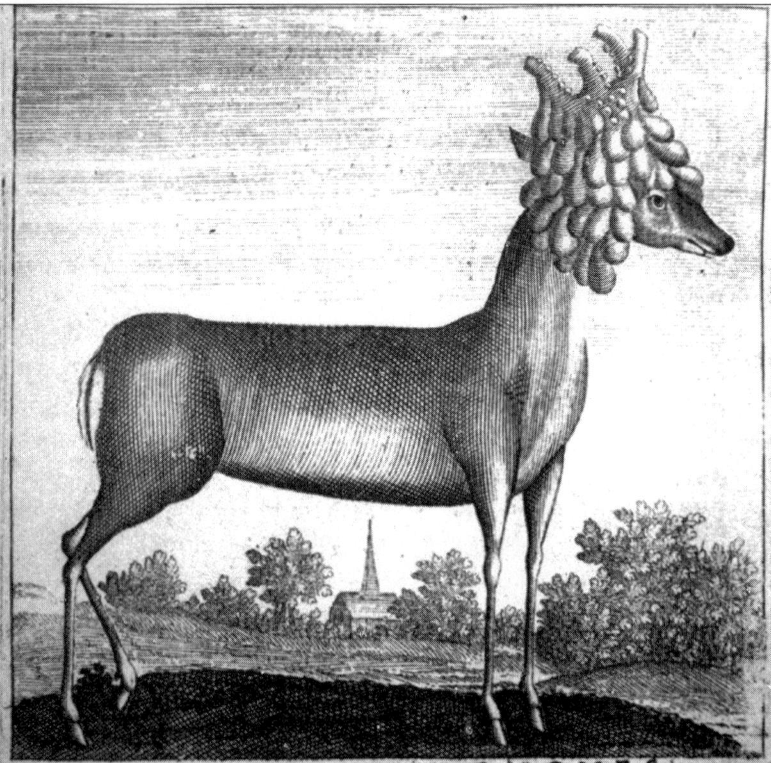

AVIS AUX ASTRONOMES,

L'Astronomie n'a rien de plus difficile que de fixer quel est le veritable mouvement de Mercure. C'est pourquoy les Astronomes sont avertis qu'ils pourront observer par la Lunette sa conjonction avec le Soleil. Elle arrive tres-rarement, & sert à perfectionner les Tables du mouvement de cette Planete, qu'on verra le 7. Novembre prochain entierement dans le Disque du Soleil. Le commencement de cette conionction paroistra sur l'horison de Paris dés les onze heures du matin, & ne finira qu'à Soleil couchant. Ceux qui feront leurs observations hors de Paris auront égard à la difference de leurs Meridiens. &c. Cet avis vient de M. Comiers Prevost de Ternant.

A PARIS,

Chez JEAN CUSSON, ruë S. Jacques, à l'Image S. Jean, proche S. Yves. *Avec Privilege du Roy.*

Figure 4.7. Figure d'un chevreüil coiffé d'une manière fort extraordinaire. Extrait d'une lettre de M. de Leibniz. (*Journal des Sçavants,* XIV, 1677).

laws that are always and everywhere the same, manifests itself in what is common, repetitive, and reproducible. When, on the other hand, we recognize no rule, then that which is rare appears as the only thing capable of representing nature" (Pomian 1990, 63).

In the example in question, the law comes from the *Journal*. On the one hand, its explanations confine the chains of causality inside the body of the deer; on the other, its case can be deployed as a more general illustration of physiology, chemistry, and physiognomy. The marvellous has not disappeared, but it has been pushed to one side by the explanatory power of the scientific perspective.

The image of the roe, we are told, is the "faithful and exact copy in miniature" of a "lifelike" portrait that was done of it after its death. In passing successively from a living curiosity to what seems to have been a painting and, finally, to an illustration in the *Journal*, the roe seems not only to have changed the register of knowledge to which it was assigned, but also its social space: from that of the simple amateur of curiosities, it has moved into the realm of the scientist. This passage, however, is not Manichean; it passes subtly from shade to shade of grey. Not only does it demonstrate the hybrid nature of the sciences, but also that of the scientific public in the mid-seventeenth century. French scientists often were dependent on the king and the court, just as their English counterparts were *amateurs* who fully subscribed to their society despite their radical choice of the experimental approach (Hooke 1987).[13] Before the laboratories (in which scientific facts were constructed) and the reviews (where they were then published for public consumption) became the preserve of a certain socially specific category, the protagonists of science and their readership, while sharing the same social universe as the new intellectual middle class, deployed their diversity at the heart of scientific writing.

4. The Domestication of Perspective

It is significant that the image of the deer is to be found in the *Journal des Sçavants* and not in what is traditionally thought of as being its English equivalent, *The Philosophical Transactions*. This is revealing of two distinct editorial approaches that, despite their proximity, diverge from the very beginning with regard to the importance assigned to experimental reporting. If the *Journal's* first editorial states its intention of publishing the "experiments of Physics and Chemistry that can serve to explain the effects of nature," the space actually allotted to such reports does not occupy more than a relatively modest portion of the early issues, if any (*Journal des Sçavants* 1665).[14]

In the unequivocal affirmation of their attachment to experimentalism, the members of the Royal Society marginalized a whole field of knowledge. In particular, the *rarities of nature*—the curiosities—were not to be discussed unless they were found useful from an experimental point of view. The first lines of the society's statutes (1661) are very clear on this point:

> The business of the society in their Ordinary Meeting shall be to order, take account, consider, and discourse on philosophical experiments and observations; to read, and discourse upon letters, reports, and other papers concerning philosophical matters; *as also to view, and discourse upon rarities of nature and art; and there upon to consider, what may be deduced from them, or any of them; and how far they or any of them, may be improved for use or discovery.* (Hall 1991, 1; italics ours)[15]

We find the Royal Society's approach reflected in *The Philosophical Transactions,* which was published in London by Oldenberg.[16] Through the intermediary of this publication, the person who was at once both one of two secretaries of the prestigious institution and the editor of the review aimed to disseminate the society's philosophy to a wider public beyond the circle of the men of science. From the experimentalists' perspective, curiosities were not, therefore, worthy of consideration unless it was possible to fix them somewhere within the regularly ordered system of nature. The attitude, both purist and militant, of the representatives of this new manner of practicing science is discernable at the level of the images they employed. When the curiosity was unique, as in the case of Aldrovandi's stag, its image could be considered as antithetical to the ideal-type representations promoted by the experimental approach. Such ideal-type images had to be the fruit of the sedimentation of various perspectives or ways of seeing that, layer upon layer, had specifically obliterated the particular at the expense of what—in the most literal sense of the word—was essential:

> "I had a large number of dispersed observations," explained Albinus in his *Tabulae.* "I started by digesting them . . . choosing principally those that I found the most frequently and which, I thought, best corresponded to the intentions of nature. (Albinus 1747, cited in Kemp 1993, 85)

We should recall at this juncture the fact that the greater the detail in which an image is rendered, the further it is distanced from ordinary visual experience. As Ackerman suggests with regard to a watercolor of a crab by Dürer, "that which permits us to say of a given image that it is 'objective' is that the artist has established the equivalent of what we might term 'laboratory' conditions, conditions which reduce the variables that affect the image's appearance and that permit the artist to standardize his procedures . . . The internal variables—psychological and social—are limited by the choice of an object that generates a minimum of emotional and conventional 'noise,' such as feeling, fear, respect, or contempt" (Ackerman 1985, 5).[17]

The images that inhabit the pages of *The Philosophical Transactions* from its earliest editions onward bear a strikingly homogeneous witness to this neutralization of perspective. In the synoptic tables of the figures published between 1665 and 1700 (Lowthrop 1749), amid all the detailed anatomical drawings, the mechanical diagrams, and zoological illustrations, we indeed do find monstrous creatures that recall the teratological tradition that existed up to Aldrovandi's day (figure 4.8). Here, however, their monstrosity has been domesticated by a description that deploys a depersonalized perspective of the image, without any emotional overtones, one

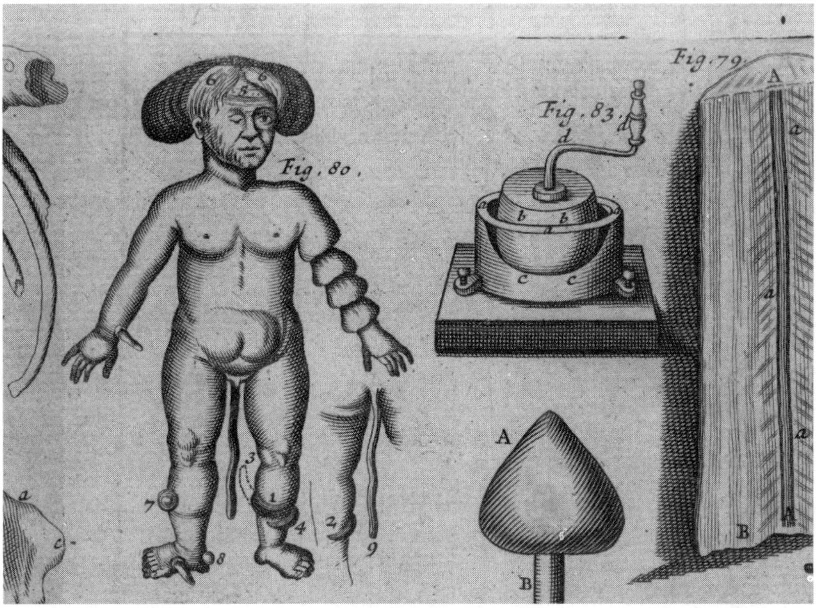

Figure 4.8. Monster, John Lowthrop, ed., in *The Philosophical Transactions and Collections . . . , Abridged, and Disposed under General Heads* [1665–1700], Vol. 3, *Anatomical, Medical and Chymical* (London: Royal Society of London Ed., 1749 [5th ed.]). [Synoptic plate of the figures published in the Phil. Trans. from 1665 to 1700]).

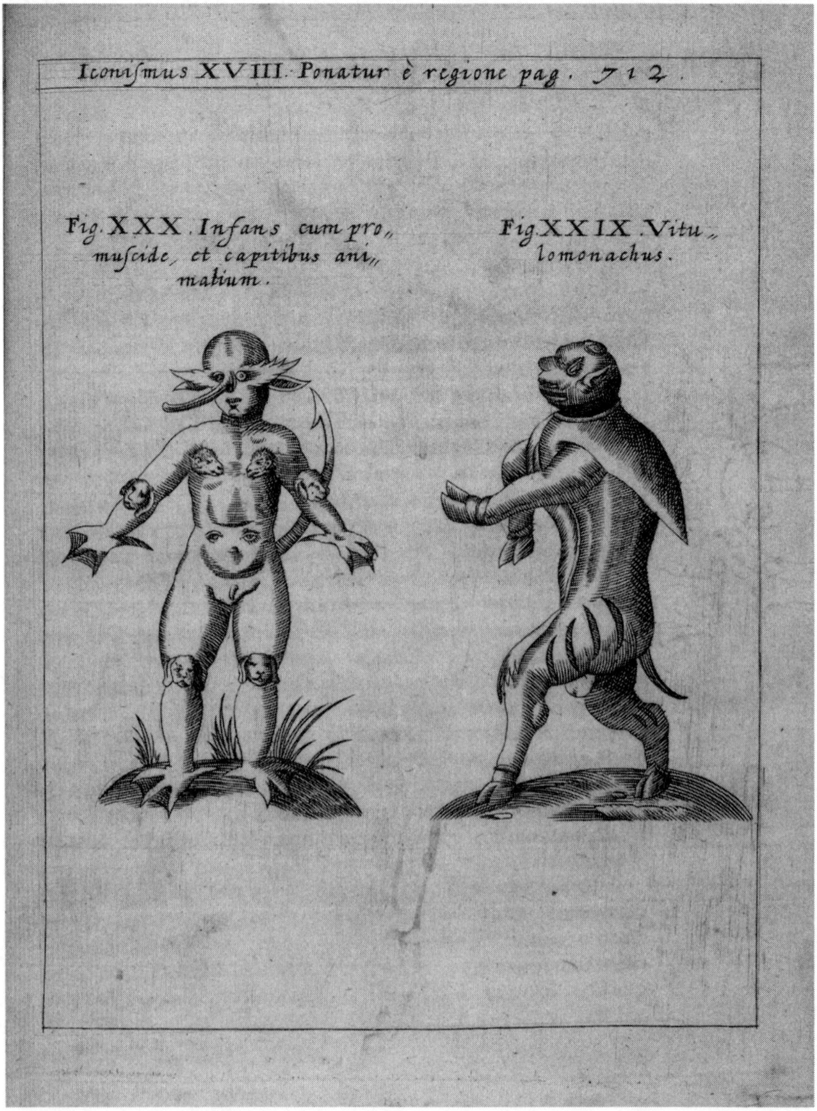

Figure 4.9. Monster, Infans cum promuscide et capitus animatum, in Caspar Schott, *Physica curiosa, sive mirabilia naturae et artis libris XII comprehensa: quibus pleraque quae de angelis, daemonibus, hominibus, spectris, etc. rara, arcana, curiosaque circumferuntur, ad veritatis trutinam expenduntur* (Herbipoli, 1662, 2 vol. Vol. 2, p. 617, plate 18).

that seems to move away from the old models. With truly geometric rigor, the creature has been broken up into duly noted parts. The deformity of an arm is translated into a succession of stackable cones; an excrescence is figured as a little sphere or a small point inserted into a disc of flesh. A detail drawing allow us to view this clinical case study from every angle, for it is necessary to see everything and to note every detail so that the figure may emerge as the faithful ally of the written commentary that must, like the figure, suffice by and to itself.

Everything differentiates the monster displayed in *The Philosophical Transactions* from the one revealed to us by the Jesuit Gaspar Schott in his *Physica Curiosa* (1662) (figure 4.9). In fact, Schott's monstrosity is a replica of the one published by Sebastian Münster in his *Cosmographiae universalis,* which was edited in Basle in 1559, more than one hundred years earlier (figure 4.10). Born in Krakow in February 1547, this poor child's deformities are rendered as the heads of sheep fixed to his breasts and the heads of dogs to the joints of his arms and legs. His nose is a horn; his ears are extended to double their usual dimensions by an excess of hair or flesh. We find two large eyes situated on either side of his navel, which itself takes the form of a nose. Finally, webbed hands and feet place this strange being on the boundary between the human and the bestial.

Obviously, we will never know in any of these cases what these unhappy progeny *really* looked like. The image from *The Philosophical Transactions* is opposed to that of Münster or of Schott as the different perspectives of two distinct eras, but which both refer to registers of translation that were rendered obsolete by the emergence of modern science. The English seventeenth century does not mourn the passing of the curiosity—neither as a way of thinking, nor as a rare object offered up to the imagination—but it does provide it with access to a new legitimacy (Blumenberg 1983, part 3, chap. 9). In the tradition established by Francis Bacon, the aim is to isolate it, to subject it to induction, and to place it within the rules of *natural philosophy.* The heterogeneity structured by similarity is then replaced by the homogeneity of measurement. Our own contemporary perspective, our way of looking at the world, is still so marked by this order that we often need to make a conscious effort to efface this perspective in order to be able to consider it as just one particular way of rendering credible the visible. For that way of looking at things is also but a part of the social history of the *truth of the image.*[18]

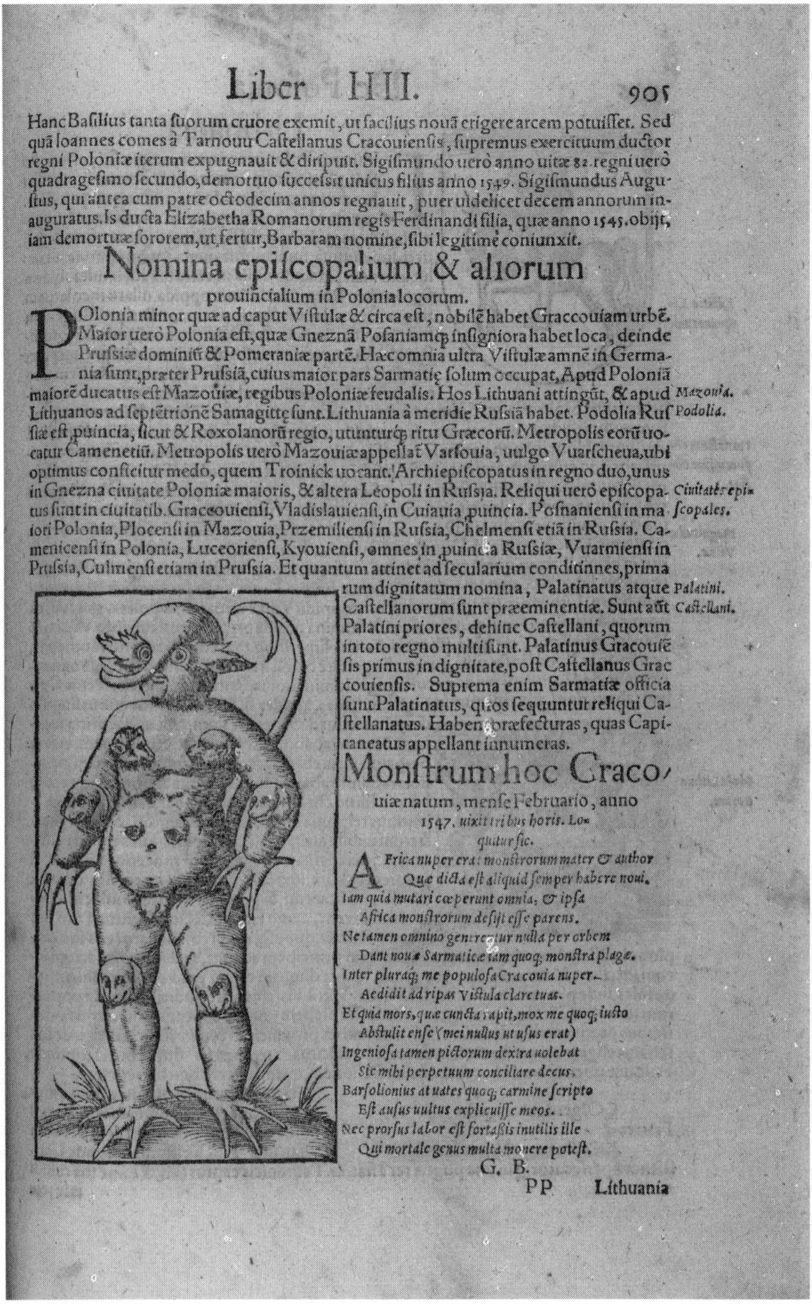

Liber IIII. 905

HancBasilius tanta suorum cruore exemit, ut facilius nouā erigere arcem potuisset. Sed quā Ioannes comes à Tarnouu Castellanus Cracouiensis, supremus exercituum ductor regni Poloniæ iterum expugnauit & diripuit. Sigismundo uerò anno uitæ 82. regni uerò quadragesimo secundo, demortuo succeßit unicus filius anno 1549. Sigismundus Augustus, qui antea cum patre octodecim annos regnauit, puer uidelicet decem annorum inauguratus. Is ducta Elizabetha Romanorum regis Ferdinandi filia, quæ anno 1545. obijt, iam demortuæ sororem, ut fertur, Barbaram nomine, sibi legitimè coniunxit.

Nomina episcopalium & aliorum prouincialium in Polonia locorum.

POlonia minor quæ ad caput Vistulæ & circa est, nobilē habet Graccouiam urbē. Maior uerò Polonia est, quæ Gneznā Posaniamq; insigniora habet loca, deinde Prussiæ dominiū & Pomeraniæ partē. Hæc omnia ultra Vistulæ amnē in Germania sunt, præter Prussiā, cuius maior pars Sarmatię solum occupat. Apud Poloniā maiorē ducatus est Mazouiæ, regibus Poloniæ feudalis. Hos Lithuani attingūt, & apud Lithuanos ad septētrionē Samagittę sunt. Lithuania à meridie Russiā habet. Podolia Russiæ est, puincia, sicut & Roxolanorū regio, utunturq; ritu Græcorū. Metropolis eorū uocatur Cameneciū. Metropolis uerò Mazouiæ appellat̄ Varsouia, uulgo Vuarscheua, ubi optimus conficitur medo, quem Troinick uocant. Archiepiscopatus in regno duo, unus in Gnezna ciuitate Poloniæ maioris, & altera Leopoli in Russia. Reliqui uerò episcopatus sunt in ciuitatib. Graccouiensi, Vladislauiensi, in Cuiauia, puincia. Posnaniensi in maiori Polonia, Plocensi in Mazouia, Przemiliensi in Russia, Chelmensi etiā in Russia. Camenicensi in Polonia, Luceoriensi, Kyouiensi, omnes in puincia Russiæ, Vuarmiensi in Prussia, Culmensi etiam in Prussia. Et quantum attinet ad secularium conditiones, primarum dignitatum nomina, Palatinatus atque Castellanorum sunt præeminentiæ. Sunt aūt Palatini priores, dehinc Castellani, quorum in toto regno multi sunt. Palatinus Gracouiēsis primus in dignitate, post Castellanus Gracouiensis. Suprema enim Sarmatiæ officia sunt Palatinatus, quos sequuntur reliqui Castellanatus. Habent præfecturas, quas Capitaneatus appellant innumeras.

Mazouia. Podolia. Ciuitates episcopales. Palatini. Castellani.

Monstrum hoc Cracouiæ natum, mense Februario, anno 1547. uixit tribus horis. Loquitur sic.

Africa nuper erat monstrorum mater & author
Quæ dicta est aliquid semper habere noui.
Iam quia mutari cœperunt omnia, & ipsa
Africa monstrorum desijt esse parens.
Ne tamen omnino generentur nulla per orbem
Dant nouæ Sarmaticæ iam quoq; monstra plagæ.
Inter pluraq; me populosa Cracouia nuper
Acdidit ad ripas Vistula clare tuas.
Et quia mors, quæ cuncta rapit, mox me quoq; iusto
Abstulit ense (mei nullus ut usus erat)
Ingeniosa tamen pictorum dextra uolebat
Sic mihi perpetuum conciliare decus.
Barsolionius at uates quoq; carmine scripto
Est ausus uultus explicuisse meos.
Nec prorsus labor est fortaßis inutilis ille
Qui mortale genus multa monere potest.

G. B.

PP Lithuania

Figure 4.10. Monstrum hoc Craciviae in (Sebastian Münster, *Cosmographiae universalis libri VI* . . . (Basileae: apud Henrichum Petri, 1559).

5. Conclusion

"I admit nothing but on the faith of the eyes." So affirmed Francis Bacon. His faith in the visible would seem to be tenable if and only if all other possible explanations of the world are replaced with the one based solely on *natural causes*. In the perspective of the project of *scientific* knowledge inaugurated by the English philosophers, to see the world is necessarily to give oneself up to idols of reason of a kind that limit the way things may be looked at:[19]

> The human understanding resembles not a *dry light*, but admits a tincture of the will and passions, which generate their own system accordingly: for man always believes more readily that which he prefers. He, therefore, rejects difficulties for want of patience in investigation; sobriety, because it limits his hope; the depths of nature, from superstition; the light of experiment, from arrogance and pride, lest his mind should appear to be occupied with common and varying objects; paradoxes, from a fear of the opinion of the vulgar; in short, his feelings imbue and corrupt his understanding in innumerable and sometimes imperceptible ways. (Bacon 1854a, 348)

Thus, everything occurs as though it is only at the cost of relinquishing those passions and that will that the objects of knowledge may be deposited upon the smooth and ordered surface of images that henceforth must parade under the scientific eye like the turning pages of nature's book.

We have attempted to reconstruct, in the light of these few examples, just a part of this episode in the history of seeing. However, we are obliged to own that the episode's general context is largely absent from our narrative. Following Shapin's invitation formulated in his remarkable *Social History of Truth*, it now remains for us to reconstruct the *cognitive and moral conditions* in which this domestication of perspective took place, so as to understand the ways in which it developed as the new regime of *credibility* emerged: that is to say, as a social agreement about natural facts that are established, among other things, as these universes of paper that are images. In doing so, the question might then be posed whether the ideals of order and of measurement proper to the scientific approach were not rooted wholly in the puritan ethics of the gentlemen-philosophers who first posited them. If they were, the domestication of the way the world is looked at would be the child of a new form of iconoclasm, the avatar of English Protestantism.

Notes

1. Upon the pedestal of the stele, Psalm 104 recalls the grandeur of God: "Quam ampla sunt Opera tua, o Jehoava! Quam ea omnia Sapienter Fecisti! Quam plena est terra possessione Tua."
2. On the relation between the two instruments in intellectual history, see Mario Vegetti (1987).
3. As is indicated by the work's title, the classes in question are: *"terrestrium variorum, volatilium, aquatilium, serpentum, insectorum, ovurumque structuram naturalem."* There are no departures from the alphabetical order of species except in the case of very close proximities, as for example that between beef, bull, cow, and calf, which follow one another in the book.
4. These technologies comprise Boyle's notion of experimental method: a material technology (for Boyle, the development and production of the air pump apparatus), a linguistic technology (for seventeenth-century science, the simultaneous invisibility and authority of the scientist-as-observer), and a social technology (the development of a scientific community, guilds, and affiliations in the production of a legitimacy for modern science).
5. This diversity recalls that which Tongiorgi Tomasi identified as the "green medicine" of the Middle Ages, which was a type of learning about the interior body that mixed the commentaries of old *auctores* and—with a more *applicable* intent—a description of the efficacy of the natural substances that enter into the pharmacopoeia of the era. See Lucia Tongiorgi Tomasi (1984).
6. A fascinating selection of these images can be found in the bestiaries of the Middle Ages that frequently were integrated into new systems. On this question, see Jurgis Baltrusaitis (1988).
7. Note that this mandragora was published for a courtly audience that already actively participates in the culture of the *curiosity.* It appears in Nicolas Robert (1701).
8. We must mention here the most celebrated work in French naturalism: Pierre Belon (1555).
9. One later, yet remarkable example of this practice is the mandrake of Abraham Bosse that we cited earlier, which also proudly boasts the caption *au naturel.*
10. A connection might be drawn between Linné's project of botanic systematization and certain projects of *perfect language* that flourished in the seventeenth century and in which there was also the question of giving things names that would synthesize ideographically their intrinsic *qualities.* We are thinking particularly here of the hypothesis formulated by John Wilkins—a member of the Royal Society—(1984 [facsimile of London edition 1668]). For a broad overview of these projects see Mary M. Slaughter (1982) and Umberto Eco (1995).
11. Author's translation from the French edition. Michel Foucault, *Les mots et les choses: Une archéologie des sciences humaines.* (Paris: Gallimard, 1966), pp. 141, 144.
12. Notice that this practice would pose serious problems of classification and denomination. It could happen, for instance, that an animal literally would be *invented* at the end of a chain of copies, names, and commentaries.
13. About the tensions between amateurs and scientists with regard to methodological choices of experimentation, we return to the preface of the well-known *Micrographia* by Robert Hooke (1987 [photocopied reproduction of the London edition 1665]), in which the curator of the Royal Society attacks *curiosities.*

14. "The aim of this Journal" states the first editorial, "is to make known what is happening in the Republic of Letters. It will firstly be composed of an exact Catalogue of the principal books published in Europe. And we shall not be content simply to cite their titles as is the current practice among most Bibliographers, but we will further tell of their subject matter and in what manner they may be useful. . . . Secondly, whenever some personage celebrated for his opinions or his works should come to die, we shall publish his eulogy along with the catalogue of his achievements and the principal circumstances of his life. Thirdly, we will make known the experiments of Physics and Chemistry that may serve to explain the effects of nature, along with the new discoveries made in the Arts and in the Sciences such as the useful or curious machines and inventions that may be furnished by Mathematics, by the observation of the Heavens, by those of the Meteors & that which Anatomy may discover in the realm of animals. Fourthly, the principal decisions of the secular and Ecclesiastical Courts, the edicts of the Sorbonne and the other Universities, whether in this kingdom or in foreign Countries. Finally, we will endeavour to ensure that nothing worthy of the curiosity of men of letters should takes place in Europe that cannot be learned of in this Journal" (*Journal des Sçavants* 1665, III–IV).
15. Marie Boas Hall further posits that if the experimental philosophy of the Royal Society was always an ideal, the very fact that they continually were referring to it suggests just how difficult it was for them to adhere to this ideal, especially given the cultural diversity of its members.
16. Contrary to popular belief, the journal was private and, even if it seems to echo the aspirations of Royal Society work, the fact remains that the journal was published personally by the secretary in Oldenburg until 1677. See Maria Boas Hall (1991).
17. Albrecht Dürer, *Crabe*, Rotterdam, Musée Boymans-van Beuningen, n M. B. 1958/T 30.
18. For a particularly clear reconstruction of this historical and methodological point of view regarding the sociological conditions of the production of *truth* in the field of English experimentalism, see Steven Shapin (1994).
19. "In order that we might be better understood, we have thought it necessary to distinguish these idols which obsess the human spirit into the four denominations that follow: The first species are the *Idols of theTribe;* the second, the *Idols of the Cave;* the third, the *Idols of Marketplace;* and the fourth, the *Idols of the Theatre*" (Francis Bacon 1854a, 348).

References

Ackerman, James S. 1985. Early Renaissance, "naturalism" and scientific illustration. In *The natural sciences and the arts,* edited by Allan Ellenius. Uppsala: Acta Universitatis Upsaliensis.

Albinus, Bernardus Siegfried. 1747. *Tabulae sceleti et musculorum corporis humani.* Lugduni Batavorum.

Ashworth Jr., William B. 1985. The persistent beast: Recurring images in early zoological illustration. In *The natural sciences and the arts,* edited by Allan Ellenius. Uppsala: Acta Universitatis Upsaliensis.

Bacon, Francis. 1854a. Novum Organum. In *The works,* edited and translated by Basil Montague. Vol. 3. Philadelphia: Parry & MacMillan.

———. 1854b. The great instauration. In *The works,* edited and translated by Basil Montague. Vol. 3. Philadelphia: Parry & MacMillan.

Baltrusaitis, Jurgis. 1988. *Réveils et prodiges, le gothique fantastique.* Paris: Flammarion.

Bauhin, Johannes. 1650. *Historia plantarum universalis nova et absolutissima cum consensu et dissensu circa eas.* Ebroduni. 3 vol. in-fol.

Belon, Pierre. 1555. *L'histoire de la nature des oiseaux: Avec leurs descriptions and naïfs portraits retirez du naturel.* Paris: s.n.

Blaes, Gerhard. 1681. *Anatome animalium terrestrium variorum, volatilium, aquatilium, serpentum, insectorum, ovorumque structuram naturalem, ex veterum recentiorum propriisque observationibus proponens.* Amstelodami. 1 vol. 4°.

Blumenberg, Hans. 1981. *Die Lesbarkeit der Welt.* Frankfurt am Main: Suhrkamp.

———. 1983. *The legitimity of the modern age.* Cambridge: MIT Press.

Braun, Lucien. 1990. *Conrad Gesner.* Genève: Slatkine.

Daston, Lorraine, and Peter Galison. 1992. The image of objectivity. *Représentations* 40: 81–128.

Daston, Lorraine, and Katharine Park. 1998. *Wonders and the order of nature, 1150–1750.* New York: Zone Books.

Defert, Daniel. 1987. Collections et nations au XVIe siècle. In *L'Amerique de Théodore de Bry: Une collection de voyages protestants du XVIe siècle,* edited by Michel Duchet. Meudon-Bellevue: CNRS.

Della Porta, Giambattista. 1650. *Phytognomonica, octo libris contenta.* Rothomagi.

Eco, Umberto. 1995. *The search for the perfect language.* Oxford: Blackwell.

Editorial. 1665. *Journal des Sçavants* 1: 3–4. Paris: De Boccard.

Foucault, Michel. 1973. *The order of things: An archaeology of the human sciences.* New York: Vintage Books.

Gesner, Conrad. 1551–1958. *Historiae animalium.* Tiguri: apud C. Froschoverum.

———. 1553. *Icones animalium.* Tiguri: C. Froschoverum.

Hall, Maria Boas. 1991. *Promoting experimental learning: Experiment and the Royal Society 1660–1727.* Cambridge: Cambridge University Press.

Hoeniger, F. David. 1985. How plants and animals were studied in the middle sixteenth century. In *Science and the arts in the Renaissance,* edited by John W. Shirley and F. David Hoeniger. Washington: The Folger Shakespeare Library.

Hooke, Robert. 1987. *Micrographia, or some physiological descriptions of minute bodies made by magnifying glasses with observations and inquiries thereupon.* Lincolnwood, Ill.: Science Heritage.

Jay, Martin. 1988. Scopic regimes of modernity. In *Vision and visuality,* edited by Hal Foster. Seattle: Bay Press.

———. 1993. *Downcast eye: The denigration of vision in twentieth century French thought.* Berkeley: University of California Press.

Journal des Sçavants, no1. 1655. Paris: De Boccard.

Kemp, Martin. 1990. Taking it on trust: Form and meaning in naturalistic representation. *Archives of Natural History* 17 (2): 127–188.

———. 1993. The mark of truth: Looking and learning in some anatomical illustrations from the Renaissance and eighteenth century. In *Medicine and the five senses,* edited by William F. Bynum and Roy Porter, 85–123. Cambridge: Cambridge University Press.

Knight, David. 1985. Scientific theory and visual language. In *The natural sciences and the arts,* edited by Allan Ellenius. Vol. 22. Uppsala: Acta Universitatis Upsaliensis.

Leibniz, Gottfried Wilhelm. 1677. Lettre de M. de Leibniz à l'Auteur du Journal, écrite d'Hanovre le 18. Juin 1677 contenant la Relation & la figure d'un Chevreüil coiffé d'une manière fort extraordinaire. *Journal des Sçavants,* 12 juillet 1677. Paris: De Boccard.

Lowthrop, John, ed. 1749. *The philosophical transactions and collections . . . , Abridged, and disposed under general heads* [1665–1700], vol. 3, *Anatomical. Medical and Chymical.* London: Royal Society of London Ed. [5th ed].

Panese, Francesco. 2003. Doctrine des signatures et technologies graphiques au seuil de la modernité. *Gesnerus: Swiss Journal of the History of Medicine and Sciences* 60 (1/2): 6–24.

Pomian, Krzysztof. 1990. *Collectors and curiosities: Paris and Venice 1500–1800.* Cambridge, U.K.: Polity Press/Cambridge, Mass.: Basil Blackwell.

Robert, Nicolas. 1701. *Recueil de plantes gravées par ordre du Roi.* Paris: Imprimeries Royales.

Rossi, Paolo. 1983. *Clavis Universalis: Arti della memoria e logica combinatorial da Lullo a Leibniz.* Bologne: Il Mulino.

Shapin, Steven. 1994. *A social history of truth: Civility and science in seventeenth century England.* Chicago: The University of Chicago Press.

Shapin, Steven, and Simon Shaffer. 1985. *Leviathan and the air-pump: Hobbes, Boyle, and the experimental life.* Princeton, N.J.: Princeton University Press.

Slaughter, Mary M. 1982. *Universal languages and scientific taxonomy in the seventeenth century.* Cambridge: Cambridge University Press.

Tomasi, Lucia Tongiorgi. 1984. Dalla "medicina verde" al naturalismo: L'immagine botanica e zoologica nei manuscritti e nei primi testi a stampa sul finire del' 400. In *Immagine e natura: l'immagine naturalistica nei codici e libri a stampa delle Biblioteche Estense e Universitaria. Secoli xv–xvii.* Modena: Edizioni Panini.

Tomasi, Lucia Tongiorgi, and Paolo Tongiorgi. 1984. I "teatri del mondo": La nascita del collezionismo scientifico. In *Immagine e natura: l'immagine naturalistica nei codici e libri a stampa delle Biblioteche Estense e Universitaria. Secoli xv–xvii.* Modena: Edizioni Panini.

Vegetti, Mario. 1987. *Il coltello e lo stilo: Animali, schiavi, barbari e donne alle origini della razionalità scientifica.* Milano: Il Saggiatore.

Wilkins, John. 1668. *Mercury, or the secret and swift messenger: Showing how a man may with privacy and speed communicate his thoughts to a friend at any distance.* Facsimile of the first, London edition, Amsterdam: J. Benjamins, 1984.

Yates, Frances Amelia. 1999. *The art of memory.* New York: Routledge.

5

Images of Science in the Classroom

Wall Charts and Science Education, 1850–1920

Massimiano Bucchi

> One of the educational by-products of German botanical scholarship was the publication of sets of large "wall diagrams" *(Wandtafeln)* for use in the lecture-room. Most British University Departments of Botany dating from the period before the first world war probably had at least one of these sets. In my own department I have used these excellent diagrams occasionally, realizing that they combined clarity, size and accuracy to an unrivalled extent.—S. M. Walters (1989)

This passage from a recent essay by S. M. Walters forms an appropriate introduction to the topic of this paper. From the start, it should be noticed that the use of such diagrams (referred to as "wall charts" in the rest of the paper) was by no means limited to botany nor to university studies.[1] As will become clear, they were one of the most important media for teaching and learning at different levels of education and within different fields.

It is not difficult to trace historical antecedents for this form of visual communication with educational purposes. In 1658, the Moravian pedagogue Comenius (Jan Amos Komensky, 1592–1670) published his *Orbis sensualium pictus,* a textbook with 150 woodcuts designed to teach pupils how to read by relating words to objects. About one century later, in 1774, the German pedagogue Johann Bernard Basedow (1732–1790) enriched

Originally published in the *British Journal of History of Science* 31 (1998): 161–184. Reprinted with the permission of Cambridge University Press.

his *Elementarwerk* (Elementary school handbook) with a companion atlas reproducing more than 100 copper engravings by the artist Daniel Nieolas Chodowiecki (1762–1801), one of the leading European engravers of the time. According to Basedow, each of these illustrations should be given to the pupil "as a gift, each time that appreciation for his conduct is to be expressed" (Basedow, quoted in Schmidt 1990, 65).[2] He also suggested gluing some of the engravings to a cardboard support, covering them with glass and then hanging them on the walls of the classroom. The subjects depicted included familiar settings (the house, the countryside) as well as arts and crafts, geographic landscapes, and natural history subjects. Together with the Philantropinum (the school he founded in Dessau in 1764), the *Elementarwerk* was part of Basedow's project to put Rousseau's educational theories into practice. It was very successful all over Europe and was reprinted up to 1909.

When asked to give a series of lectures about educational theory in 1776–1777, Kant chose to discuss this book and named his lecture series *Pädagogik über Basedows Methodenbuch*. Another important work in this line was the *Bilderbuch für Kinder* (Children's picture book) published by Friedrich Justin Bertuch (1747–1822) between 1790 and 1830 with illustrations portraying not only plants and animals but also exotic items such as Indian fakirs and dragons. The illustrations appeared in no systematic and ordered sequence but they were marked with a number and a reference topic that allowed readers to assemble them and bind them together.

The first wall charts, that is, images explicitly designed to be displayed in the classroom, appeared in the 1820s. They initially were produced in small format (20 × 30 cm; 7.8 × 11.7 inches), presenting scenes and objects from familiar settings for primary school teaching: pets, kitchen utensils, activities such as hunting, the cycle of the seasons. With the help of such images, the young pupils were supposed to learn to recognize and name animals, objects, and professions. But the "golden age" for such a genre of didactic communication is to be located between about 1870 and 1920, a period during which several different kinds of wall charts were produced and sold in large quantities. A wide range of themes were covered, from the historical to the religious, with titles such as "The creation of the world," "Greek and Roman war machines," and "The process of beer production." Most of the wall charts were published in series, and they were advertised extensively and reviewed regularly in school magazines such as *Schulwart* (founded in 1903) and *Natur und Schule* (founded in 1902); their purchase also was recommended strongly by school authorities. (See, for instance, the recommendation by the Düsseldorf school authority as reported in

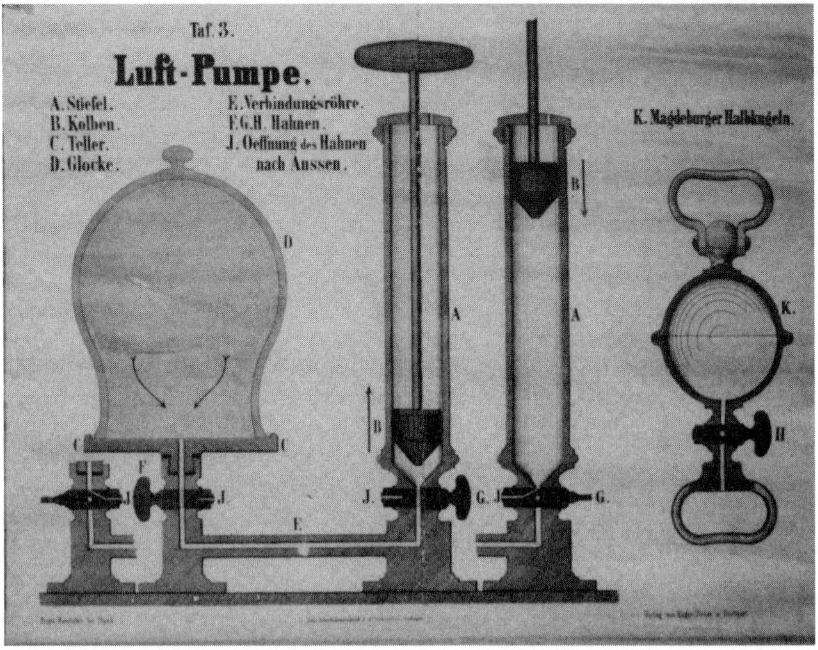

Figure 5.1. Tab. 3: Luft-Pumpe [Airpump], in C. Bopp, *Wandtafeln für Physik* [Wall charts for physics], Stuttgart, 1868. Chromolithography on cloth, 46.5 × 65 cm (18 × 25 in). (Collection of the State School Liceo G. Prati, Trento, Italy)

Lehnemann 1994, 14.) Their use gradually spread from primary school teaching to almost any field of study, and they began to be employed at other levels of education also (from middle schools to university).

To give just a few figures: In geography alone, more than thirty series appeared between 1880 and 1915; in only five years, seven editions of the *Wandtafeln für Physik* by Carl Bopp (1833–1904) were published following the first edition in 1866 (figure 5.1); two thousand copies of the *Zoologische Wandtafeln* by Paul Pfurtscheller (1855–1927) were sold quickly, while the *Schädlingstafeln* (Wall charts of noxious insects) edited by the entomologist Curt Schlüter (1881–?) were translated into Italian, Polish, Turkish, Arabic, Bulgarian, and Russian.[3] Initially limited to German-speaking countries, wall charts later became common in other countries such as France and Italy, where the German series were sold (sometimes in a translated versions) and where at a later stage original series also were published, such as one edited by the famous French botanist Gaston Bonnier (1853–1922) (Bonnier 1905) and that by the Italian botanist Pietro Voglino (d.1933) (Voglino et al. 1917). At the turn of the century, wall charts no longer were confined exclusively to the classroom, but were used to address the public on a variety of themes. This can be seen, for instance, in the case of a wall chart prepared in 1904 by the anatomist Anton Weichselbaum (1845–1920, director of the Pathological-Anatomical Institute of the University of Vienna and later rector of the university) for the Österreichischer Verein gegen Trunksucht (the Austrian Association against Alcoholism) (figure 5.2).

Long series of wall charts became the chief vehicle for information campaigns about health and hygiene (Schmidt 1995). Later, the Nazi propaganda machine made extensive use of wall charts; for instance, they were used to highlight the (supposed) differences between German and Jewish people (Müller 1983).

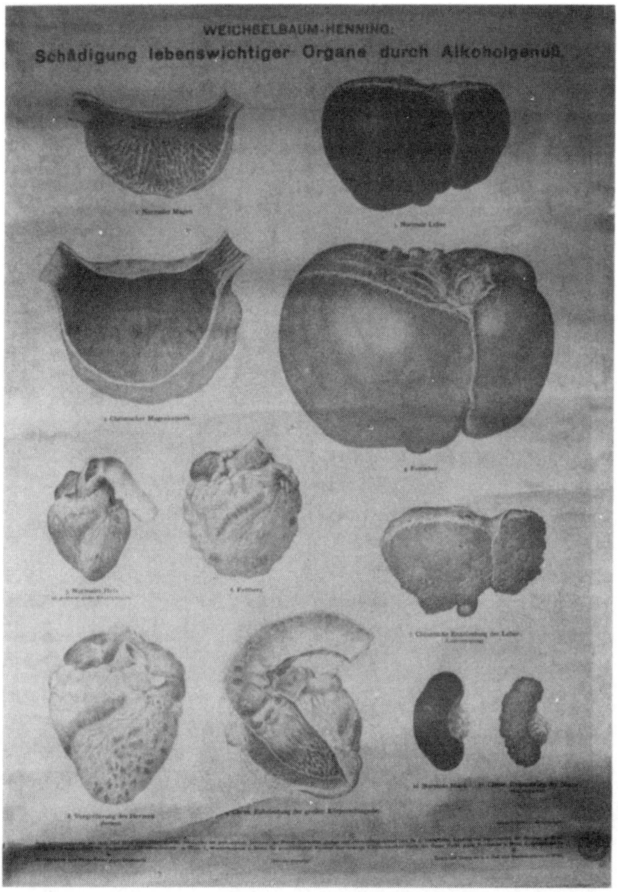

Figure 5.2. Schädigung lebenswichtiger Organe durch Alkoholgenuss [Damage caused by alcohol to vital organs], in A. Weichselbaum, Vienna, 1904. Chromolithography on cloth, 104 × 75 cm (40.5 × 29 in). (Collection of the State School Liceo G. Prate, Trento, Italy)

1. Wall Charts and Science Education

With their rich images and colors, wall charts soon proved to be a particularly suitable medium for the teaching of different classifications in botany and zoology. One of the earliest series circulating was about fungi, *Die essbaren und giftige Schwämme in ihren wichtigsten Formen* (Edible and poisonous fungi in their most important forms), by the illustrator Anton Hartinger (1806–1890) and M. A. Beeker, first published in 1855 (figure 5.3).

Another significant series was the *Zoologische Wandtafeln* (Zoological wall charts) edited between 1877 and 1892 by the German zoologists Kari Rudolf Leuckart (1822–1898) and Heinrich Nitsche (1845–1902) consisting of more than 100 images (figure 5.4).

Unlike biological specimens and wax models, wall charts provided a quick and relatively inexpensive means of permitting all the students in a classroom to form an overview of a class of animals or plants and, at the same time, take a closer look at external and internal details. Therefore, it was soon acknowledged that there was a need to provide not just classification schemes but a comprehensive view of an object in its entirety ("Einer ganzheitlichen Gesamtschau") (Schmidt 1990, 69). Otto Schmeil (1860–1943), himself a biologist, director of a *Volksschule* (primary school) in Magdeburg, published an important essay in 1896, which was reprinted eleven times, concerning the need to reform teaching in natural history: *Über die Reformstrebungen auf dem Gebiete des naturgeschichtlichen Unterrichts* (Schmeil 1896). He argued that teachers should not fill the heads of their pupils with "dry descriptions and classifications" but rather "lead them to a true understanding of nature." The morphological-systematic approach should be replaced by a biological approach, considering in the first place organisms as living bodies (Seybold 1986) and accounting for the strong causal relationship existing between their structure and their way of living. In a chapter entitled "On Observation," he reported the results of an interesting survey that he had conducted among 150 pupils from his school, aged twelve to fourteen, asking them questions like "Have you ever seen a beehive?" or "Have you ever seen mushrooms in a wood?" From the number of negative answers recorded, he concluded that the teaching of biology was lacking the equivalent of laboratory experiments in physics or chemistry, that is, real, live encounters with its objects. He wrote:

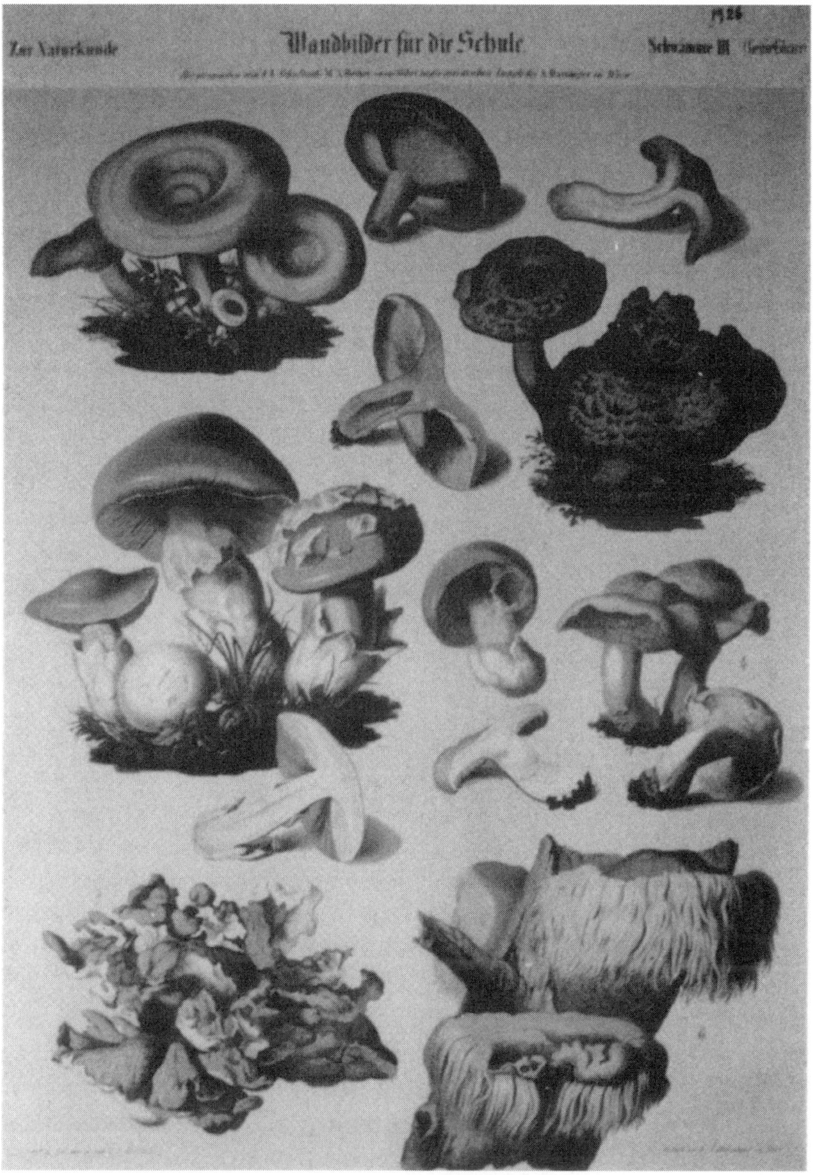

Figure 5.3. Tab. 3: Schwamme III (Geniessbare) [Fungi III (edible)], in A Hartinger and M. A. Becker, *Die esbaren und giftige Schwamme* [edible and poisonous fungi], Vienna, 1872. Chromolithography on paper, 71 × 22.5 cm (27.5 × 8.5 in). (Collection of the State School Liceo G. Prati, Trento, Italy)

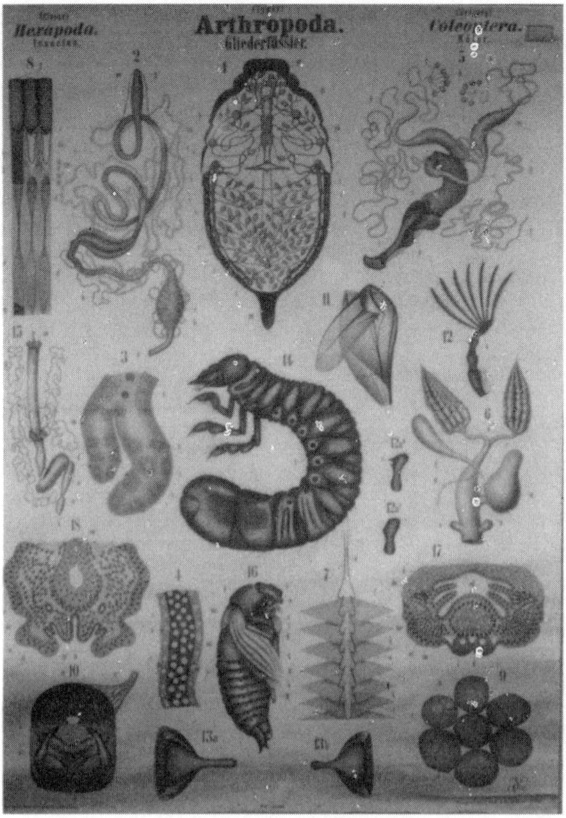

Figure 5.4. Tab. 84: Arthropoda-Hexapoda-Coleoptera, in K. Leuckart and H. Nitsche, *Zoologische Wandtafeln* [Zoological wall charts], Cassel, 1877–1892. Chromolithography on cloth, 133 × 102 cm (52 × 40 in). (Collection of the State School Liceo G. Prati, Trento, Italy)

> Nothing is in the understanding that has not before been in the senses! . . . Only through diligent observation, personal viewing and personal investigation, is it possible to banish the worst enemy of any formative teaching: verbalism. (quoted in Seybold 1986, 243)[4]

Schmeil and other teachers of his time were influenced strongly in this by the work of biologists such as Karl August Möbius (1825–1908), founder of the first German sea aquarium in Hamburg (1863) and later director of the Zoological Museum of Berlin University. From 1870 on, Möbius had been delivering lectures for school teachers in Kiel in which he emphasized the importance of considering animals and plants as part of a "living community" *(Lebensgemeinschaft)* rather than as isolated objects.[5]

> All living plants and animals that the child sees belong to living communities, either natural or shaped by human action . . . Gardens, meadows, field and

> woods offer plenty of opportunities for teaching pupils—whatever their age—that all living beings are members of living communities. The child feels he himself belongs to a living community that includes his parents, his siblings, his playmates, servants and pets. (Möbius in Schmidt 1984, 81)[6]

Series of wall charts such as Adolf Lehmann's *Zoologischer Atlas* clearly demonstrate how these ideas were reflected in the production of educational materials (figures 5.5 and 5.6). In his book about the teaching of biology, to which I already have referred, Schmeil himself praised the use of illustrations of living communities *(Lebensgemeinschaften)* in inducing pupils to see that "nature should be understood not as a conglomerate but as an organism" (in Widar Lehnemann 1994, 72).[7] These principles were to be applied by Schmeil in his own series of wall charts, the *Botanische Wandtafeln* (botanical wall charts). For instance, in the wall chart shown in figure 5.7, the plant is not just displayed in its different parts; one of its most important living functions, reproduction, is presented in a separate figure. This form

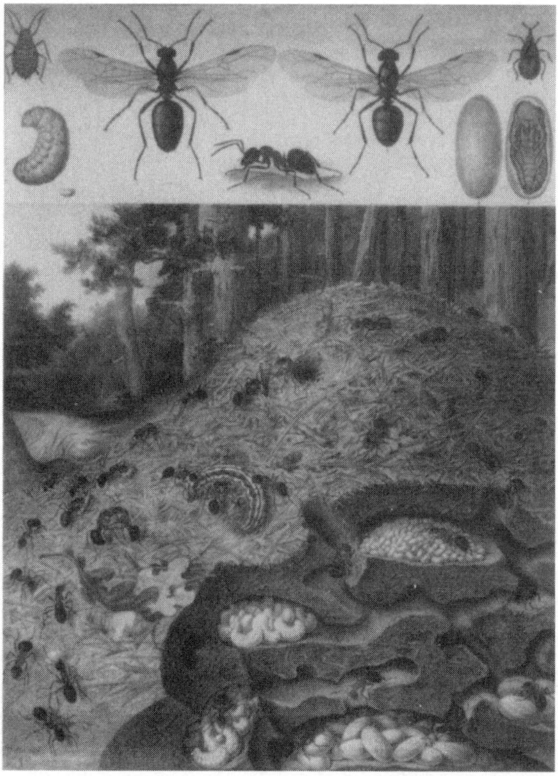

Figure 5.5. Tab. 47: Ameise mit Bau [Ants with their anthill] in A. Lehmann, *Zoologischer Atlas* [Zoological atlas], Leipzig, 1895. Chromolithography on cloth, 83 × 64.5 cm (32 × 25 in). (Collection of the State School Liceo G. Prati, Trento, Italy)

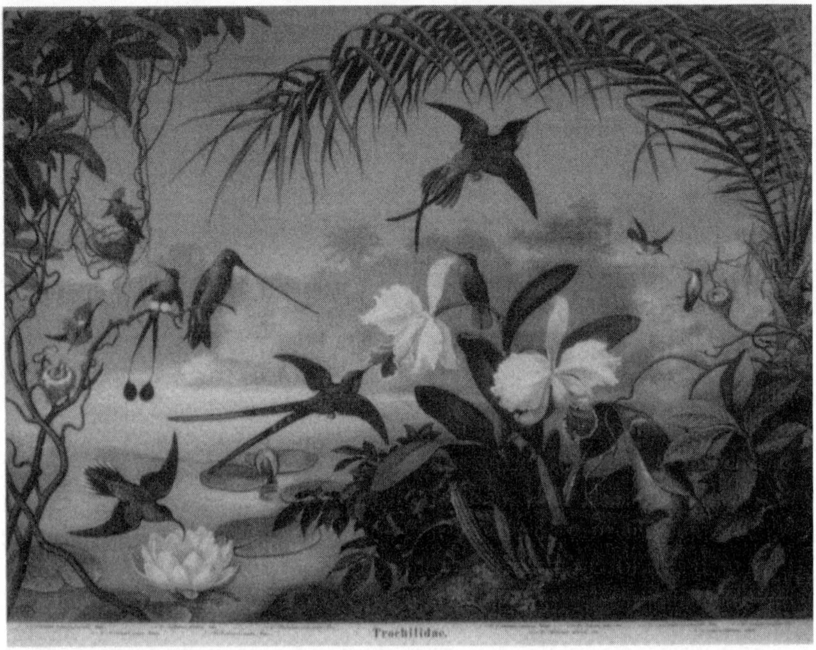

Figure 5.6. Tab. 54: Kolibri [Hummingbird], in A. Lehmann, *Zoologischer Atlas* [Zoological atlas], Leipzig, 1895. Chromolithography on cloth, 63 × 85.5 cm (24.5 × 33 in). (Collection of the State School Liceo G. Prati, Trento, Italy)

of representation can be observed in many other images of plants, such as the one in figure 5.8, which is from the *Anatomisch-physiologischen Atlas der Botanik* (Anatomical-physiological atlas of botany) written by the Swiss botanist Arnold Dodel-Port (1843–1908, professor at the University of Zürich) with the help of his sister Carolina. In this wall chart, the role of the pollinator in reproduction is the most prominent element.

Thus wall charts were expected to provide not only descriptions, but also an explanation of processes and, in some cases, even prescriptions of the conduct to be adopted in regard to certain natural phenomena, as in the wall charts of the *Schädlingstafeln* series (figure 5.9). The caption contained information about the life and behavior of each parasite, as well as practical suggestions on how to combat it.

This impulse directed to a more *lively* presentation of biological subjects corresponded perfectly with the two main, and often overlapping, strategies used in wall charts to facilitate the pupil's encounter with plants and animals. One consisted in presenting the object not as such, but in its natural environment (see figures 5.10 and 5.11).

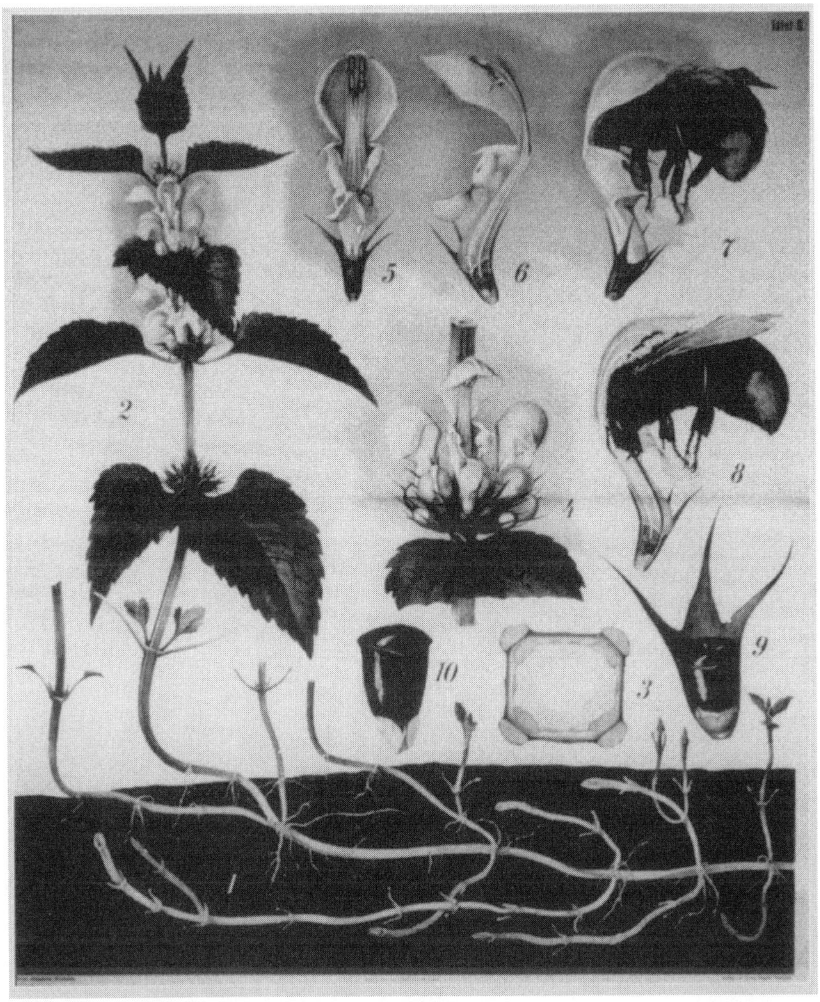

Figure 5.7. Tab. 2: Weisse Taubnessel [White nettle], in O. Schmeil, *Botanische Wandtafeln* [Botanical wall charts], Stuttgart, 1903. Chromolithography on cloth, 133 × 115 cm (52 × 45 in). (Collection of the State School Liceo G. Prati, Trento, Italy)

The use of this form of presentation was one of the cardinal rules of *Anschauungsunterricht*, teaching through images: to proceed from the known (familiar animals and plants) to the unknown (more exotic subjects). The other strategy was based on the aesthetic appeal of images. In reviewing the *Zoologische Wandtafeln* edited and drawn by Paul Pfurtscheller (a high school teacher—*Gymnasial Lehrer*—from Austria), Schmeil wrote: "In

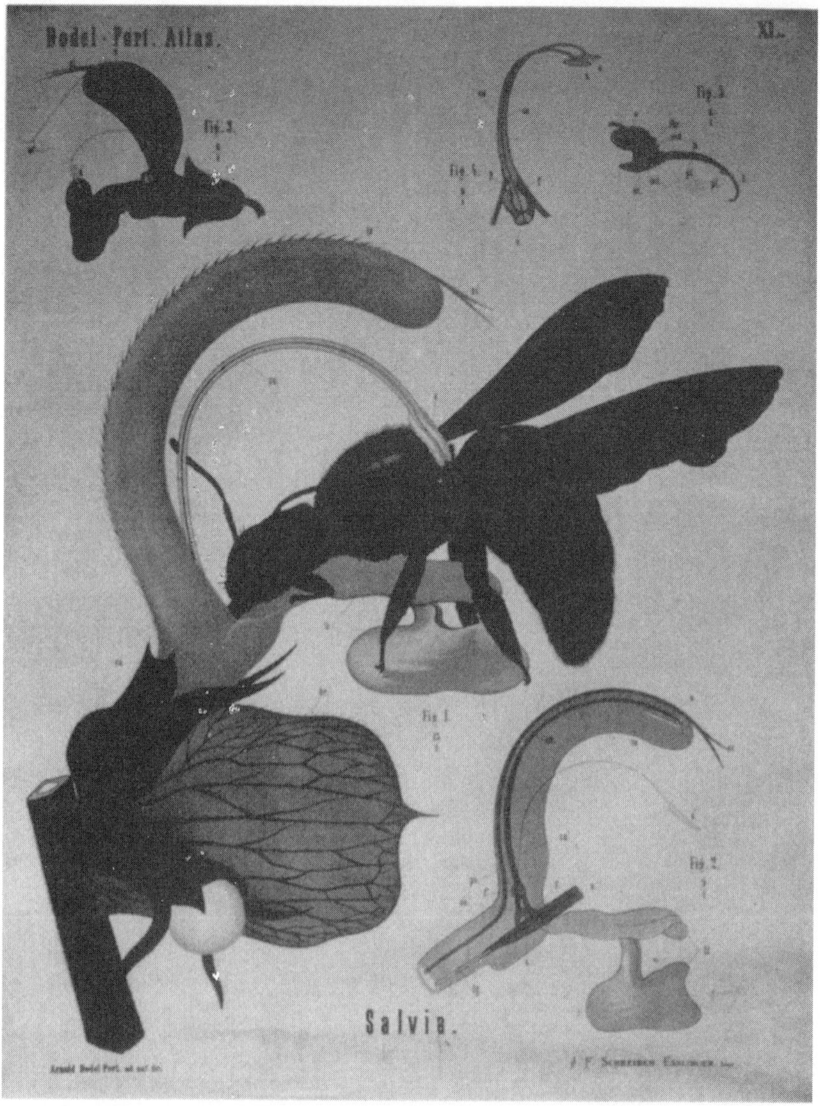

Figure 5.8. Tab. 40: Salvina [Sage], in A. and C. Dodel-Port, *Anatomisch-psychologischen Atlas der Botanik* [Anatomical-psychological atlas of botany], Esslingen, 1878–1883. Chromo-lithography on cloth, 82 × 64 cm (32 × 25 in). (Collection of the State School Liceo G. Prati, Trento, Italy)

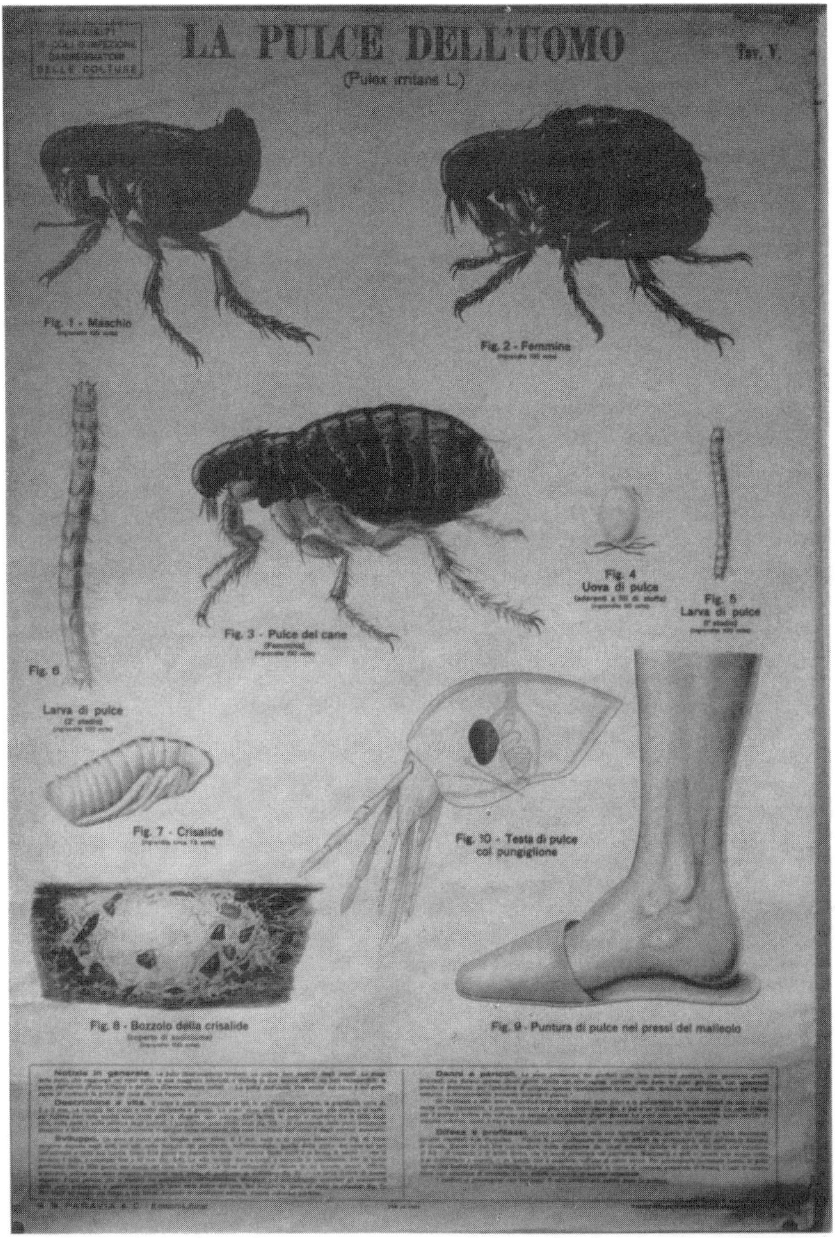

Figure 5.9. Tab. 5: La pulce dell'uomo [Human flea] (Pulex irritans), in A. Hase, *Parassiti—Veicoli d'infezione—Danneggiatori delle colture* [Parasites—Carriers of infectuous diseases—Pests of crops], Turin, Italian translation of Der Menschfloh, from the Schädlingstafeln series originally published in Germany between 1917 and 1925. Chromolithography on cloth, 85 × 68 cm (33 × 26.5 in). (Collection of the State School Liceo G. Prati, Trento, Italy)

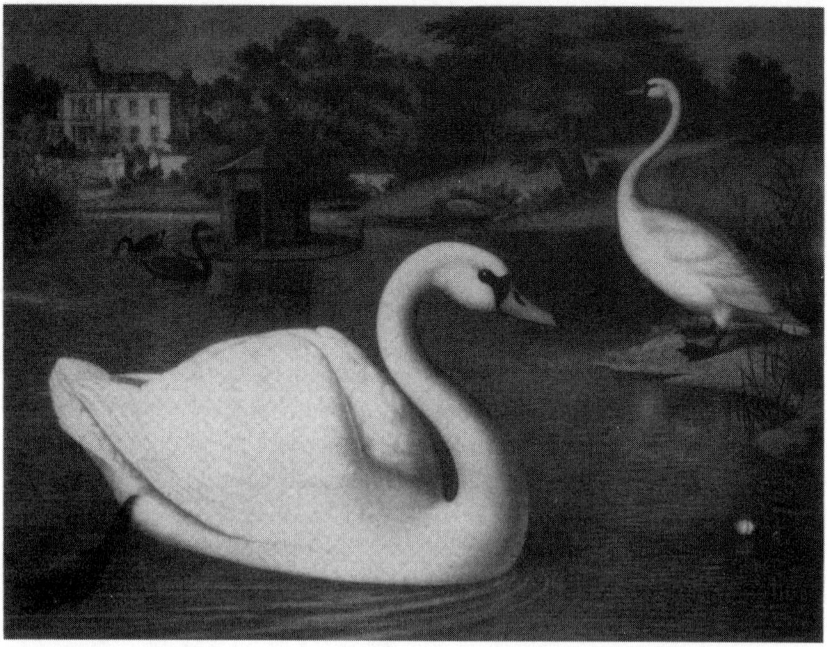

Figure 5.10. Tab. 45: Schwan [Swan], in A. Lehmann, *Zoologischer Atlas* [Zoological atlas], Leipzig, 1895. Chromolithography on cloth, 63 × 85.5 cm (24.5 × 33 in) (Collection of the State School Liceo G. Prati, Trento, Italy)

my opinion, they are the most artistically accomplished wall charts that have appeared so far" (Schmeil 1903, 308).[8] Other reviewers made similar comments:

> These images, being "art works for the wall" *[künstlerischer Wandschmuck]* in the school, will accordingly also exercise a favourable aesthetic influence upon pupils. (Fuss 1903, 5)[9]

> [These wall charts] are moreover different from those by Leuckart in giving greater importance to the formal elements in the image *[Habitusbild]*, and at the same time try to display the internal organization of the structure, while Leuckart's wallcharts neglect such formal elements. (Fedde 1903, 5)[10]

> Living nature is the best teacher and pedagogue; an artistic medium of representation *[Anschauungsmittell]* tries to replace nature and this can be possible in practice only if the images are true to natural objects. (Dodel-Port and Dodel-Port 1883)[11]

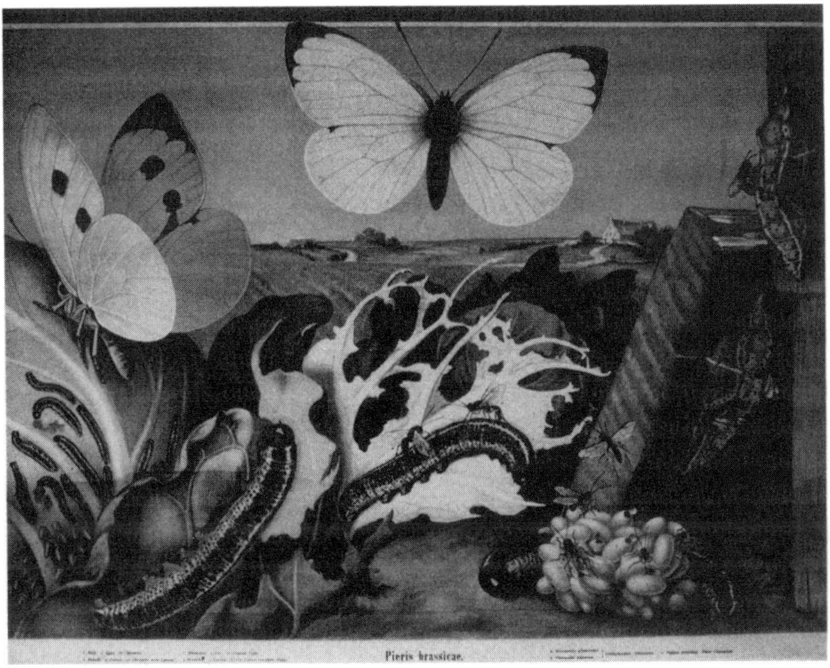

Figure 5.11. Tab. 53: Kollweissling [Cabbage butterfly], in A. Lehmann, *Zoologischer Atlas* [Zoological atlas], Leipzig, 1895. Chromolithography on cloth, 62 × 85 cm (24 × 33 in). (Collection of the State School Liceo G. Prati, Trento, Italy)

It is also a matter of interest that illustrators like Heinrich Leutemann (1824–1905) came to draw scientific wall charts after a long experience in drawing animals for children's books of tales and fables. Gottlieb von Koch (1849–1914), who participated in the production of Jung's *Neuen Wandtafeln für den Unterricht in der Naturgeschichte* (New wall charts for teaching natural history), was a painter with a strong background in biology; he also had been an assistant of Ernst Haeckel (1834–1919) in Jena and later was appointed professor of zoology at a *Hochschule* in Darmstadt (Thienie and Becker 1972, 74; see figures 5.12 and 5.13). Given the importance of their contribution to the design of wall charts, illustrators were often the most well known and well paid among those involved in the production of such series (such as authors of texts or editors). Leutemann, for instance, one of the biggest celebrities of the wall chart market, did not simply execute the drawings for the *Zoologischer Atlas* (Zoological atlas) edited by Carl Adolph Lehmann (1842–1927)

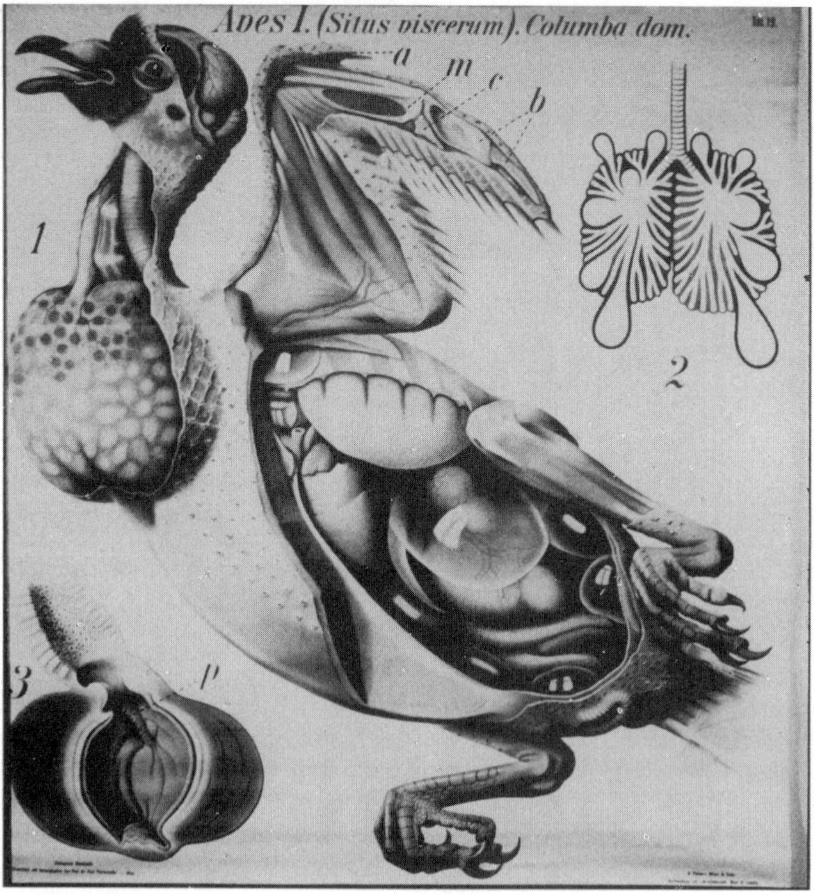

Figure 5.12. Tab.19: Aves I (Situs Viscerum) Columba Dom [Birds I (intestine) domesticated dove], in P. Pfurtscheller, *Zoologische Wandtafeln* [Zoological wall charts], Vienna, 1907. Chromolithography, 137 × 126.5 cm (53 × 49 in). (Collection of the State School Liceo G. Prati, Trento , Italy)

but actually was able to negotiate and shape its content (Bertram 1985); Hartinger generally was regarded as the author of the series he illustrated *(Hartingers Wandtafeln)*, despite the involvement of other contributors in terms of texts and scientific supervision. The dominant role played by illustrators is confirmed by the fact that in several cases (including the first editions of Hartinger's *Schwämme* series and the first editions of Pfurtscheller's *Zoologische Wandtafeln*), the illustrators showed a lively entrepreneurial spirit by acting themselves as publishers of the series.

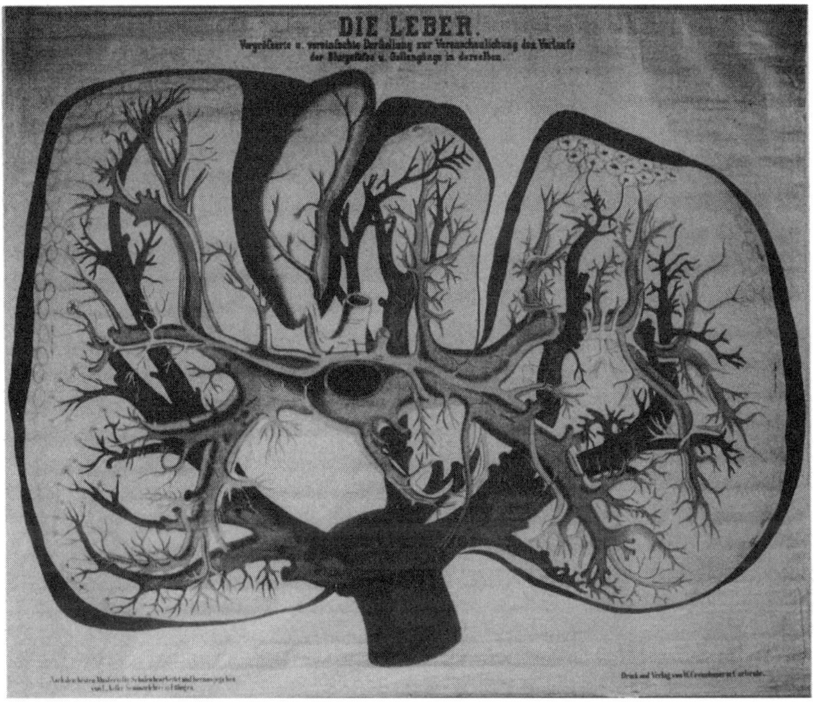

Figure 5.13. Tab. 4: Die Leber. Vergrösserte und vereinfachte zur Veranschaulichung des Verlaufs der Blutgefass und Gallegänge in derselben [The liver. Enlarged and simplified to display the vessels for the circulation of blood and of bile in the same figure], in L. Keller and R. Zilles, *Anatomische Schulwandtafeln* [Anatomical wall charts], Carlsruhe, 1883. Chromolithography, 66.5 × 79.5 cm (26 × 31 in). (Collection of the State School Licco G. Prati, Trento, Italy)

2. The Origin and Success of Wall Charts

Why did this particular medium for the visual communication of science enjoy such great diffusion and popularity? I will suggest three main sets of reasons.

Technical Reasons

Wall chart diffusion clearly was related to important advances in printing technology. The invention of lithography in 1798 by Aloys Senefelder (1771–1834), which in Germany was developed on an industrial scale by the Englemann workshops from 1816 on, laid the basis for the production of large color prints at a reasonably low price.

Educational Reforms

Processes of reform had taken place in the organization of education in the German states. To quote the historian of education Karl Ernst Jeismann, "In the nineteenth century the State put into effect, what it had announced in the eighteenth century and it became a school state. After the military and the tax state, the school state came into being with the state educational system" (1987, 4).[12] As a result, the number of schoolchildren grew enormously, especially in primary school. In the 1840s, there was a 40 percent increase in the number of teachers, while the number of schoolchildren increased by 108 percent in the same period. The average teacher had 103 pupils in 1833 and 136 pupils in 1852 (Schleunes 1989, 113; see also Leschinsky and Roederer 1976, 122). It became increasingly difficult to show images by simply circulating picture books or small engravings. There were similar difficulties in permitting every student to have a good view of a biological specimen or a chemical experiment. As the Dodel-Ports noted: "There are many objects which it is impossible to show through the microscope when teaching in a classroom" (Dodel-Port and Dodel-Port 1883).[13] It soon was recognized that these difficulties could be overcome by using large wall images that could be seen reasonably well from every part of the classroom.

In 1866, in the text accompanying his *Wandtafeln für Physik,* Bopp cautioned:

> In order not to increase production costs too much, a format has been chosen for the wall charts that should fit the majority of needs; however, should an even larger format become necessary for extraordinarily large schools, it would not be very difficult to enlarge one image or another by some graphical technique. (Bopp 1866)[14]

Presenting his series of botanical wall charts, Heinrich Jung explained, "These images are so executed as to allow them all to be recognized from any seat in the classroom" (Jung 1894).[15] Paul Wossidlo made a similar remark about Pfurtscheller's zoological wall charts: "The format has been chosen to be large enough to enable the details of illustrations to be visible even in very crowded classrooms" (1903, 5).[16]

Visualization was assigned a fundamental role by the educational thought of the time. Reformers of the educational system such as Karl Wilhelm von Humboldt (1767–1835) were influenced heavily in their practice by the ideas of Heinrich Pestalozzi (1766–1827). Central to Pestalozzi's pedagogical theory was the idea that children learn by transforming their own original "vague sense impressions" into "distinct and definite ideas." As Pestalozzi put it in *Wie Gertrud ihre Kinder lehrt* (How Gertrud teaches her children, 1801), "the world lies before our eyes like a sea of confused sense impressions, flowing into one another" (in Boyd 1921, 323).

In order to achieve this transformation process, Pestalozzi considered it crucial that pupils be provided with what he called *Anschauung*. *Anschauung*, that is "intuition," in the Kantian sense, given by sense data within the transcendental forms of space and time, implied that the pupil should become acquainted directly with an object by seeing it or handling it instead of merely hearing about it. Soon an *Anschauungsunterricht* was developed, designed for young pupils to learn to recognize animals or environments and their names. But *Anschauung* was most of all a method that could be applied to any kind of teaching, as we have seen, including that of scientific topics. This was recognized by the same researchers, whose commitment to extend and reform the teaching of biology culminated in 1901 in the so-called "Hamburger Theses," which declared biology to be "an empirical science *[Erfahrungswissenschaft]* . . . with no interest in metaphysical speculation," and capable of teaching "the so often neglected art of observing solid reality." (Gutzmer 1905, 533).[17] A few years later, the Education Commission of the German Society of Scientists and Physicians recommended that "pupils should be given an overview, grounded in intuition *[Anschauung]* and observation, of the richness of forms existing within the worlds of plants and animals." Authors of and commentators on wall charts strongly emphasized this aspect. For instance, in introducing the *Schädlingstafeln,* to which we have already referred, Frickinger wrote:

> Youth requires one thing, it is not satisfied with the spoken word, it wants visual material *(Anschauungsmaterial),* good visual material. The use of such material is our foremost means of awakening the abiding interest of the young. (Frickinger 1917, 2)[18]

An anonymous reviewer of Pfurtschelier's wall chart (1971) gave it as his opinion that "in teaching, words alone are not enough. To reinforce them we really need adequate images to be available."[19] And Carl Bopp judged that, "No explanations by the teacher can bring understanding when they are not made on the basis of elementary images" (Bopp 1866),[20] while the Dodel-Ports stated, in the preface to their botanical atlas: "Natural, scientifically reliable wall charts can replace a natural object in classroom teaching and in lectures; they are more enlightening than the spoken word" (Dodel-Port and Dodel-Port 1883).[21]

3. Wall Charts and Scientific Communication

One of the most striking aspects of wall charts is that they are not instantly intelligible. For instance, in the series by Heinrich Jung et al., *Neue Wandtafeln für den Unterricht in der Naturgeschichte* (New wall charts for teaching Natural History, 1894), figures are simply displayed on a black background without any caption or explanation (figure 5.14).

In several cases, description was provided by the teachers using their specialized knowledge and the explanatory booklet *(Begleitheft)* that accompanied the wall charts. This method had two significant implications. First, it called for greater involvement and participation on the part of both the teacher, who was usually supposed to know the text by heart, and the pupils. If we go back to Hartinger's series of fungi (figure 5.2), it is interesting to observe that the names of the different types of fungus are not printed on the wall chart. In fact, there was a transparent overlay, with the names, to be placed over the wall chart itself. By virtue of this and other similar arrangements, the pupils first could be shown images together with their names and salient features, and later asked to recall such elements by looking only at the wall chart (May 1909, 132).[22]

The second implication was that the same image could be used for different types of student, at different levels of education. The Dodel-Ports noted of their botanical atlas:

> We had in mind not only the needs of *Hochschule*, but also those of the *Mittelschule*. Pupils of different age mainly have been served badly in regard to schematic representations of all kinds, so that it is actually difficult for them to gain a correct understanding of natural living things . . . Accordingly, the *Anatomisch physiologische Atlas der Botanik* will be designed to be used at every level of botanical teaching and in every branch of botanical knowledge. (Dodel-Port and Dodel-Port 1883)[25]

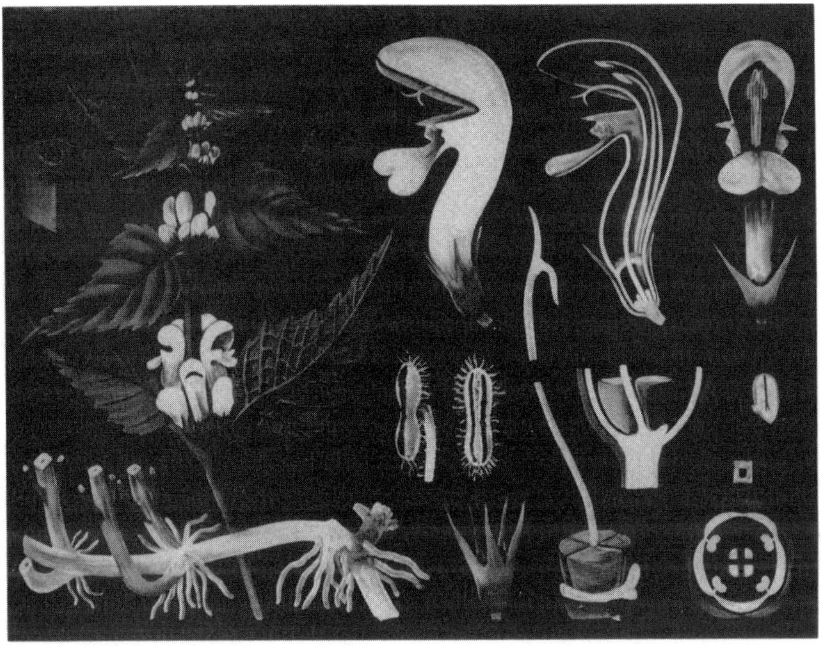

Figure 5.14. Tab. 14: Weisse Taubnessel [White nettle], in H. Jung, G. Von Koch, and F. Quentell, *Neuen Wandtafeln für den Unterricht in der Naturgeschichte. Botanik.* [New wall charts for teaching natural history and botany], Darnstadt, 1894. Chromolithography, 70 × 98.5 cm (27 × 38 in). (Collection of the State School Liceo G. Prati, Trento, Italy)

This use of images as a *universal* means for communication across different areas and levels of knowledge was also meant to bridge the gap between teaching and research. Among the authors of series of scientific wall charts, there were not only school teachers but also distinguished scientists such as the zoologist and paleontologist Karl Leuckart (professor of zoology at the University of Leipzig), the geologist Eberhard Fraas (1862–1915, Konservator at the Naturalien-Kabinett of Stuttgart), the geologist and palaeontologist Kari Alfred von Zittel (1839–1904, the "leading teacher of palaeontology in the nineteenth century and the only encyclopedist of the subject") (Hölder) and Gaston Bonnier (1853–1922, professor of botany at the Sorbonne) (see figures 5.15, 5.16, and 5.17).

As Otto Schmeil wrote: "It is without question one of the first duties of a teacher to keep abreast with the present state of research; because otherwise he would be teaching ideas that have had their day or things unfounded in science, and either erect a crumbling ruin or an imaginary

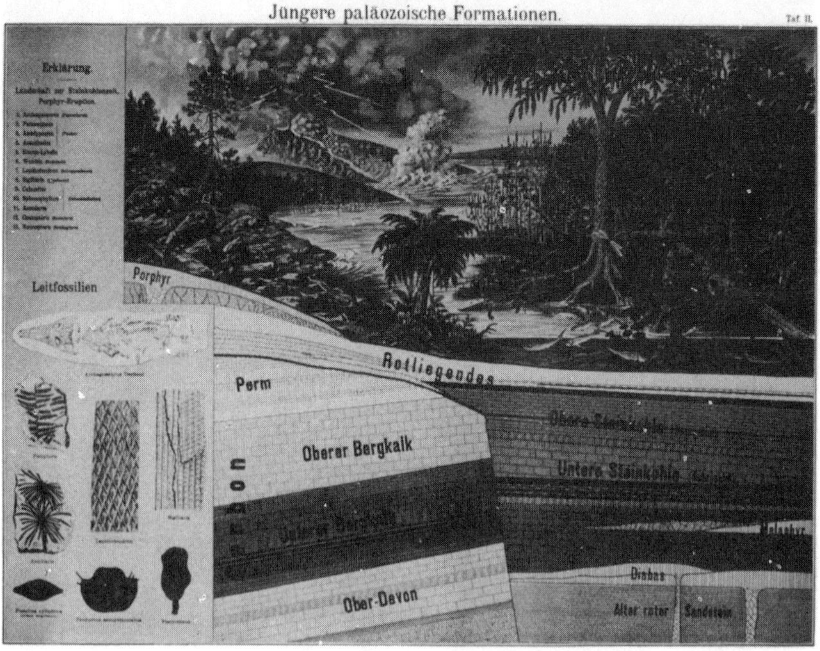

Figure 5.15. Tab. 2. Jüngere Paläozoische Formationen [Late Palaeozoic formations], in E. Fraas, *Die Entwicklung der Erde und ihrer Bewohner* [The development of earth and its inhabitants], Stuttgart, 1906. Chromolithography on paper, 30 × 163 cm (12 × 63.5 in). (Collection of the State School Liceo G. Prati, Trento, Italy)

castle in the souls of his pupils" (quoted in Seybold 1896, 233).[24] A similar point was also made by Arnold Dodel-Port: "[School teachers,] who have the first word on what should be included in or excluded from teaching, have requested the authors of this work not to hesitate to include in the charts, without reservation, everything that should be taken into account from the point of view of science" (Dodel-Port and Dodel-Port 1883).[25]

It is certainly interesting, from this perspective, to look at a wall chart about Coelenterates by Leuckart (figure 5.18). It was in fact Leuckart himself who introduced in 1847 the division of Cuvier's Radiata into Cocienterata and Echinodermata. (Schadewaldt n.d.). Clearly, the degree of actual involvement of scientists in wall chart production was not always as high as in the case of Leuckart (his series of wall charts was specialized and prepared mainly for the later years of the gymnasium and for university studies), or A. Dodel-Port (who discussed the selection of images for his *Anatomisch-physiologisch Atlas der Botanik* with a committee of botanists

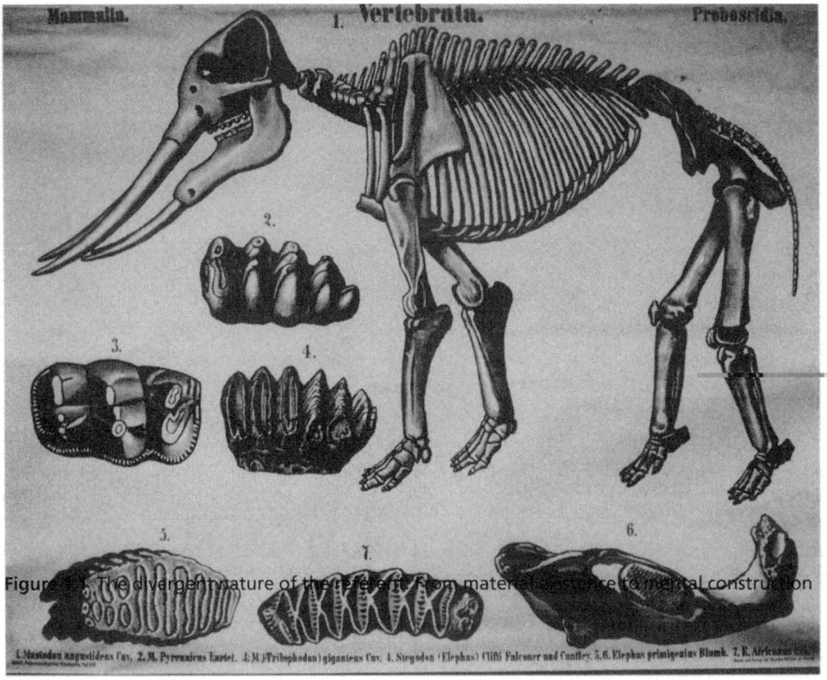

Figure 5.16. Tab. 57: Vertebrata-Mammalia-Proboscidea, in K. A. von Zittel and K. Haushofter, *Paleontologische Wandtafeln und Geologische Landschaften* [Paleontological wall charts and geological landscapes], Cassel, 1879–1901. Chromolithography on cloth, 108 × 143.5 cm (42 × 56 in). (Collection of the State School Liceo G. Prati, Trento, Italy)

during the Fiftieth Meeting of German Scientists and Physicians in Munich), and one legitimately may entertain the suspicion that some of the researchers were simply hired by publishers to act in a supervisory capacity or even merely to confer scientific legitimacy on a series of charts.

Later wall charts such as those presented in figure 19 (1920) and figure 20 (1963), also include visual metaphors, for example a man as a factory and the liver as a storehouse. It is worth quoting Dodel-Port, an enthusiastic follower of Charles Darwin (see for instance Dodel-Port 1889), who also considered educational tools such as wall charts an important channel for spreading ideas and theories: Our greatest gratitude goes to a person who took active part in our work at its beginning and who is unfortunately no longer among the living: "Charles Darwin . . . He was very much pleased by some of these wall charts—may they be passed on to future generations of his disciples" (Dodel-Port and Dodel-Port 1883).[26]

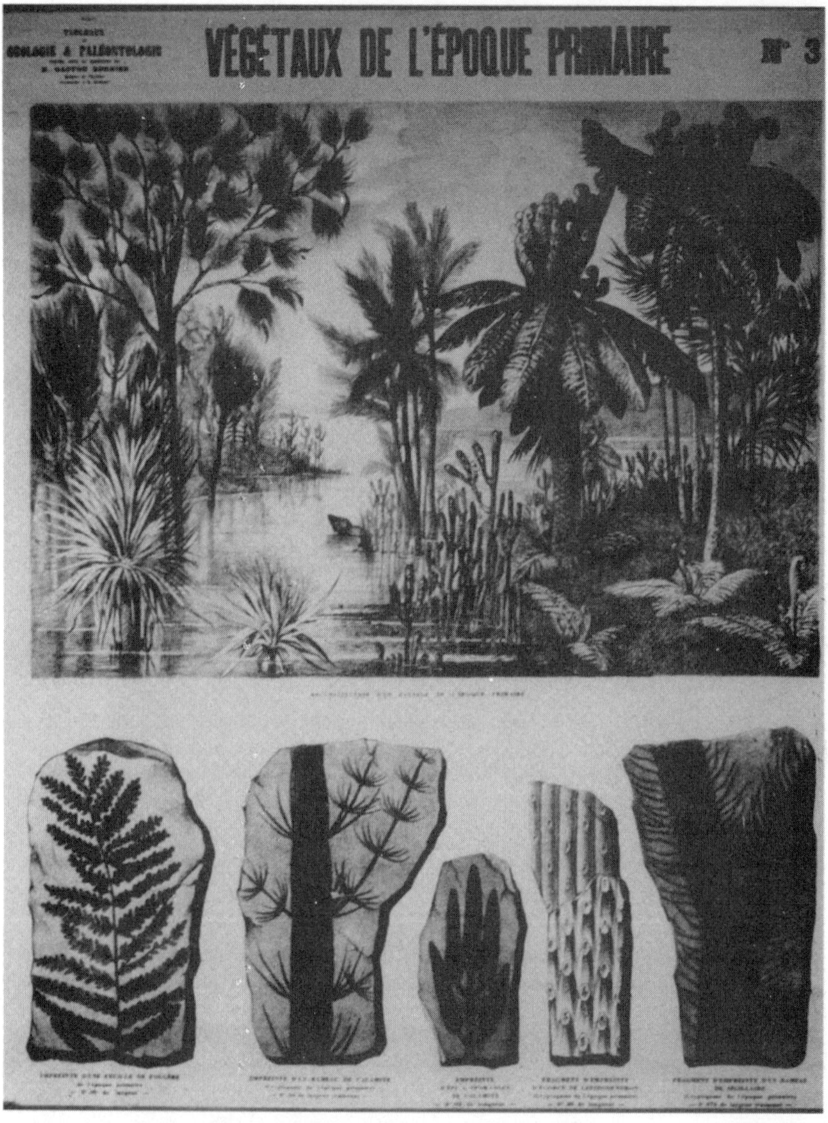

Figure 5.17. Tab. 3: Végétaux de l'Epoque primaire [Plants from the Primary era], in G. Bonnier, *Tableaux de Géologie et Paléontologie* [Geology and palaeontology wall charts], Turin, c. 1905. Lithography on cloth, black and white, 113 × 92.5 cm (44 × 36 in). (Collection of the State School Liceo G. Prati, Trento, Italy)

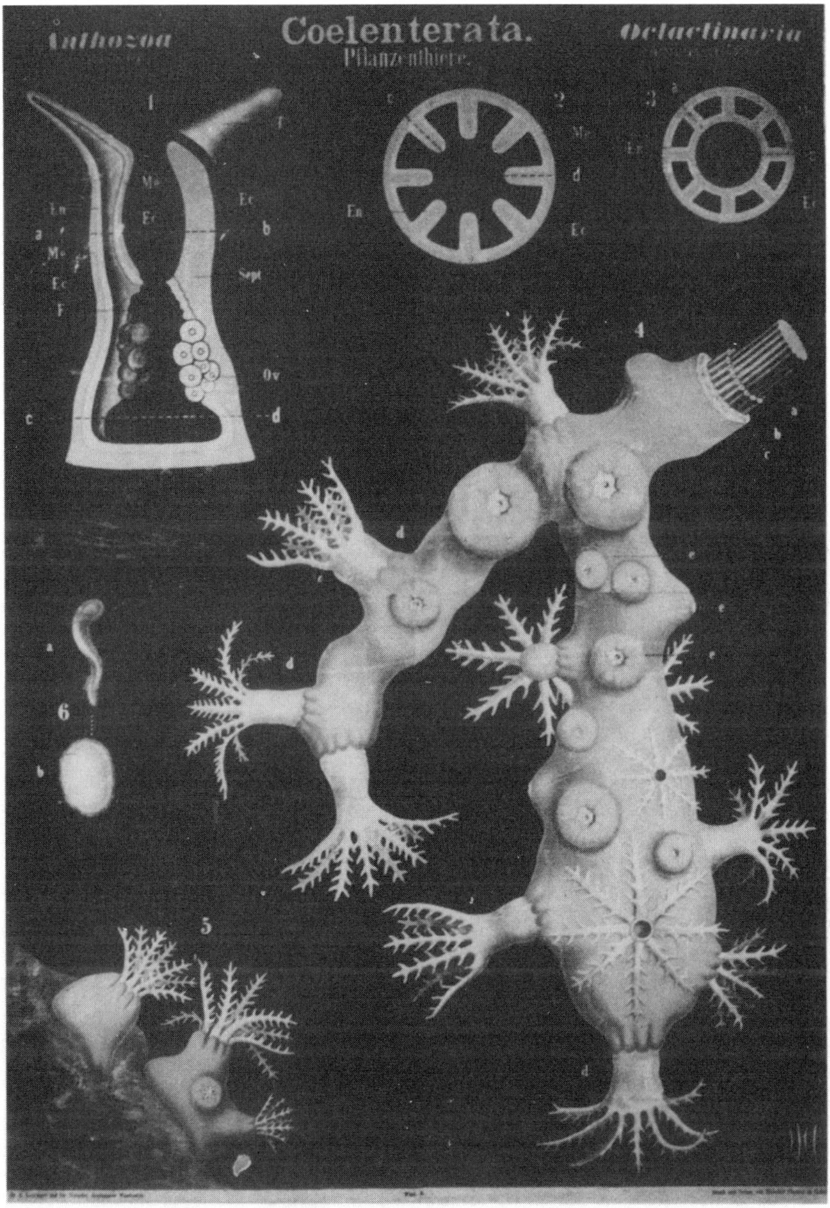

Figure 5.18. Coelenterata (Pflanzenthiere)—Anthozoa (Corallenthiere)—Octactinaria (achtstrahlige Korallen) [Coelenterata (plants)—Anthozoa (corals)—Octactinaria (eight-rayed corals)], in K. Leuckart and H. Nitsche, *Zoologische Wandtafeln* [Zoological wall charts], Cassel, 1877–1892. Chromolithography, 133 × 102 cm (52 × 40 in). (Collection of the State School Liceo G. Prati, Trento, Italy)

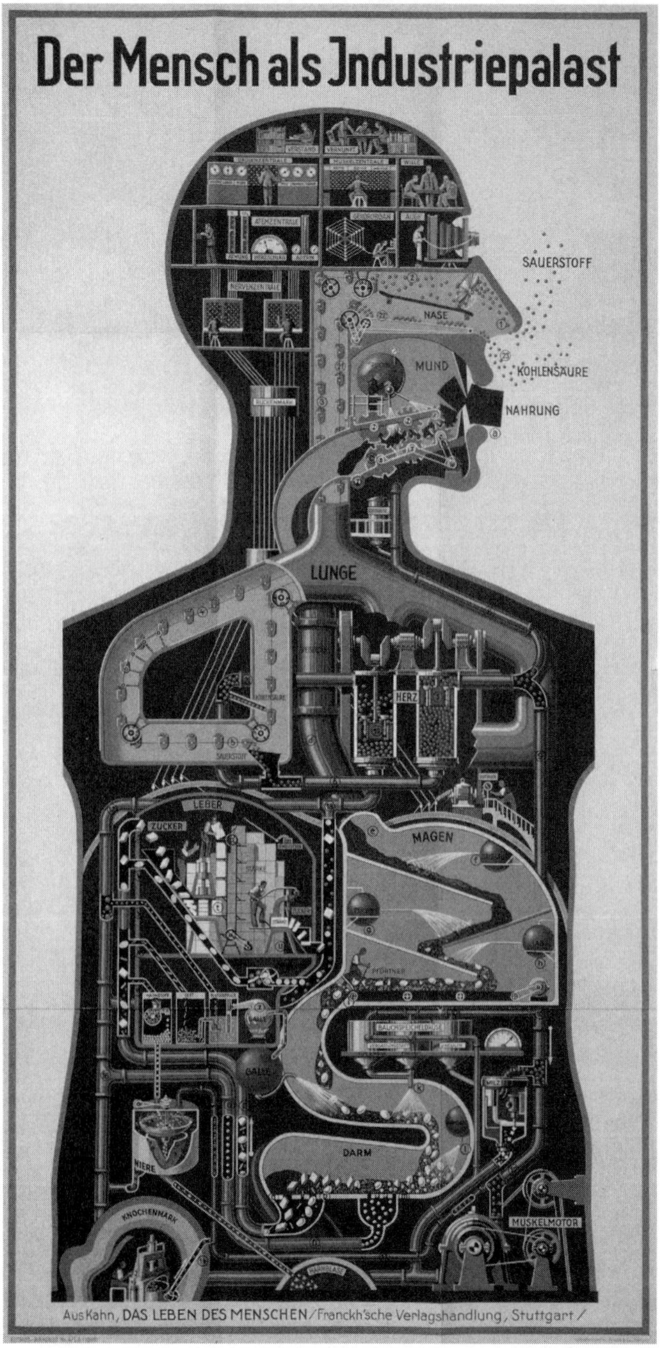

Figure 5.19. Der Mensch als Industriepalast [Man as a factory], in Fritz Kahn, Stuttgart, 1926. Chromolithography (National Library of Medicine, Germany)

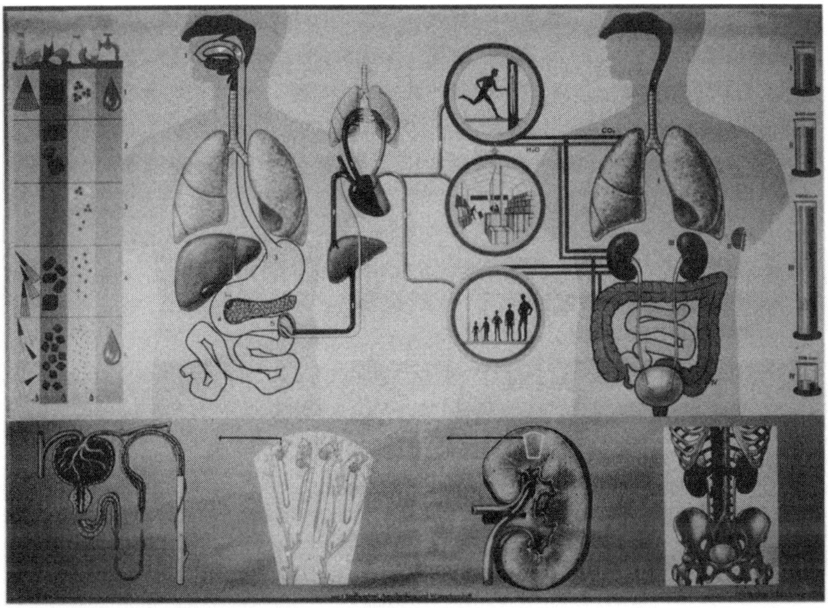

Figure 5.20. Tab. 9: Stoffwechsel, Ausscheidung und Wasserhaushalt [Metabolism, secretion and balance of fluids], in K. Rosenbauer and F. von Hagen, *Unser Körper, Lehrtafeln für menschenkundlichen Unterricht* [Our body. Wall charts for the teaching of human anatomy and physiology], Düsseldorf, 1963. Chromolithography on cloth, 117 × 165 cm (46 × 64 in). (Collection of the State School Liceo P. Prati, Trento, Italy)

4. Concluding Remarks

The emergence and flowering of wall chart production in late nineteenth-century Germany was made possible by a unique combination of several elements: structural transformations within the educational system, trends in scientific practice and pedagogical thought, and artistic skills and reproduction technologies. Thus wall charts, and in particular scientific wall charts, represent a valuable source for the study of the historical development of these different areas and for their reciprocal interrelations, particularly since some of their authors and editors operated simultaneously within more than one area. Ideas about the most appropriate forms and content for the teaching of biology in school and universities were derived from theories about the nature of a discipline, its place in society, and its relations to other practices (such as fine art). A group of scholars has even gone so far as to suggest that wall charts might be seen as mirrors of a *Zeitgeist,* since they capture through their style and content the intellectual *spirit* of a certain historical period (Müller 1984; Stach et al. 1988).

In any case, wall charts undoubtedly remind us of the crucial role that visual media play in scientific education and in scientific communication in general. Together with scientific apparatus and laboratory practice, images are an important element in teaching science and, given their nonverbal, more direct appeal, they also can communicate to different kinds of audience and different types of practitioner.

Notes

1. This was the English expression generally used to translate the original German terms for such products, *Wandtafeln* or *Schulwandbildern.*
2. "Ein jedes Blatt muss als ein neues Geschenk an die Kinder, und zwar zu einer solcher Zeit sein, da man ihnen vorzügliche Zufriedenheit mit ihrem Verhalten bezeugen will." J. B. Basedow, quoted in Schmidt (1990, 65).
3. The *Lehrmittel-Netto Katalog* of 1911 listed more than six thousand different wall charts (Wingolf Lehnemann 1994). Schlüter did not write or actually design individual wall charts but conceived of the series and coordinated the work of several other specialists.
4. "Nichts ist im Verstande, was nicht zuvor in den Sinnen war! . . . Nur durch fleissiges Beobachten, durch Selbstschauen und Selbstuntersuchen ist es möglich, den schlimmsten Feind alles geistbildenden Unterrichts aus der Schule zu verbannen: den Verbalismus" (Schmeil in Seybold 1986, 243).
5. Friedrich Junge, a teacher from Kiel, had a key role in translating Möbius's principles into pedagogical practice (F. Junge 1917).
6. "Alle lebenden Pflanzen und Tiere, die das Kind sieht, gehören natürlich oder durch den Menschen veränderten künstlichen Lebensgemeinschaften an . . . Gärten, Wiesen, Felder und Wälder bieten reichliche Gelegenheit dar, Schüler aller Alterstufen zu belehren, dass alle lebenden Wesen bedingte und mitbedingende Glieder von Lebensgemeinschaften sind. Das Kind fühlt sich selbst als Glied einer Lebensgemeinschaft mit seinem Eitern, Geschwistern und Gespielen, mit dem Hausgesinde und den Haustieren" (K. A. Möbius, quoted in Schmidt 1984, 81).
7. "Die Natur nicht als ein Konglomerat, sondern als ein Organismus erfasst werde" quoted in Widar Lehnernann (1994, 72).
8. "Dabei sind die Bilder künstlerisch in einer Weise vollendet, wie dies m. A. nach in keinem Tafelwerk bisher erreicht worden ist" (Schmeil 1903, 308).
9. "Diese Bilder werden daher auch als 'künstlerischer Wandschmuck' eigener Art in Schulen aufs vorteilhafteste ästhetisch auf die Jugend einwirken" (Fuss 1903, 5).
10. "Sie unterscheiden sich von den Leuckartschen Tafeln allerdings dadurch, dass sie mehr Wert auf das Habitusbild legen, zugleich aber auch am habitusbilde selbst die innere Organisation zu zeigen suchen, während die Leuckartschen tafeln das Habitusbild vernachlässigen" (Fedde 1903, 5).

11. "Ist die lebendige Natur die beste Lehrerin und Erziehrin; ein künstliches Anschauungsmittel will aber die Natur ersetzen und das ist nur dann in zweckdienlicher Weise möglich, wenn die Bilder von Naturgegenständen wahr sind" (Dodel-Port and Dodel-Port 1883, ii).

12. "Im 19. Jahrhundert verwirklichte der Staat, was er in 18. Jahrhundert angekündet harte: er wurde zum Schulherrn. Nach dem Militär- und dem Steuerstaat entstand mit dem 'staatlichen Unterrichtswesen' (Lorenz von Stem) der Schulstaat" (Jeismann 1987, iii, 4).

13. "Das vorweisen und demonstriren von zahlreichen Obiekten am Microscop selbst ist für den Klassenunterricht eine Unmöglichkeit" (Dodel-Port and Dodel-Port 1883, ii).

14. "Sollte für ausnahmsweise grosse Schulen ein noch grösseres Format nöthig sein, so ist es nicht sehr schwer, sich die eine ober andere Tafel durch einen des Zeichnen Kundigen in der nöthigen Vergrösserung nachbilden zu lassen" (Bopp 1866, v.). At this time, a variety of apparatus were available to assist in copying drawings in different sizes.

15. "Die Abbildungen sind so auszuführen, dass jede einzelne deutlich von allen Plätzen einer Klasse erkannt wird" (Jung 1894, iii).

16. "Der Massstab ist so gross gewählt, dass auch in stark besetzten Klassen die Details der Abbildungen zur Anschauung kommen" (Wossidlo 1903, 5).

17. "1. Die Biologie ist eine Erfahrungswissenschaft, . . . Für metaphysische Spekulationen hat die Biologie als solche keine Verantworten . . . 2. . . . Im besondern lehrt die Biologie die sonst so vernachlässigte Kunst des Beobachtens an konkreten" (Gutzmer 1905, 533).

18. "Aber die Jugend . . . will eines: sie gegnügt sich nicht mit dem gesprochenen Wort, sie will Anschauungsmaterial, gutes Auschauungsmaterial. Erst an Hand von diesem wecken wir ihr nachhaltiges Interesse!" (Frickinger 1917, 2).

19. "Das belehrende 'Wort' allein genügt jedoch nicht. Zu seiner Unterstützung sind vielmehr geeignete Bilder erforderlich" (Review of Pfurtscheller's wall chart 1971).

20. "Aller Erklärung des Lehrers nie recht zum Verständnis kommen, weil die zu Grund liegenden elementaren Anschauungen fehlen" (Bopp 1886, iii).

21. "Naturgetreue, wissenschaftlich zuverlässige Wandtafeln können beim Klassen-Unterricht und bei Vorlesungen die Vorweisung der Naturkörper selbst am besten ersetzen; sie belehren mehr als das lebendige 'Wort'" (Dodel-Port and Dodel-Port 1883, ii).

22. "Die Wandtafeln sind nicht nur bei der Neudurchnahme, sondern auch bei der Wiederholung zu benutzen" (May 1909, 132).

23. "Wir harten nicht allein das Bedürfnisse der Hochschule, sondern auch diejenigen der Mittelschulen ins Auge gefasst. Die Schüler aller Unterrichtsstufen werden häufig so sehr mit schematische Darstellungen aller Art misshandelt, dass sie gerade deswegen das richtige Verständniss der lebenden Naturkörper nicht erlangen . . . So wird der *Anatomisch physiologische Atlas der Botanik* auf allen Stufen des botanischen Unterrichtes und in allen zweigen des botanischen Wissens verwendbar sein" (Dodel-Port and Dodel-Port 1883, i–ii).

24. "Der Lehrer . . . so ist es fraglos eine seiner ersten Pflichten, sich über den derzeitigen Stand der Wissenschaften zu orientieren; denn im anderen Falle würde er ja wissenschaftlich längst überlebte oder wissenschaftlich nicht feststehende Tatsachen lehren, entweder eine morsche Ruine oder ein Luftschloss in dem Geiste seinder Schüler aufbauen" (Schmeil quoted in Seybold 1896, 233).

25. "Die Lehrern, welche in Sachen des pädagogischen Taktes das erste Wort über das zu reden haben, was vom Unterricht auszuschliessen sei und nicht, haben die Herausgeber des Werkes gebeten, alle Bedenken fallen zu lassen und in den Tafeln ohne irgend welche andere Rücksichten alles das darzustellen, was vom Standpunkt der Wissenschaft aus in erster Linie zu berücksichtigen sei" (Dodel-Port and Dodel-Port 1883, i).
26. "Der verdientesten Einer, welche am Fortgang unseres Werkes regen Antheil genommen, ist leider nicht mehr unter den Lebenden: Charles Darwin . . . Er hat an manchem dieser Blätter seine helle Freude gehabt—möge diese nun an seine kommende Schüler übergehen" (Dodel-Port and Dodel-Port 1883, iii).

References

Bertram, A. 1985. Heinrich Leutemann als Illustrator von Schulwandbildren. Dargestellt am "Zoologischen Atlas." Thesis, University of Duisburg.

Bonnier, G. 1905. *Tableaux de geologie et paléontologie.* Turin.

Bopp, C. 1866. *Text zu den acht Wandtafeln für Physik.* Ravensburg.

Boyd, W. 1921. *History of Western education.* London.

Dodel-Port, A., and Dodel-Port, C. 1883. *Erläutender Text zum Anatomisch-physiologischen Atlas der Botanik.* Esslingen.

Dodel-Port, A. 1889. *Moses oder Darwin? Eine Schulfrage.* Zürich.

Fedde. F. 1903. Urteile über Professor Dr. Paul Pfurtschellers Zoologische Wandtafeln. In *A. Picklers Witwe und Sohn Katalog, Buchandlung und Lehrmittel-Anstalt.* Leipzig.

Frickinger, H.W. 1917. *Die Schädlingstafeln der Deutschen Gesellschaft für angewandte Entomologie.* Halle.

Fuss, K. 1903. Urteile über Professor Dr. Paul Pfurtschellers Zoologische Wandtafeln. In *A. Picklers Witwe und Sohn Katalog, Buchandlung und Lehrmittel-Anstalt.* Leipzig.

Gutzmer, A. 1905. Reformvorschläge für den mathematischen und naturwissenschaftlichen Unterricht. *Zeitschrift für mathematische und naturwissenschaftlichen Unterricht* 36.

Hölder, H. *Karl Alfred von Zittel.* DSB.

Jeismann, Karl Ernst. 1987. Zur Bedeutung der Bildung in 19. Jahrhundert. In *Handbuch der deutschen Bildungsgeschichte.* Munich.

Jung, H. 1894. *Begleitheft zu den Neuen Wandtafeln (Zoologie und Botanik) für den Unterricht in der Naturgeschichte.* Darmstadt.

Junge, Friedrich. 1917. *Der Dorfteich als Lebensgemeinschaft.* Kiel.

Lehnemann, Wingolf. 1994. Schulwandbilder—Bilder für den Unterricht. In *Von Tieren und Pflanzen. Schulwandbilder für die Naturkunde.* Exhibition Catalogue. Münster.

Lehnemann, Widar. 1994. Einheimische Wildtiere auf Schulwandbildern. In *Von Tieren und Pflanzen. Schulwandbilder für die Naturkunde.* Münster.

Leschinsky, A., and P. M. Roederer. 1976. *Schule im historischen Prozess.* Stuttgart.

May, F. 1909. *Methodik der Naturkunde auf Grunde der Reformbestrebungen.* Düsseldorf.

Müller, W. 1983. *Anschauung und Ideologie. Schulwandbilder der NS-Zeit aus Duisburg.* Duisburg.

———. 1984. Schulwandbilder als Spiegel des Zeitgeistes? In *Die weite Welt im Klassenzimmer. Schulwandbilder zwischen 1880 und 1980.* Exhibition Catalogue. Cologne.

Review of Pfurtscheller's wall chart. 1971. *Zeitschrift für Lehrmittelwesen und pädagogische Literatur* 7.

Schadewaldt, H. *Karl Georg Friedrich Rudolph Leuckart*. DSB.

Schleunes, K. A. 1989. *Schooling and society: The Politics of Education in Prussia and Bavaria, 1750–1900*. Oxford.

Schmeil, Otto. 1896. *Über die Reformstrebungen auf dem Gebiete des naturgeschichtlichen Unterrichts*. Stutgart.

———. 1903. Review of Paul Pfurtscheller's *Zoologische Wandtafeln*. *Natur und Schule* 2: 308.

Schmidt, J. 1984. Die "biologischen" Wandbilder von Otto Schmeil. In *Die weite Welt im Klassenzimmer. Schulwandbilder zwischen 1880 und 1980*. Exhibition Catalogue. Cologne.

———. 1990. Naturgeschichtliche Wandbilder im Wandel didaktischer Theorien. In *Schule halten heisst Welt gestalten*, edited by N. Just and A. Kleckers. Essen.

———. 1995. *So Bleibt man gesund! Gesundheitserziebung in visuellen Medien von 1900 bis 1950*. Essen.

Seybold, A. 1986. Otto Schmeils Lebenswerk. In *Otto Schmeil: Leben und Werk eines Biologen—Lebenserinnerungen*. Heidelberg.

Stach, R. et al. 1988. *Schulwandbilder als Spiegel des Zeitgeistes zwischen 1880 und 1980*. Opladen.

Thienie, U., and F. Becker. 1972. *Allgemeines Lexikon der bildenden Künstler von der Antike bis zu Gegenwart*. Leipzig.

Voglino, P., G. Roda, and V. Bongini. 1917. *Organografia vegetale*. Turin.

Walters, S. M. 1989. Introduction to *Botanical atlas. A guide to the study of plants*, by D. McAlpine. London.

Wossidlo, P. 1903. Urteile über Professor Dr. Paul Pfurtscheller's *Zoologische Wandtafeln*. In *A. Picklers Witwe und Sohn Katalog, Buchhandlung und Lehrmittel-Anstalt*. Leipzig.

6

Representing Moving Cultures
Expression, Multivocality and Reflexivity in Anthropological and Sociological Filmmaking

Luc Pauwels

The scientific significance of the photographic imaging technique already was underlined in 1839, when it was first presented to the public, and indeed it was not long before the method was put into practice by ethnographers and anthropologists. Likewise, the development of the moving image toward the end of the nineteenth century was driven by scientific rather than commercial motives, even though non-fiction cinema since has stood in the shadow of fictional film production.

The initial appeal of the camera to anthropological research was that it made traditional fieldwork lighter and more interesting for researchers, who previously had to try to capture the complexity of the cultures under study by pencil and paper. Whatever escaped their attention or was hard to put into words therefore inevitably was lost to the research (Mead 1963). In consequence of technological innovations, shooting pictures in the field soon became less cumbersome, and an altogether more relaxed approach could be taken. Fewer technical obstacles meant that researchers could engage more actively and more variedly in a dialogue with their field of research, and to an extent this is what hap-

An earlier and shorter version of this chapter has been previously published as: Pauwels, L. 2004. Filmed science in search of a form. Contested discourses in anthropological and sociological film-making. *New Cinemas* (2) 1: 41–60, doi: 10.1386/ncin.2.1.41./0.

pened. Yet, technological progress alone could never have brought about social and scientific innovation. After all, the integration of this far-reaching visual approach into scientific practice implies the development of a theoretical framework and an adapted methodology, issues that only took shape very slowly and indeed continue to be debated today. The emphasis that anthropologists have put on the documentation of often disappearing societies quite often has been at the expense of efforts to underpin the visual productions theoretically and to put the new means of research at the service of new insights. The implicit schemes that were adopted were often indicative of a naïve-realistic and ethnocentric outlook on the multicultural meeting that anthropological fieldwork essentially is.

1. The Multiformity of Anthropological and Sociological Filmmaking

Besides providing a description or (an inevitably imperfect) reflection of certain aspects of the reality studied, social scientific filmmaking represents a specific form of scientific communication that is directed toward specialized practitioners of the trade, aspiring social and cultural scientists or, as the case may be, a broader public (although in each of these cases it may concern very different kinds of film). The usefulness of film for data collection—both in observation and in experimentation—is now beyond question, but the possibilities it offers in terms of the integration of images into a scientific discourse and the precise rules that should be followed in this respect are still very much debated.

Many authors have tried to distinguish between films that are little more than raw footage and films as independent expressive scientific constructions. According to some, the scientific film can consist only of *pure data* that has been collected under meticulously controlled circumstances, while others argue that such products cannot possibly be scientific because, being primary databases or pure descriptions, they cannot offer a *processed* and scientifically sound perspective of the phenomenon under study. Many visual researchers have tried to represent this polarity (i.e., between emphasizing mimesis or scientifically motivated expression) in concrete terms. Lajoux (1975), for example, refers to a dichotomy between "scientific film" on the one hand and "expressive film" on the other. De Heusch (1988) calls the pure-data form "research film" as opposed to the more processed "sociological film." Leroi-Gourhan makes a distinction between "filmic notes" and "documentary" (Chiozzo 1989,

11) and Collier (1967) uses the term "cultural documentation" for the more mimetic variant, which he distinguishes from the "anthropological film" and "cultural drama for a broader public." Wolf (1967) feels that one should differentiate between "research film" ("Forschungsfilm") and "documentation film" ("Dokumentationsfilm"), arguing that these two genres are fundamentally divergent in terms of their truth claims, scientific significance, and use of specific filmic means of expression. Prost introduced a similar division between "representational film" and "illustrative film," in which the former term refers to cinematographic material that can be used as a scientific data source, while the latter can fulfil merely an entirely illustrative purpose, much like documentaries and educational films do. However, Prost does concede that it is much harder to make this distinction in practice and that, at best, there is a gradual difference in the manner in which an effort is made to guard the scientific integrity of whatever is depicted. Moreover, a single film (sequence) may be "representative" for one research purpose and merely "illustrative" for another (Prost 1975, 325–327).

If the divergent purposes of film in a scientific or educational context are taken into account, as well as their different target audiences, it soon transpires that this crude dichotomy (mimesis/expression, or raw data versus elaborated view) only reflects part of the existing diversity.

In an early report by the Department of Scientific and Industrial Research, entitled *The Film in Scientific Research,* the notion of a "research film" is described rather restrictively as "film taken in the course of research for the purpose of analysis" (FSR 1963, 8). However, the report also sums up broader or related descriptions of scientific film:

a) *Research film.* Film material assembled in logical sequence by the addition and arrangement of isolated record shots, made for analysis.

b) *Concept film,* comprising short lengths of silent film taken in the course of research and illustrating events analysed by slow-motion cinephotography or speeded up by time-lapse techniques, which can be used as "moving lantern slides" in post-graduate teaching.

c) *Research film report.* A film designed to communicate information about research in much the same way as does a paper in a specialized journal, with orderly presentation of the problem, method of experimentation, results, and interpretation.

d) *Research information film.* Film taken during research, which is subsequently prepared for showing to scientists or others for the purpose of information.

e) *Research educational film.* Film, possibly including research material, which is prepared for the purpose of describing a piece of research.

f) Film prepared specifically or incidentally to *demonstrate* new applications of photographic *techniques.*

A more recent attempt to grasp the multiformity of the phenomenon of film in an ethnographic context by means of a number of distinguishing criteria is that by Crawford. This categorization, too, provides elements for further debate on the distinguishing marks of scientific film (including outside the ethnographic sphere) and its subgenres:

1. Ethnographic *footage* is unedited film material, which may be used in its unedited form for research purposes or eventually be edited into a film.

2. *Research* films are edited films made specifically for research purposes and hence not intended for public screening or an audience other than a highly specialized academic audience.

3. The ethnographic *documentary* is a film which has a specific relevance to anthropology but which is in one way or another part of documentary filmmaking in general.

4. The ethnographic *television* documentary is a film made for, and very often by, a television company, with the intention of reaching a wide non-specialized audience.

5. *Education* and *information* films are made for educational purposes and meant for classroom audiences or the general audiences.

6. Other *non-fiction* films include journalistic reports, newsreels, travelogues, and so on.

7. *Fiction* films and drama documentaries may be labelled ethnographic because of their subject matter. (Crawford 1992, 74)

Both attempts exemplify that there is an urgent need for a more qualified approach and specification with regard to the notion of film as a component or an end product of scientific endeavor in either the broad or the narrow sense. The implicit or explicit truth claims made in such films and their scientific usefulness cannot be detached from specific goals and target groups, and indeed a production methodology that (hopefully) is adapted to this purpose. Ultimately, we need to ask what *science* is or should be, and we need to ascertain what may be regarded as legitimate scientific objectives and how they are best attained.

2. Filmic Codes and Controversies: Expressiveness Up for Debate

The narrative structure of a scientific film may tie in closely with the inherent structure of the events recorded or it may provide a very manifest and scientifically founded and processed perspective on these events, using typically filmic means. According to Wright (1992, 276), controversies over this issue also are related to the fact that there are diverging opinions on the so-called "'natural state' of the medium—subjective constructed meaning versus an objective 'transparent' recording."

The strongly dramatized exoticism of a number of early films that were labelled as ethnographic even though they were not necessarily made by ethnographically schooled filmmakers, soon made way for a striving toward a more "scientifically sound" camera use. This resulted in a very sober, functional style of shooting, with the intention of the most accurate possible collection of data in their broader context. Many researchers in this rather positivistic and often naïve realistic tradition used to have (and indeed still have) almost blind faith in the automatic objectivity of the camera, in which they saw (or see) the long hoped-for elimination of subjectivity and selectivity of field notes.

The creative or constructive editing of images into a film was considered to be incompatible with scientific praxis—or at least it was felt that it should be kept to a minimum—because it entailed a moment of subjective selectivity that would destroy the truth value of the image. This is a somewhat short-sighted view that nevertheless was shared by theoreticians of the fictional genre (see, for example, Bazin 1975). Young asserts about this rather stubborn view that "it was looking at only a very small part of the problem that led to anthropologists in the early days to ask filmmakers to preserve all the footage exposed and not 'edit' it into a 'film'" (1975, 66). The choice of what is filmed, at what moment in time, and in which manner is an equally selective process; it is already a form of *editing* (selection and rearrangement) of reality. The decision not to edit recorded material merely has an influence on the retention of the integrity of the recorded film—the original chronology and context—but in itself it offers insufficient guarantees that the visual document is complete and objective (which is and will always be an unattainable ideal). Nevertheless, it remains useful to preserve all recordings in their original sequence, even after the material has been edited into a film. In the course of time, or after repeated viewing, certain details can lead to new insights or questions that necessitate a close review of the original material.

Editing remains a crucial and controversial aspect of scientific filmmaking, probably because this genre is very much focused on the legitimacy of the construction of film. It is indisputable that careless editing according to non-scientific standards can seriously compromise the research value of the recorded material (e.g., the application of cinematographic principles for purely aesthetic or dramatic purposes: rapid succession of shots, alternating frame size in order to create variation or artificial tension and dynamism). On the other hand, a careful selection and ordering of shots according to the technical principles of scientific research and in accordance with the purpose of the research (accurate description or communication of insights) need not result in a less scientifically sound product. Similar interventions (selection of moments and ordering of time and space) also occur during the recording, at shot level, so that generally one needs to respond much more quickly and impulsively to unexpected circumstances, while often one is still unaware (certainly in the case of spontaneous action) of the precise unfolding of events. The possibility of ordering the recorded material in a carefully considered manner and according to a motivated structure largely determines the strength of the film as a form of scientific communication.

Besides the ordering of the images, the relation between the visual representation of an event and the synchronically recorded sound is usually of great significance. Synchronic sound—be it vocal or non-vocal—offers an important complementary dimension. Sound film in varying genres rarely will contain soundless moments. Besides synchronic sound, most film productions also make extensive use of sounds that are recorded later and that tie in with the images (so-called wild sound), off-screen commentary, and background music. These kinds of additions should be handled with great care and circumspection from a social scientific perspective. Music, unless it actually is performed during the recording and is thus an integral part of the reality conveyed (i.e., synchronic sound), indeed can have a disrupting effect on the process of knowledge transfer. Sound that is added afterward can set a particular mood and thus impose a certain interpretation upon the viewer, for which there may be no scientific grounds whatsoever. Or it will at least distract some of the viewer's attention, which otherwise would be focused on the actual material. Likewise, the use of commentary, be it *guiding* or not, remains a hotly debated aspect of audiovisual productions by social scientists. The viewer needs to receive the required information in order to be able to put whatever is shown in its proper context. For that matter, this need not necessarily happen (entirely) in the film itself, but also may be achieved through a verbal introduction, written information, or other means.

The application of motivated punctuation, a scientifically functional rearrangement and selection of raw material need not imply a loss of research value, but may in fact be enriching from a scientific point of view. To claim that the use of filmic means is simply incompatible with scientific goals holds a danger that a number of specific possibilities of the image as a means of knowledge transfer or communication will remain unused. Moreover, this conviction is based on the faulty premise that there is such a thing as unprocessed data. Even in the most sober verbal or numeric representation of facts lies hidden a particular way of seeing things, a perspective or belief. In other words, a theory. To try to banish this from scientific endeavor is simply impossible. Wright, too, subscribes to this conviction when he asserts quite resolutely that "even in the act of pointing we cannot escape some degree of construction. So, if narrative is unavoidable, this gives an important emphasis to the need to understand the terms in which narrative has been established. It cannot simply be regarded as the unfolding of events before the camera" (1992, 276). It therefore clearly is unrealistic to demand that the producer of images should merely be an intermediary and must never express a point of view.

For that matter, the unconscious conveyance of a view can be much more dangerous than expressing it consciously and explicitly. It therefore is paramount for the scientific filmmaker to make assessments on the basis of scientific insights and to use the appropriate expressive devices without succumbing to the temptation to apply much more attractive means of expression in an uncontrolled and unmotivated way.

It is clear that scientific criteria should always come first, not the striving toward a smooth production according to the standards of the entertainment industry. Rollwagen stresses quite emphatically the role of scientific theory in the film production process. He sharply criticizes the filmic approach that does not put scientific theory central and that relies instead on other frames of reference (e.g., purely cinematographic ones). Among other reasons, he claims this because the filmmakers involved are insufficiently trained in the research discipline (1988, 290–293). At the same time, opportunities are missed because researchers are unable to integrate the two areas of expertise and mental frameworks. Integration implies that one is capable of translating scientific views and insights into visual information and, conversely, that one can distill scientific information appropriately from visual manifestations.

On the other hand, lack of fluency and visual aplomb is not synonymous with "being scientific." Science, too, stands to benefit from effective communication. The point is that another set of rules must be obeyed; rules, for that matter, that are often yet to be formulated and implemented. More concretely, this boils down to a well-considered selection and acceptance of existing, possibly adapted (visual) codes and the development of specific conventions for a visual scientific discourse. It is clear that the audience of sociological films (and by extension "scientific films" in general) will need to get used to the audible and visible absence of familiar filmic devices and slowly come to accept other norms and conventions with regard to the application of such devices. Also, the intended target audience (experts, prospective scientists, non-experts) should be made aware of precisely what they are watching, of the constitutive rules of whatever has been visually represented.

In some circles, an understandable but exaggerated response to the naïve realism of many filmmakers and consumers has been that visual observation as such is considered contaminated while the word is seen as the source of salvation. Consequently, more contemporary, often postmodern films unfortunately are offering (even) less visual approaches. Such films consist primarily in verbal exchanges with and within the research field, and the verbalization of the persons under study is promoted to a new kind of anthropological or sociological truth.

Yet the utterances by participants remain views of individuals, rather than of entire cultures, which subsequently are edited in accordance with the insights of the researcher or the dramatic instinct of the editor. In essence, one naïve perspective is replaced with equally naïve or one-sided views, in particular the conception that research subjects are truthful, that groups under study always have full and accurate insight into their own situation, and that a pure representation of dialogues highlights the underlying reality and makes it understandable. Such approaches contribute very little to the further development of a theoretically and methodologically sound form of visual data collection and communication whereby the complementary nature of word and image can be exploited to the full. Of course, the views and perceptions of those involved remain important sources of information for the acquisition of a qualified scientific understanding of a situation.

In a much debated publication, the anthropologist/filmmaker Jack Rollwagen sharply criticized ethnographic filmmakers who claim that the structure of reality can be uncovered through pure observation ("observational cinema") and that the viewer thus can acquire full insight into events merely by watching the film. At the same time, however, he attacks the previously mentioned post-modern, altogether language-oriented, approach that purposely foregrounds the (often entirely verbalized) view of the informants of the culture and the author of the audiovisual text, but that offers very little in the way of theoretical underpinning or guidance.

> Anthropological filmmaking is not simply the recording of what human beings say or do but the interpretation of those recordings within the disciplinary framework of anthropology throughout the total filmmaking process from conception to execution. The incorporation of anthropology into anthropological filmmaking as a conscious process is a statement by the anthropological filmmaker that he/she is willing to have his/her ideas and conclusions subjected to the same rigorous evaluations by colleagues as a written work receives. There is no room in such framework for evasion of responsibility as there is in an approach which suggests that what is recorded speaks for itself. (Rollwagen 1988, 295)

3. The Meeting of Researcher and Researched: Towards Multivocality and Mutual Promotion of Interests

In the past, many ethnographic filmmakers tended to minimize both the impact of the camera on exhibited behavior and the impact of the specific styling possibilities and choices in the various phases of the production process. They often believed that their productions were extremely objective and systematic observations. Goldschmidt's definition of ethnographic film is characteristic of this view: "Ethnographic film is film which endeavors to interpret the behavior of people of one culture to persons of another culture by using shots of people doing precisely what they would have been doing if the camera were not there" (Goldschmidt 1972 in MacDougall 1975, 114). MacDougall rightly remarks that conceptions (or misconceptions) such as those expressed by Goldschmidt often go hand in hand with another illusion, namely that the camera can grasp the totality of events. The invisibility of the camera is a fiction (with the exception of truly hidden cameras). To pretend that the presence of the camera (or the camera crew) has no influence whatsoever on exhibited behavior actually entails a denial of an existing situation. As MacDougall asserts, "no ethnographic film is merely a record of another society: it is always a record of the meeting between a filmmaker and that society" (1975, 119). The pursuit of objectivity, and a befittingly sober and meticulous filmic style, are nevertheless very legitimate goals. However, an almost exclusive attention toward these aspects easily can result in other important methodological questions being ignored. Furthermore, an attempt to restrict filmmaking efforts to making the most accurate representation of perceivable reality leaves unused a number of new possibilities for scientific communication.

The ethnographic ideal of invisibility still persists in certain circles. We can distinguish between at least two camps among researchers/cinematographers: On the one hand, those who feel that the impact of the camera on events is negligible or at least controllable and, on the other, those who clearly recognize the reactivity of those observed to the presence of the camera and who try to apply this insight (with varying degrees of success), so that scientifically valuable data can still be collected and made into a film. The underlying belief among the latter is that the explicit entry by the camera crew into the world of the research subjects and the ensuing possibilities for the acquisition and verification of information through the participation of the subjects examined in the research project enhance the authenticity of the acquired material rather than create an unreal, disrupted, and therefore scientifically invalid situation. Rouch (1975, 100) typifies this important shift in thought as follows: "The anthropologist is no longer an entomologist observing his subject as if it were an insect ('putting it down') but rather as if it were a stimulant for mutual understanding." The striving of the scientist to record and observe natural and representative behavior is not affected by the introduction of the camera. Yet, the camera's presence should be recognized as an integral part of the real situation, an element that may have a great or a small impact on other elements in this unusual setting (as is the presence of researchers themselves).

In addition to the fact that the contribution and cooperation of the group studied is necessary to the acquisition of knowledge, another important human driving force is involved in participative or interactive forms of visual research, namely the desire to reshape the traditional research situation into a more egalitarian relationship. Those who are being researched should not be kept under tutelage or subjected to arbitrariness on the part of the researcher. Moreover, if at all possible, they should take something away from their cooperation with the research team, for example in the shape of knowledge or means to address an undesirable situation (such as economic deprivation or political oppression). Although these propositions are essentially commendable, certain conceptions (such as the notion of "shared" production) need to be approached with some circumspection. Sometimes, they appear to be rationalizations that hardly tie in with research practice. The participatory idea may be rooted in an inarticulated feeling of guilt among anthropologists occasioned by a colonial past and the often-dubious role that their discipline played in it. The camera, in particular, proved to be a very useful instrument for creating and sustaining a particular image of the homeland that supported a certain political doctrine or scientific experimentation with an outspokenly Western perspective.

Despite this understandable and legitimate response to a colonizing use of the camera and the ensuing rearrangement of the relation between the researcher and the researched, it makes little sense to conceal that researchers ultimately play a steering role, and must continue to do so in order to fulfil their scientific mission adequately. But clearly there can be various degrees of imbalance between researcher and researched and there are, real alternatives for the relationship between the two. This relationship used to be a matter of the researcher needing something from the researched and subsequently disappearing with the loot.

The necessary reformulation of the meeting between the researcher and the researched may begin to take shape through the skillful use of visual feedback methods. These make it possible to give a voice to the research subjects (both visually and verbally), certainly if this contribution—this new or different story or perspective—is also given a place in the visual end product.

4. Visual Feedback Techniques: Possibilities for Constructive Interaction between Researcher and Researched

Although working with cameras and visual material sometimes can invoke resistance, spontaneous participation and enhanced interest among respondents are more common reactions. The researcher can use this interest to uncover new perspectives and insights and to help close the gap between researcher and researched. By encouraging the respondents to provide verbal or visual feedback (i.e., commentary on images or image production), the interpretation of certain aspects of the production of images is, to an extent, entrusted to the respondent.

While the notion of acquiring an overall view of *the* truth vis-à-vis cultural events is unquestionably an illusion, feedback techniques can help adjust the story of science or the scientist, so that it becomes a multivocal narrative of a meeting between cultures. This in turn may result in more ethical and, most probably, scientifically more valuable products and practices.

Interview with Camera Images (Photo and Film Elicitation)

The technique whereby camera images are used as a stimulus in interviewing originally was applied in psychological research. It subsequently was adopted by a number of social scientists. The confrontation of the interviewee with the reality of camera images can provide two kinds of information for the researcher. First and foremost, the interview with visual materials can offer the researcher a simple and quick way to acquire information about whatever has been recorded. A properly informed respondent (e.g., in the case of a visual report on a ceremonial or festive event) can tell very accurately who or what has or has not been captured by the images, which actions are performed, and what the significance is of certain signs and symbols.

The purpose of the interview with photo or film elicitation, however, is not restricted to the collection or explanation of a series of concrete facts about whatever has been recorded. The technique also allows the researcher to elicit deeper, more abstract perceptions and values of respondents as individuals who are involved in the depicted world. Carefully chosen visual material of a photographic or non-photographic nature (such as prints or drawings) combined with a good interview technique can broaden the interview from an information review about what has been recorded to a data collection session about the significance of the recorded material to the respondent. Thus, the focus of attention shifts from external manifestations to an *experience*, to an *interior* perspective, as it were. The possibility of eliciting crucial information (including emotions and perceptions) in interviews through camera images is valued by researchers who have used the technique intensively. Krebs (1975, 284) asserts that "if the film elicitation technique is employed skillfully, the researcher may obtain some of the most exciting data of anthropology—how members conceptualize and structure the world in which they live." Collier (1967, 49) goes one step further, claiming quite resolutely that "methodologically, the only way we can use the full record of the camera is through the projective interpretation by the native."

In comparison to the purely verbal interview, the visual interview certainly offers a number of specific benefits. While the interviewer often will not find it easy to convince his respondents to participate, visual material—especially if it concerns the immediate environment or field of interest of the interviewee—often can serve as a door opener. Visual material may provide the kind of structure to an interview that will not be considered irritating or compulsory by the respondent, so that it rarely forces the latter into a defensive or artificial position. It always provides a concrete talking point, while the stimulus provision that is characteristic of interview situations is experienced as less direct. The visual material often evokes spontaneous and unpredictable answers from respondents.

While the verbal interview often is characterized by a typical role differentiation between interviewer and interviewee (Segers 1983, 216), this relationship can assume an entirely different form in the visual interview. Rather than being pushed into the role of the person who is being questioned, the respondent takes on the role of an expert. This encourages him or her to speak freely. When commenting on images, interviewees tend to feel much less as if they are talking about things that they should not be talking about. Instead, they feel as if they are simply explaining things that already have been captured by the recording. Collier found that purely verbal interviews tend to become unproductive much more quickly than interviews with visual stimuli. If new images are made available perpetually, then the respondent's attention and interest can be retained for long periods of time. Irrelevant digressions on the part of the respondent can be prevented elegantly by providing a new image that focuses attention on a new fact. Visual material thus jolts the memory of the respondent, who will tend to become more easily irritated during a purely verbal interview because of the lack of interesting new stimuli from the researcher and through the confrontation with the failure of his own memory (Collier 1967, 48).

This comparison between the verbal interview and the interview with visual stimuli, which is based largely on Collier's findings, may create the impression that the latter type of interview is superior and effortlessly resolves all problematic aspects of the former. However, the interview, be it purely verbal or with visual aids, remains a complex interaction process that is subject to a whole range of influences. In the visual interview, the result ultimately depends not only on the quality of the interviewer and the perception and the attitude of the respondent, but also on the adequacy of the selected visual material. This visual material—like the interviewer himself—can disrupt or distort the research process by being irrelevant to the issue studied, by being ill-adapted to the respondent or by creating an all too suggestive, one-sided, or incomplete picture of the phenomenon or event considered. Moreover, an additional and often challenging expertise is required from interviewers: They are expected to be familiar with the visual material used, the manner in which it was obtained, its research possibilities, and interpretations. Even if the visual material used by the interviewers has been collected and produced with the greatest possible care, they need to avoid suggesting certain interpretations based on their own perspective or ignoring the conceptual frameworks of the respondents. Both Krebs (1975, 297) and Wagner (1979, 91) believe that this specific problem can be resolved by avoiding naming or interpreting depicted phenomena and by posing open questions ("What's happening here?" "What's that?" "Tell me about . . .").

The significance to the respondents that lies enclosed in the image cannot be approached in a strictly standardized manner. By always remaining alert for respondents' reactions to the visual material, and by adopting a flexible attitude toward unpredictable turns, researchers often can collect the most significant information.

Cultural Self-portrayal or Native Image Production

The interview with visual elicitation can offer a wide range of relevant information about how an informant perceives his world, but the clearest visual feedback is obtained by allowing the members of the group or culture studied to produce their *own* images. The underlying premise is that, in this way, significant patterns of the culture of the makers will be expressed in the images and thus be communicated to the researcher and to the other members of the group.

A comparable presupposition constitutes the basis of sociological research of static or dynamic familial images (photos, home video, family Web sites, etc). In a sense, these are also forms of "native image production" (Pauwels 1993, 2002a; Chalfen 1987). An essential difference between the two, however, is that in the case of elicited self-portrayal, the image production occurs in a research context rather than in a social setting. Unlike the family photo album, elicited image production is not part of normal social life. After all, the researcher initiates and steers the production of camera images in places and at moments that may be very unusual for the groups studied.

This form of visual feedback typically has little structure: The researcher has only limited control of the manner in which certain aspects are portrayed (which makes it more difficult to compare to other data of similar events). In fact, a basic assumption of this approach is that the influence of the researcher should be minimal. Still, the question remains to what extent one can prevent the cultural outlook (or, in this case, *bias*) of the researcher from affecting or disrupting the visual practice of the group studied, even in the teaching of the most elementary techniques for the production of camera images. This bias may manifest itself in specific choices with regard to filmic parameters (framing, duration and structure of recordings, selection of topics, etc.). Although this issue does not compromise the applicability of the technique, it is quite clear that the researcher must be very wary of possible ways of influencing the research process, as well as its scope and effects.

One of the best-known studies in which the visual feedback technique is applied is *Through Navajo Eyes* by Sol Worth and John Adair (1972). These two researchers taught a number of Navajo Indians to operate a 16 millimeter camera. In an amazingly brief period of time, they succeeded in making a series of short films about their culture. This convinced Worth and Adair that teaching basic filmmaking techniques to members of other cultures could provide insight into typical cultural and mental patterns among such groups:

> A working hypothesis for our study was that motion picture film, conceived, photographed, and sequentially arranged by a people such as the Navajo, would reveal aspects of coding, cognition and values that may be inhibited, not observable, or not analyzable when the investigator is totally dependent on verbal exchange—especially when such research must be done in the language of the investigator. We were interested not only in studying the general nature of the cognitive processes involved in this visual mode of communication itself, but were searching for specific pattern, code, and rules for visual communication within a cultural context. It was the interdependence between a mode of communication and its context as expressed in patterned, interrelated behavior that was sought. (Worth and Adair 1972, 27–28)

The researchers noticed, for example, that the Navajo filmmakers used a significant amount of footage to record people who were moving from one place to another (e.g., in order to find the right wood for making a fire in a film on the activities of a silversmith). Later, they found that being "on the move" also plays an important role in Navajo songs, poems, and everyday language. Moving from one place to another is considered a significant event in itself, and not just as a means to an end (Worth and Adair 1972,

146). They also felt that Western filmmakers tend to work more "elliptically": they usually try to shorten the process by leaving out any moments that are perceived to be unimportant. They would for instance use a shot of a man who walks out of the frame, possibly a number of short shots of someone picking up some twigs and then moving out of the picture again, and finally a shot of him back home with a load of wood.

Today, visual feedback techniques in various forms have become favored approaches for many visual sociologists and visual anthropologists. Douglas Harper, who frequently has used the photo-elicitation method, provides a comprehensive overview of such research, primarily in the sociological field. Photo-elicitation techniques often have been applied to subfields, such as social class research, family studies, identity research, community research, historical ethnography projects, and various culture-related topics (Harper 2002, 16). However, the potential area of application is much broader, and the types of imagery and visual media that can be employed are manifold.

Also, these techniques have proven particularly useful when combined with other visual or more commonly applied social science methods and approaches. In a recent study, social researchers Gloor and Meier (2000) investigated people's verbal and visual reactions to a river revitalization project in Switzerland using a visual approach that merges "native image making" with photo elicitation, sensibly combined with more established social science methodologies (survey, interview). Visitors to the area under study were asked to take Polaroid photographs of what they liked and what they disliked about the locality. Afterward, they were invited to explain the content of their pictures and their reasons for taking them. This approach yielded unique insights into the different significances of the riverside area and into the many facets of people's attitudes and demands with regard to that particular social space.

Interestingly, three distinct *frames* were employed consistently by the respondents: a nature-oriented or ecological perspective emerged as just one of many ways in which visitors experienced the area. Social and aesthetic viewpoints were more prominent in the photographs and in the verbal evaluations of the revitalization project. All three perspectives or frames had an appreciative (positive) and a more critical (negative) variant (2000, 119–134).

Scholars from disciplines other than sociology or anthropology also have embraced and further developed such visual feedback techniques. Horst Niesyto (2000), for example, focuses in his research on the media-ethnographical exploration of symbolic milieus and in particular uses techniques of self-representations through video in the context of media-educational youth research in Germany. His theoretical work deals with important questions of conceptualization and methods when working with adolescent video self-productions and reflects upon the quality, the validity, and the hermeneutics of self-produced videos.

Psychoanalysts continue to use projective drawing techniques as a therapeutic device. Drawings or other types of patient-produced visual materials (still pictures, video) in a controlled therapeutic setting are understood to be ways of allowing individuals to express their inner world, the preconscious and unconscious motives, and their relationship with the analyst. Child psychoanalyst Diem-Wille (2001) strongly believes that children's drawings express their fantasies, wishes, and experiences in a symbolic way and she uses projective drawing techniques mainly as a means to overcome inhibitions and self-censoring mechanisms that tend to occur routinely in purely verbal exchanges with the therapist. This particular field of expertise—perhaps more than any other—proves how much the interpretation of visual materials remains a daunting enterprise, in spite of more or less clear procedural preparations and models. While some minor methodological guidelines have been developed and agreed upon (e.g., "Look at what is drawn first," as this is believed to reveal what is most important to the creator), analysts ultimately have to interpret the numerous features of the drawing and its often highly metaphorical and symbolic nature, relying on their experience, informed guesses, and intuition, which they can then try to substantiate with other pieces of evidence.

These examples also illustrate that visual techniques should not be thought of as a short-cut or an undemanding method of data gathering, and that they do not lead to uncomplicated analysis and interpretation of such data.

5. Reflexivity, or the Metadiscourse of Filmed Science

Increasingly, the (visual) representation of cultures is seen as an extremely complex meeting of cultures: first and foremost, between those of the researchers and those of the researched, but subsequently also with the culture of the viewers or users of the resulting visual product. A scientist is not an objective and emancipated subject who observes, but a "positioned" and "repositioned" subject (Renato 1989), with personal beliefs, preferences, experiences, characteristics, cultural backgrounds, etc. From this position, the scientist approaches a field of study, influences it, and is influenced by it. The audience of scientific products is also inevitably positioned and actively introduces its own world into the interpretation of whatever it is presented with.

Moreover, it has become quite clear that every medium, even one that looks as realistic as film, inevitably has mimetic limitations and cannot but reduce or reshape (be it in a consciously or an unconsciously expressive manner). And finally, awareness is growing that the primary research object itself (an aspect of the culture studied) is extraordinarily multiformal and dynamic and can therefore never be grasped in a single or even a whole series of representations.

Reflection upon the generative process of science and the role of the scientist as an individual is not new in this respect. However, by gradually abandoning the persistent mental image of the objective scientist, contemporary scientific writings and visual productions have come to focus more on the notion of *reflexivity*. Yet, it often remains unclear what this notion means precisely.

In defining the concept of reflexivity, some authors restrict themselves to attaching new labels to existing methodological principles and praxis. Moreover, they tend to claim that past research projects and visual productions never considered that perhaps their approach contained a degree of subjectivity. In addition, however, reflexivity does concern broadened insights, principles, and focal points that should be appropriately incorporated into scientific praxis.

Ruby provides a translucent and qualified description of what the term "reflexivity" may signify in an anthropological context: "To be reflexive, in terms of a work of anthropology, is to insist that anthropologists systematically and rigorously reveal their methods and themselves as the instruments of data generation and reflect upon how the medium through which they transmit their work predisposes readers/viewers to construct meaning of the work in certain ways" (2000, 152).

How the various aspects should be dealt with more concretely is not so easily resolved. Indeed, this leads to new misunderstandings. Closer attention to one's own influence need not gain the upper hand over the actual topic of a visual production. While reflexivity should result in self-relativization and qualification of the results, some researchers and filmmakers sometimes use it as a means to put themselves in the foreground in an almost self-glorifying manner. Such strongly autobiographically oriented explorations, paradoxically enough, can be at the expense of insight into the broader cultural meeting that the researchers believe they should strive for. The same holds for exaggerated reflexive approaches that in turn may quickly degenerate into a repetitive metadiscourse on filmmaking and scientific work, against an alternating background of given cultural contexts. There are no quick or simple procedures for a reflexive approach: occasionally filming oneself or the crew (to say "hey, this is only film not reality!"), or a merely verbal acknowledgement of reflexivity, provides little insight into the driving forces behind a scientist and the scientific usability of the product. A fierce advocate of a thoughtful and well-balanced reflexivity, Ruby (2000, 155) does admit that it is extremely difficult to determine exactly how much knowledge and precisely which aspects of the producer and the production process are useful to reveal, and, conversely, what should be considered an excess or superfluous and thus detrimental to the end result.

In any case, the conclusion that much subjectivity inevitably is involved in the creation and the further use of visual representations in a scientific context does not mean that all approaches are equally (in)valuable for developing an insight into the phenomena that surround us and for the conceptual constructions that scientists weave around them. To a scientist, reflexivity or reflexive action does not mean liberation from a rigid and explicit methodology, but for each scientific activity, a necessary and demanding expansion of it. After all, non-recognized subjectivity or the denial of subjective influences and the unmistakable epistemological consequences of all choices inevitably lead to much less scientifically useful insights, although this might not always appear to be the case at first sight. Ruby does draw attention to the paradox that as scientists become better at this demanding assignment, their scientific status appears increasingly to be undermined: "their methodological statements then begin to appear to be more and more personal, subjective, biased, involved, and culture bound—in other words, *the more scientific anthropologists try to be by revealing their methods, the less scientific they appear to be*" (2000, 163).

The concept of reflexivity encompasses a clear recognition that all scientific knowledge is *provisional,* perspectivistic, and incomplete, but it should go beyond that recognition. A justification of the procedures followed in research—including the subjective influences and influencing- must assume the concrete shape of a carefully considered metadiscourse that is developed partly in the film (internal contextualization) but usually also outside it (external contextualization, e.g., through verbal elucidation, study guide, book). The audience of a visual end product must be allowed to become as conscious as possible of the *status* of what is shown: Why were certain choices made? Is one looking at naturally occurring phenomena, elicited behavior, or (re)staged events? Are the materials resulting from covert observation or co-operation with the subjects? What difficulties have been experienced during the shooting and editing? To what extent can what is shown be generalized? These and other questions should be made explicit.

6. Conclusion and Applications

The debate on scientific filmmaking remains very complex, because it is at the same time a debate about the status and the potential of a particularly diverse and culturally embedded medium (which includes formal aspects, visual and verbal codes of expression, the impact of technology, truth claims) and a debate about the essence and the possibilities of science as a culturally specific practice.

Film as a medium of representation clearly is an exceptionally rich one, both in a mimetic and an expressive sense. Compared to many other types of visual representations in science, it has a particularly strong mimetic and expressive potential because of its visual diachronic nature (enabling development over time) and its powerful multimedia character (incorporating sound, image, text). A complex visual medium in its own right, film also can include and emulate other types of visual representations (charts, drawings, computer simulations). Thus film is not limited merely to reflecting a pro-filmic reality, in other words not confined to depicting naturally occurring, elicited, or re-enacted phenomena, but also is able to reproduce more conceptual representations. Many recent anthropological films tend to take advantage of this capability to mix representations and discursive strategies of a very different kind.

However, film also can become part of a larger multimedia structure. A good and fairly early example is the CD-ROM titled *Yanomamö Interactive,* made by Peter Biella, Napoleon Chagnon, and Gary Seaman (1997). This multimedia product is built around a complete digitized version of Timothy Asch and Napoleon Chagnon's complex and multi-layered anthropological film *The Ax Fight* (1975). Users of this cleverly designed multimedia product can explore this rich anthropological event in a non-linear fashion at their own pace and have access to illuminating genealogical charts, maps, biographies, analytical texts, annotations, diagrams, and audio narration. They also can replay scenes, freeze frames, or use slow motion for added understanding. Additional contextual information is provided on a Web site that permits easy updating in contrast to the CD-ROM format.

Thus current digital media technology allows the further expansion of the discursive potential of film and makes it into an even more hybrid product with more possibilities and challenges. Digitized images, texts, and sounds in non-linear digital base formats offer numerous possibilities to bypass the sometimes too rigid linearity of the moving image and may accommodate various audiences (according to their level of knowledge and specific interests). New media, rather than merely supplanting the older media, are both emulating and extending many of their features in addition to producing completely new ones. If well thought through and skillfully applied, these new opportunities may enrich the science discourse dramatically.

However tempting and promising it may seem in its own right, technology should remain the servant of ideas and insights, and not become the driver of a scientific project. This is not an unqualified plea for a limited use of the growing—and largely unexplored—capabilities of visual media, but a certain degree of sobriety will remain essential to the creation of valuable scientific products. New opportunities should always be measured by their added value as well as by their costs since *more* may still mean *less*.

By way of closure, the following film reviews provide concrete applications and elaborations of the aspects and issues of anthropological and sociological filmmaking discussed in the present chapter, and, by extension, of the possibilities that film offers as a means of scientific communication.

FILM REVIEW 1

A Country Auction: The Paul V. Leitzel Estate Sale

A visually expressive and reflexive scientific film

A Country Auction: The Paul V. Leitzel Estate Sale, an anthropological film produced by Robert Aibel, Ben Levin, Chris Musello, and Jay Ruby (1984), is about the social, personal, and economic aspects of an auction of the last general store in a rural community in Pennsylvania. The film follows the relatives of the deceased manager on the eve of the public auction, as they go through all the objects and comment on them. Then the day of the auction is shown. Members of the community turn up in large numbers to look for bargains, to purchase a memento, or to swap memories and take a final look at the store. After the big auction, the camera follows a number of sold items to their new owners. At the basis for the production lies the belief that this kind of public auction serves various social purposes: A household is physically dismantled, so that the next of kin of the deceased person experience the disappearance of their relative very intensely; the members of a community are given an opportunity to acquire certain objects (with a primarily symbolic or emotional value); antique

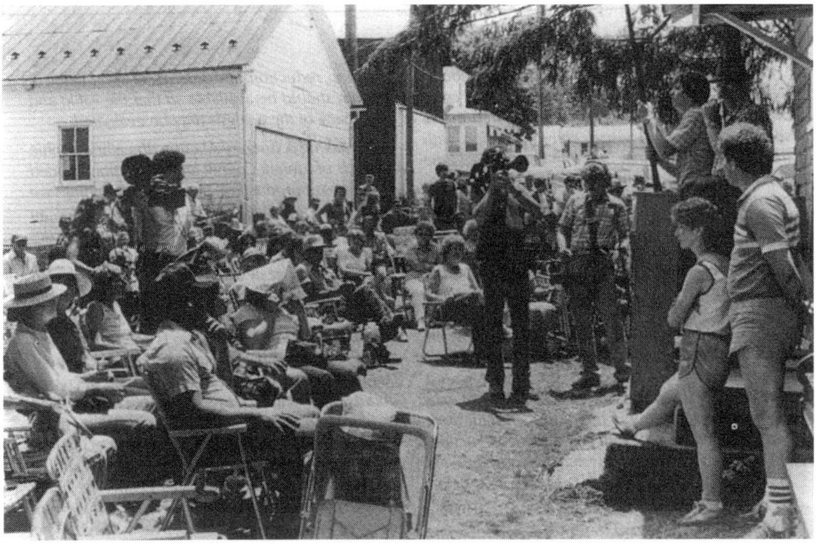

Figure 6.1. The film crew of *A Country Auction* at work. Besides the excellent study guide that provides verbal image-external contextualization, photos such as these can be conceived as a form of *visual* image-external contextualization of the project. (Photo: Frances Cox)

dealers can make a living and may distribute a considerable share of the goods over large areas. The film uses filmic techniques and conventions very thoughtfully to construct a scientific argument. Both the intentions of the makers and the various methods employed are dealt with explicitly in this production. The actual production process of the film and the filmmakers are also a topic of this visual study, so that the role of the researchers and the research process are not concealed.

Having designed their production as social science, the authors subjected their film and its accompanying materials to similar (though not necessarily the same) and equally rigorous standards as those that are imposed on written forms of science. The study guide that accompanied the film provided the necessary context. It also elucidated the theoretical framework surrounding the production and revealed and substantiated each of the cinematographic choices made. An extract of the study guide illustrates the very thoughtful approach that typifies this important anthropological film:

> The "look" of this film is not the result of a belief in any one particular aesthetic of documentary realism, but due to a set of pragmatic decisions

about how best to convey social science knowledge in film. Narration and a brief methodology statement were included near the beginning of the film in order to establish the guiding analysis and to make the filmmakers' purposes clear. The interviewer was often left in the frame so that viewers were reminded of the interactive quality of the scene. Dissolves were used reflexively to avoid "artificial" continuity in editing. Fades to black were used to distance viewers from the narrative surface of the film and to divide theoretically distinct segments of the event. An editing structure was developed which periodically diverged from the chronology of the auction in order to develop particular concepts. (Aibel et al. 1984, study guide).

FILM REVIEW 2

Tobelo Marriage

The mimetic tradition in ethno-cinematography

Tobelo Marriage, by the anthropologist and filmmaker Dirk Nijland (1985) from Leiden, depicts a marriage ceremony in the village of Paca (Northern Moluccas, Indonesia) and the somewhat difficult negotiations between the two groups of relatives. The distinctive actions are elucidated and placed in a broader cultural context. The gradual progress in the process of negotiation and exchange is explained: Situations and problems are introduced; sometimes they are depicted or summarized in diagrams; the key figures and entourage are identified clearly for the benefit of the viewer; conversations in Tobelo and Indonesian have been subtitled. Still, few of the expectations that Western viewers typically have, both formally and in terms of content, actually are fulfilled. On the whole, the framing is broad for the benefit of the image's internal context. The long shots tie in with the slow rhythm of unfolding events. The subject, which contains hardly any dramatic moments (e.g., rows) is dealt with in a detached and didactic manner.

Tobelo Marriage is a representative product of the "ethnocommunication" department at Leiden. This term refers not only to the field of study, but also to a specific visual approach, which insists that phenomenon should be shown in a transparent and descriptive manner and that the *source value* of a recorded event must be retained, and in which showing is more important than suggesting. In addition, much attention is devoted to ethnographic foundations and to the establishment of a good rapport with the community studied. Film is not seen here as a substitute for a written

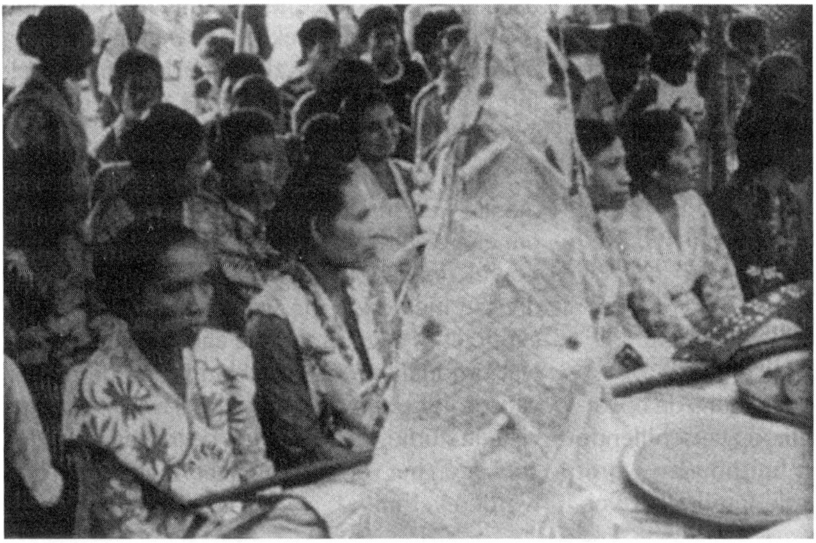

Figure 6.2. Reaction of a Tobelo after viewing this image: "Ah, this . . . in my experience, in my eyes . . . this is interesting to me, because at this moment the bride-takers have already taken a decision with . . . together with the bride-givers. This is apparent from what is on the table and what is given in return is standing in between. Because this is the shield and in the middle is a *peroso higi* (a tower-shaped wicker basket filled with uncooked rice, several of which are contributed by the bride-givers)" (Nijland 1989, 18). (Photo: Asinoellah/Rioly)

anthropological report, but as an addition of those aspects of cultural activity that are easier to depict with images and harder to put into words, and best seen in their meaningful context. Besides the use of the camera to support observations, attention is paid to the possibilities for acquiring additional data through visual feedback to participants and local experts, in order to determine the accuracy of the execution of rituals and their filmic representation as well as the reactions to specific actions by participants that are shown (Nijland 1989, 148–151).

Tobelo Marriage is a very meticulous study in which the mimesis of a cultural activity or artefact is central. Thus it is a typical example of the ethnographic tradition of retaining the integrity of the pro-filmic event as much as possible. What appears before the camera is what is important, not the camera as a mediating agent with its own lingual/expressive possibilities. It is Nijland's explicit aim to make recordings that contain as little interpretation by the researcher/cameraman as possible and that follow as much as possible the "natural unfolding of events." He continues to believe strongly in the possibilities of the representative descriptive ethnographic film that may be conceived as an "open representation," that is, as uninterpreted source material. In a sense, meticulously recorded sequences of images indeed can be considered to be "open" (Folmer 1992, 4) or at least more "open representations" than, for example, purely textual constructions, although the former inevitably concern selective constructions (even at the level of the recording, not only at the level of editing).

This very visual mimetic approach has been contested heavily for some time (among others as an emanation of "naïve realism"), yet it does have its merits, if only because it rightly contests the growing degradation of the visual in some recent approaches that appear, moreover, to propagate uncontrolled subjectivity and an emphasis on verbal dialogue. However, the sociological and anthropological film as a fully fledged scientific genre will have to take a mimetic and an expressive stance with a view to the specificity and the contribution to the field under study, and founded on solid theoretical ground, on the basis of a carefully considered and communicated reflexivity, both in the visual and in the verbal field.

FILM REVIEW 3

Polka: Roots of Conjunto Accordion Playing in South Texas

A stimulating and provocative post-modern approach

A remarkable application of the *visual feedback* technique is found in an anthropological film by Robert Boonzajer and Maarten Rens, entitled *Polka: Roots of Conjunto Accordion Playing in South Texas* (1986). It is suggested in the film that there are connections between the emergence and the development of the "Chicano Polka" in the border region between Mexico and Texas (in which the diatonic accordion played a central role) and the central European traditions of polka and accordion music. The researchers brought the Chicano musicians into contact with the direct descendants of their musical ancestors through video recordings that they had made in Salzburg just a few weeks earlier. Afterward, the recordings of the polka-performing Chicanos were shown to the Austrian musicians. The reactions of the two groups to their counterparts' performances were recorded visually and incorporated into the film.

While the Chicanos reacted very enthusiastically to the music and the playing styles of the Austrians and immediately felt an urge to imitate them, the Austrians were rather critical of the "corrupt" polka of the Chicanos and they liked to show before the cameras "how it should be done." The com-

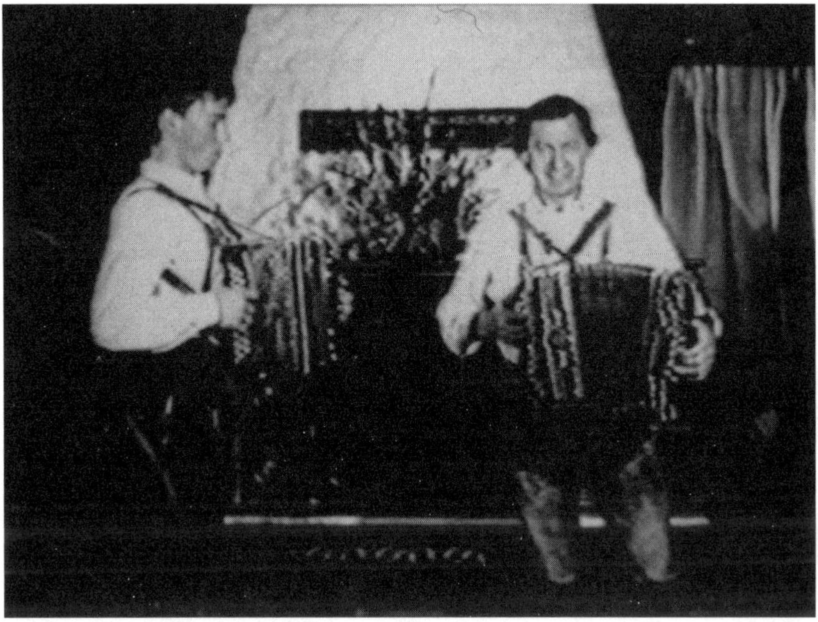

Figure 6.3. The Austrian choice of *framing.* Still from Flaes Robert Boonzajer and Maarten Rens, *Polka: Roots of Conjunto Accordion Playing in South Texas.* Video, color, 50 mins. (University of Amsterdam: Center for Visual Anthropology, 1986)

ments by the Austrians mainly concerned the (low) social position of the Chicanos ("this is gypsy music and it has nothing in common with the Austrian style"). The film also deals with the social and political functions of Chicano music. Originating in a mixture of influences, Chicano music appears to play an important role in the retention of cultural identity by a population group that is still severely discriminated against by the white community.

This production, which from a traditional perspective is somewhat provocative, visualizes a number of typical problems and points of debate with regard to field research and the role of the researcher. Unlike many other anthropological films, no attempt is made in *Polka* to conceal the role of and the actual interferences by the researcher in the situation. Especially striking in this production is the rapport that the anthropologist, himself a talented accordionist, succeeds in establishing with the research subjects, as well as the overt interventions by the anthropologist, who is almost constantly in the picture and interacts very directly with the subjects. While there is evidence of an unconcealed recognition and exploration of the influence of the filmmaker on the pro-filmic aspect, this is much less the case at the filmic level (camera handling), where the approach is expressive but not always

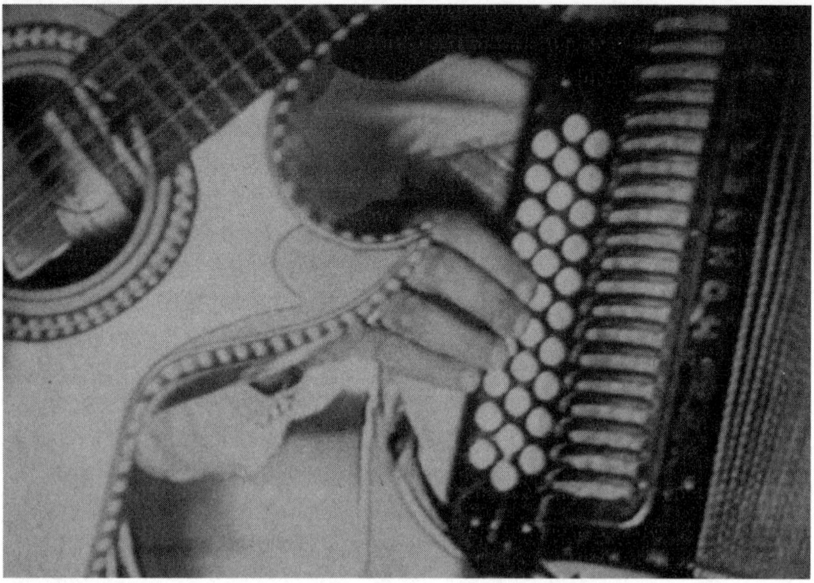

Figure 6.4. The Chicano *framing.* Still from Flaes Robert Boonzajer and Maarten Rens, *Polka: Roots of Conjunto Accordion Playing in South Texas.* Video, color, 50 mins. (University of Amsterdam: Center for Visual Anthropology, 1986)

purposeful, well-considered, and explicit. The authors honor the hyperbolic statement that "every documentary is a feature film," which often results in an abandonment of any reflexive action in the cinematographic field rather than a striving for deliberate and explicit scientific expressiveness.

In their films, Boonzajer and Rens try to emphasize "showing" rather than "telling" and, for this reason, they resolutely reject the use of explicative commentary or off-screen narration. The researchers believe that the subjects should "tell their own story." One of the consequences of this approach is that their *visual* study depends largely on *verbal* sources (albeit in their non-verbal context). Therefore, the film largely remains a *telling* of events, albeit from the perspective of the subjects rather than that of an off-screen commentator. As no attempt is made to provide an informed scientific view from the exterior as a complement, the film is largely restricted to the (scientifically valuable, yet incomplete) viewpoints of the Chicanos and the Austrians or, to be more precise, of a number of individuals from these two distinct cultures.

The rather loose and unmotivated use of such cinematographic means as editing and framing, the prominent presence and influence of the anthropologist, the very strong verbal impact and the idea that the respondents have complete insight into their own situation, but also the remarkable rapport with and participation of the field make this a *post-modern* anthropological film that provides extremely interesting food for thought regarding the further development of scientific film as a mimetic and expressive genre.

At one point, the researcher found that the Austrian accordionists were not happy with the manner in which they were being filmed: They wanted the camera to be placed further back, so that "everything would be visible." Then the Austrian father purposely positioned himself very centrally, together with his son, who was however in a slightly less favorable position than himself, surrounded by numerous signs of Austrian affluence and culture. The Chicanos, too, seemed to feel nothing for the *medium* shot. However, rather than moving the camera back, they preferred to bring it closer to the performance, so that only the most important parts of the instrument and the hands that were playing it were in the shot. Heads and limbs were regarded as interferences: all attention had to be focused on the music. This spontaneous form of feedback illustrates quite clearly how music is interwoven closely with broader cultural conceptions. In the initial version of the film, which was shown at the 1986 IVSA (International Visual Sociology Association) congress in Bielefeld, this process of reframing was incorporated in full (focusing and zooming in and out until the desired frame size was found) but this meaningful form of visual participation by the subjects (a form of cultural self-portrayal) unfortunately was not retained in subsequent versions. This probably was due to the fact that the more "professional" editor of the definitive version did not understand this "bungling" with the camera or did not want to come across as incompetent. As the authors stubbornly refuse to provide *commentary,* this event is not even mentioned verbally in the later version. There are other examples of how the second version of *Polka* uses more *typical* editing techniques (for feature film and entertainment) in order to make the production smoother and more varied (such as very rapid successions of shots or unmotivated cut-away shots, while there was clearly only one camera in the field).

FILM REVIEW 4

The Visual Representation of a Ritual

A video article as a stand-alone product of social science

An interesting sociological attempt to produce a stand-alone scientific visual product that can match and indeed surpass the traditional scholarly article

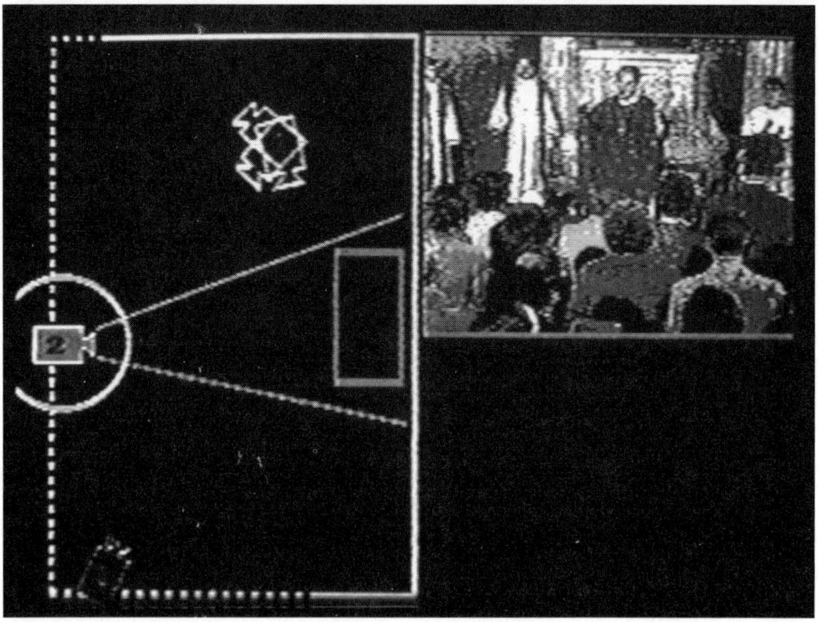

Figure 6.5. The camera's point of view *constructs* a spectator role. From Jean-Paul Terrenoire, *Mise en images d'un rituel: La messe catholique à la télévision française, un approche sociologique* [The visual representation of a ritual: The Catholic mass on French television, a sociological approach]. (Paris: CNRS, 1991)

is the video article entitled *Mise en images d'un rituel: La messe catholique à la télévision française, un approche sociologique* (The visual representation of a ritual: The Catholic mass on French television, a sociological approach), produced by the French sociologist Jean-Paul Terrenoire (1991). Of particular interest is the visual article's multi-layered structure: It is basically about a ritual (a first *representational* practice) as captured by a television crew (and thus made into a second-level representation: the television program). Moreover, the relationship between the two types of representation are discussed within a third: that of the *video article* (which is itself a media product). This video article also offers a glimpse of what may comprise a "more visually literate science," that is, a science that not only studies visual phenomena but also tries to integrate visual approaches and techniques completely into its processes of research and communication.

Terrenoire's video article provides an interesting illustration of the typical tension that exists between the real-life flow of events (which may or may not be staged or prescribed) and their *representation*, which necessarily embodies some sort of—intended or unintended, functional or dysfunctional—reinterpretation. Moreover, it points quite clearly at the *alterable* status of the implied spectator as imposed by the camera (through position, distance, movement, framing). Or, put in more general terms, it illustrates the powerful impact that the *point of view* (POV) also can have in non-fictional productions. Though this is not claimed explicitly by the author, Terrenoire's production may be considered to be a powerful statement in favor of the thoughtful development of more appropriate ways of integrating visual media and visual information into scientific practice. In other words, it strongly rejects the notion that there should or could be such a thing as a standard visual culture or grammar that is suitable for visualizing any event or phenomenon, irrespective of its purpose.

The video article's introduction and conclusion consist of considerable portions of text scrolling from the bottom to the top of the screen. While this may seem a very *non-visual* approach (and the pace of the scrolling can hardly be set to suit all readers/spectators), one should never foster the illusion that an audio-visual, supposedly self-contained article ever will be able to get around the need for some sort of verbalization (be it in the shape of on-screen text, an off-screen voice, an on-screen speaker, or an accompanying study guide).

Though there is certainly room for improvement with regard to several aspects (e.g., the balance between text, sound, and images; the choice of fonts and of background colors) the product still succeeds in transmitting scientific insight in a way that a written article could not, or at least not as well. With today's technology, one surely could overcome a number of limitations, such as by offering ways of bypassing the strictly linear construction and the imposed pace of the images and of the texts, which still make up a considerable part of the product, and by offering opportunities for revisiting portions of the analysis or to consider selected bits of images and text at one's own pace and level of interest. These enhancements are connected with increasing the level of proficiency of the producers, who are forced to integrate several different types of expertise, as well as with deploying the characteristics of the ever-evolving visual media. Today's digital media provide much easier and more intelligible ways of combining text, sound, and images compared to the rather unwieldy analog video formats.

References

Bazin, A. 1975. *Qu'est-ce que le cinema*. Paris: Cerf.

Chalfen, R. 1987. *Snapshot versions of life*. Bowling Green, Ohio: Bowling Green State University Popular Press.

Chiozzi, P. 1989. Reflections on ethnographic film with a general bibliography. *Visual Anthropology* 2 (1): 1–84.

Collier, J. 1967. *Visual anthropology: Photography as a research method.* New York/London: Holt, Rinehart and Winston. Later revised with Malcolm Collier, University of New Mexico Press, 1986.

Crawford, P. I. 1992. Film as discourse: The invention of anthropological realities. In *Film as ethnography,* edited by Peter Ian Crawford and David Turton. Manchester/New York: Manchester University Press.

De Heusch, L. 1988. Ethnographic and sociological films. *Visual Anthropology* 1 (2): 99–156. Reprint of a 1962 UNESCO series on Social Sciences.

Diem-Wille, G. 2001. A therapeutic perspective: The use of drawings in child psychoanalysis and social science. In *Handbook of visual analysis,* edited by T. van Leeuwen and Carey Jewitt. London/Thousand Oaks/New Delhi: Sage Publications.

Film in scientific research, the. 1963. Research report, Department of Scientific and Industrial Research.

Folmer, P. 1992. Het postmodernisme en de etnografische film. Paper methodologie. http://rulso1.fsw.leidenuniv.nl/-etno/pomo.html.

Gloor, D., and H. Meier. 2000. A river revitalization seen through the lens of local community members. In *Visual Cultures and Visual Literacies,* special issue of *Visual Sociology* 15: 119–134, edited by L. Pauwels.

Goldschmidt, W. 1972. Ethnographic film, definition and exegesis. *PIEF Newsletter of the American Anthropological Association* 3 (2): 1–3.

Harper, D. 2002. Talking about pictures: A case for photo elicitation. *Visual Studies* 17 (1): 13–25.

Heider, K. 1976. *Ethnographic film.* Austin/London: University of Texas Press.

Hockings, P., ed. 1975. *Principles of visual anthropology.* Den Haag/Paris: Mouton.

Krebs, S. 1975. The film elicitation technique. In *Principles of visual anthropology,* edited by Paul Hockings. Den Haag/Paris: Mouton.

Lajoux, J. D. 1975. Ethnographic film and history. In *Principles of visual anthropology,* edited by Paul Hockings. Den Haag/Paris: Mouton.

MacDougall, D. 1975. Beyond observational cinema. In *Principles of visual anthropology,* edited by Paul Hockings. Den Haag/Paris: Mouton.

———. 1998. *Transcultural cinema,* edited with an introduction by Lucien Taylor. Princeton: Princeton University Press.

Mead, M. 1963. Anthropology and the camera. In *The encyclopedia of photography,* edited by W. Morgan. New York: Greystone Press.

———. 1975. Visual anthropology in a discipline of words. In *Principles of visual anthropology,* edited by Paul Hockings. Den Haag/Paris: Mouton.

Niesyto, H. 2000. Youth research on video self-productions: Reflections on a social-aesthetic approach. In *Visual Cultures and Visual Literacies,* special issue of *Visual Sociology* 15: 135–153, edited by Luc Pauwels.

Nijland, D. 1989. Schaduwen en werkelijkheid: Cultureel bepaald kennen, waarnemen, nonverbaal communiceren en onderzoek daarvan via de terugvertoning van de etnografische film *Tobelo Marriage.* PhD diss., Rijksuniversiteit Leiden.

Pauwels, L. 1990. *Visuele sociologie? De camera en de verbeelding van de wetenschap en de samenleving.* Antwerpen: Universitas.

———. 1993. The visual essay: Affinities and divergences between the social scientific and the social documentary modes. *Visual Anthropology* 6: 199–210.

———. 1994. De betekenis van het banale, familiefotografie als sociale praktijk en sociologische databron. *Tijdschrift voor Sociologie* 15 (1).

———. 1996. *De verbeelde samenleving: Camera, kennisverwerving en communicatie*. Leuwen/Apeldoorn: Wetenschappelijke Uitgeverij Garant.

———. 2000. Taking the visual turn in research and scholarly communication: Key issues in developing a more visually literate (social) science. *Visual Sociology* 15: 7–14.

———. 2002a. Families on the web. In *Sociology: Exploring the architecture of everyday life,* by D. Newman. 4th ed. Thousand Oaks, Calif.: Pine Forge.

———. 2002b. The video- and multimedia-article as a mode of scholarly communication: Toward scientifically informed expression and aesthetics. *Visual Studies* 17 (2): 150–159.

Peters, J. M. 1989. *Het filmische denken, of de binnenkant van de beeldcultuur*. Leuven/Amersfoort: Acco.

Prost, J. 1975. Filming body behavior. In *Principles of visual anthropology,* edited by Paul Hockings. Den Haag/Paris: Mouton.

Renato, R. 1989. Grief and the headhunter's rage. In *Culture and truth*. Boston: Beacon.

Rollwagen, J, ed. 1988. *Anthropological filmmaking: Anthropological perspectives on the production of film and video for general public audiences*. Chur/London: Harwood Academic Publishers.

Rouch, J. 1975. The camera and man. In *Principles of visual anthropology,* edited by Paul Hockings. Den Haag/Paris: Mouton.

Ruby, J. 2000. *Picturing culture: Explorations of film and anthropology*. Chicago/London: University of Chicago Press.

Segers, J. 1983. *Sociologische onderzoeksmethoden: Deel 1, Inleiding tot de structuur van het onderzoeksproces en tot de methoden van dataverzameling*. 3rd ed. Assen: Van Gorcum.

Wagner, J., ed. 1979. *Images of information: Still photography in the social sciences*. Beverly Hills/London: Sage.

Wolf, G. 1967. *Der Wissenschaftliche Dokumentationsfilm und die Encyclopaedia Cinematographica*. Johann Ambrosius Barth. München: Institut für den Wissenschaftlichen Film Göttingen.

Worth, S., and J. Adair. [1972] 1975. *Through Navajo eyes: An exploration in film communication and anthropology*. Bloomington: Indiana University Press.

Wright, T. 1992. Television narrative and ethnographic film. In *Film as ethnography,* edited by Peter Ian Crawford and David Turton. Manchester/New York: Manchester University Press.

Young, C. 1975. Observational cinema. In *Principles of visual anthropology,* edited by P. Hockings. Den Maag/Paris: Mouton.

References to Visual Resources

Aibel, Robert, Ben Levin, Chris Musello, and Jay Ruby. 1984. *A country auction: The Paul V. Leitzel estate sale* (with study guide). Pennsylvania Humanities Council.

Asch, Timothy, and Napoleon Chagnon. 1975. *The ax fight*. 30 mins. University Park, Penn./Watertown, Mass.: Center for Documentary Anthropology, Pennsylvania State University/Documentary Educational Resources.

Biella, Peter, Napoleon Chagnon, and Gary Seaman. 1997. *Yanomamö interactive: The ax fight*. CD-ROM with user's guide. Forth Worth: Harcourt Brace and Company.
Boonzajer, Flaes Robert, and Maarten Rens. 1986. *Polka: Roots of Conjunto accordion playing in South Texas*. Video, color, 50 mins. University of Amsterdam: Center for Visual Anthropology.
Nijland, Dirk, in co-operation with Jos Platenkamp. 1985. *Tobelo marriage*. 16mm, color, 108 mins. ICA.
Terrenoire, Jean-Paul. 1991. *Mise en images d'un rituel: La messe catholique à la télévision française, un approche sociologique*. Paris: CNRS.

7

Arguing with Images

Pauling's Theory of Antibody Formation

Alberto Cambrosio, Daniel Jacobi, and Peter Keating

> And so we have now proved, I think, that molecular complementariness is the basis of life. So I think I shall write my book and call it, "The Nature of Life."
>
> —Linus Pauling, 1984

According to Linus Pauling (1901–1994)—one of only four individuals to have won the Nobel prize more than once, in 1954 for Chemistry and in 1962 for Peace—"complementariness" or, to be more precise, "molecular complementariness" was the basis of life.[1] In formulating this claim, Pauling had something very precise in mind, namely the hypothesis that "the extent to which certain surface regions of two molecules . . . can be brought into juxtaposition" (in less technical terms: the fit between two molecules), as determined by the size and shape of molecules, accounted for crucial

This paper was previously published as an article in *Representations* 89 (spring). It is reprinted here with the permission of the University of California Press.

Initial research for this paper was made possible by a grant from the Hannah Institute for the History of Medicine. Subsequent research on imagery by AC and PK has been funded by grants from Quebec's Fonds FCAR (now FQRSC), the Social Sciences and Humanities Research Council of Canada and the U.S. Burroughs Wellcome Fund. The French National Institute for Health and Medical Research (INSERM) has greatly facilitated the collaboration between the French and the Canadian co-authors by awarding one of us (AC) a one-year position as a senior invited researcher with its Unit 379 (Social Science Applied to Medical Innovation). We would like to thank Chris Petersen from the Ava Helen and Linus Pauling Papers at Oregon State University, who complied with our repeated requests with remarkable skill and professionalism. Eric Francoeur deserves special thanks for having introduced us to Pauling's molecular models.

biological properties, such as "the specificity of the physiological activity of substances" (Pauling 1946, 1376). Specificity—considered a defining property of living substances—was thus translated in physico-chemical terms: Three-dimensional molecular structures held the key to a scientific understanding of life (Kay 2000, 42–43, 45–46). Pauling's focus on molecular structures, in turn, betrayed a scientific style grounded in visual thinking: "I realized at the time that my original ideas in the field of science usually came or perhaps always came from my having a picture or model relating to some phenomenon" (Pauling, quoted in Laszlo 1986, 43).

To present-day readers, Pauling's claim seems somewhat uncanny, for we now tend to think of information, or, to be more precise, genetic information—rather than complementariness—as the secret of life. Indeed, as argued by the historian of molecular biology Lily E. Kay, the 1950s saw a "rupture in representations of life" whereby an informational gaze came to characterize molecular representations that, until then, had been defined in purely material and energetic terms. A flatland of instructions—a code—replaced three-dimensional shapes (Kay 2000, 41). The epigraph for this article, excerpted from an oral history interview with Pauling, exemplifies the pre-1950 approach, even though it dates from a period—the mid-1980s—when the information metaphor already had swept the field. Indeed, Pauling bitterly complained that complementariness (and thus his contribution to elucidating its role) had been lost in the transition: "I note that in modern books on biology the history of the concept of detailed molecular complementariness as the basis for life is almost entirely ignored" (Pauling, quoted in Laszlo 1986, 272).

In this chapter, we would like to re-visit Pauling's claims about complementariness, by focusing on a relatively neglected aspect of his work, namely the visual dimension—the "picture or model relating to some phenomenon"—that Pauling himself placed at the center of his heuristic and, we will argue, argumentative style. Our analysis will center on Pauling's immunochemical work, since it best exemplifies his notion of complementariness and his resort to visual imagery. Pauling's contributions in this field include empirical work—detailed chemical analyses of antigen-antibody reactions—but also a theoretical synthesis, namely his "template" or "instructionist" theory of antibody formation, first made public in 1940 (Pauling 1940).

The article introducing Pauling's template theory was accompanied by a set of striking, if highly speculative illustrations and became an immediate success, rapidly achieving the status of a classic, generating hundreds of requests for reprints (Kay 1993, 171, 175). In 1948, Pauling published a popularized, even more lavishly illustrated version of his theory, in which he emphasized the broader significance of its cornerstone notion—complementariness—for other life processes (Pauling 1948). Two decades later, however, Pauling's account had fallen out of favor, having been replaced by a new account of antibody formation—the "clonal selection theory"—grounded in cellular biology rather than structural chemistry. This emblematic episode in the history of modern biomedicine has been discussed, albeit in different terms, by several historians, including Arthur Silverstein and Lily Kay: As the saying goes, the facts are well known (Silverstein 1989; Kay 1993, 2000). Nonetheless, by focusing on the role and function of visual arguments, we offer several correctives to the standard accounts and draw a few broader conclusions concerning the role of visual arguments in scientific practice.

Kay's early work on Pauling, part of a larger account of the role of the Rockefeller Foundation and of Caltech—Pauling's home institution during the period of interest to us—in the rise of molecular biology, called attention to the increasing importance attributed to protein chemistry during the 1930s for the understanding of reproduction, growth, and other key life processes (Kay 1993). It also laid emphasis on the role played in this respect by the development of new physico-chemical tools (in particular the ultracentrifuge and electrophoresis) for analyzing substances such as viruses, enzymes, and antibodies. Kay's more recent contribution, as already mentioned, situates Pauling's work within the more general framework of the transition to the information age when a major shift in molecular representations of life saw representations centered on the informational dimensions of the genetic code, the new "Book of Life," replace representations grounded in structural biochemistry (Kay 2000). Such a gain in historical generality, however, comes with a price: an almost exclusive focus on textual metaphors that overlooks the visual culture of disciplinary traditions. Accordingly, our corrective to Kay's picture—in fact, a friendly amendment—will center on two related features of scientific practice: the interaction of disciplinary traditions with broader scientific undercurrents, and the production of visual statements as opposed to textual or theoretical pronouncements.

Silverstein's account of this episode is quite different from Kay's (Silverstein 1989). Beyond the obvious contrast between Kay's *post-structuralist* approach and Silverstein's largely descriptive, internalist account, the latter stands out because of its detailed, exclusive focus on immunology's experimental and theoretical tradition. This allows Silverstein to postulate a radical discontinuity between the beginnings of immunology, characterized by a biomedical focus (more on this below) and the "dark ages of immunochemistry," of which Pauling was a leading protagonist, and to which the rise of cellular immunology and clonal selection theory put an end in the aftermath of World War II, thereby vindicating the almost prophetic vision of the early, clinically oriented immunologists. As our comments on Kay should make clear, we cherish Silverstein's attention to disciplinary traditions. However, we grant a larger role to the material culture of immunology. In particular, our focus on visual productions uncovers the presence of continuities where Silverstein sees only theoretical discontinuities. Our own account of the Pauling episode, thus, differentiates itself from the versions provided by Kay and Silverstein, while borrowing elements from both.

Part of a broader project on the history of immunological imagery, this paper is a contribution to the analysis of *non-art* (in the present case: scientific) images. As such, it follows James Elkins' injunction "to begin writing the history of images rather than that of art" (Elkins 1999, 46). Concerning, more specifically, scientific imagery, Elkins has called attention to five research topics: the relation of non-art images to other, often quite dissimilar images that are nonetheless essential for their interpretation; the multi-layered modes of interpretation needed to make sense of how non-art images refer to the world; their different reception by successive interpretative communities; the specific relations they entertain between pictorial and linguistic markings; and, finally, the issue of how scientific images represent what is unrepresentable by resorting, for instance, to multiple, complementary pictorial strategies (Elkins 1999, 31–51).[2] Our paper will focus primarily on the first and the last of these five topics. We have cast our analysis within a thick historical description that approaches scientific images as visual arguments and that draws, in this respect, on the twin notions of "figuration" and "de-monstration."

1. Pauling's Immunological Drawings

To appreciate Pauling's contributions to immunology, readers need to recall that although proteins had been recognized as the body's building blocks by the early twentieth century, until the end of the 1920s protein biochemistry was dominated by the idea that proteins amounted to colloidal dispersions, that is, small molecules suspended within another substance. *Molecular geometry,* from this point of view, could not and did not play any role in the determination of biological function. Only in the 1930s did the idea that proteins were very large molecules, with a distinctive geometrical shape, gain gradual acceptance. Pauling made an important contribution to this transition with a landmark paper published in 1936, which proposed a "simple structural interpretation" of two phenomena—denaturation and coagulation—associated with proteins (Mirsky and Pauling 1936). The paper asserted that a protein molecule consisted of "one polypeptide chain which continues without interruption throughout the molecule (or, in certain cases, of two or more such chains); this chain is folded into a uniquely defined configuration, in which it is held by hydrogen bonds [between certain atomic groups]," adding that "the characteristic specific properties of native proteins [should be attributed] to their uniquely defined configurations," even though, at the time, very little was in fact known about the latter (Mirsky and Pauling 1936, 442–443).[3] Two important consequences of this claim were that the specificity of proteins was determined by their three-dimensional geometry or spatial folding, rather than by the chemical composition of the polypeptide chain, and that a given configuration was held together by small interatomic forces (hydrogen bonds) rather than ordinary chemical bonds, a stipulation that would have important consequences for Pauling's account of immunological reactions, as we will see.

In July 1940, Pauling followed up his previous work, this time in collaboration with the bacterial geneticist Max Delbrück, by publishing a highly speculative paper suggesting, on the basis of quantum-mechanical considerations, that the mechanism of protein synthesis was governed by the fact that molecules complementary to each other could be held together in a stable fashion by a number of weak, short-range forces (Pauling and Delbrück 1940). According to Pauling and Delbrück, this very general and somewhat abstract principle held the key to a number of living processes. In particular, it meant that substances could be created in living cells through the molding, "like die and coin," of complementarily shaped substances.

Ultracentrifugation and electrophoresis experiments recently had convinced scientists that antibodies belonged to the category of (blood serum) proteins; moreover, antibodies exhibited the kind of exquisite specificity that was the hallmark of living substances. It is thus hardly surprising that Pauling quickly turned to them as a testing ground for his theory. He did so by proposing the following solution to the process of antibody formation: Essentially, the antigen or substance against which the antibody was directed acted as a template molding the antibody protein and thus turning it into a substance specific to that particular antigen. That same year, Pauling gave a detailed description of his "theory of the structure and process of formation of antibodies" in an article published by the *Journal of the American Chemical Society* (JACS) (Pauling 1940).

The article featured a set of remarkable illustrations, the first of which—a composite drawing—depicted the "four stages in the formation of a molecule of normal serum globulin," and the "six stages in the process of formation of an antibody molecule" (figure 7.1). It is most likely that, as he had done on previous occasions, Pauling himself drafted the *JACS* illustrations, producing pencil sketches that subsequently were inked by a draftsman. A placard with a hand-made, final sketch of the *JACS* diagrams is to be found in the Ava Helen and Linus Pauling Archives (figure 7.2).[4] The annotations on the placard—consisting of instructions for the draftsman who was charged with producing the final version of the drawings—are in Pauling's handwriting, and so, presumably, are the drawings.

In 1948, Pauling published a new account of his antibody theory in *Endeavour,* a semi-popular science journal published in several languages between 1942 and 1976 in the United Kingdom by Imperial Chemical Industries (ICI). E. J. Holmyard, the editor of the journal, had contacted him in June of 1947, asking for an article and adding that "It would be an advantage if the article could be accompanied by about six illustrations, preferably hitherto unpublished."[5] Pauling obliged but, this time, he asked a

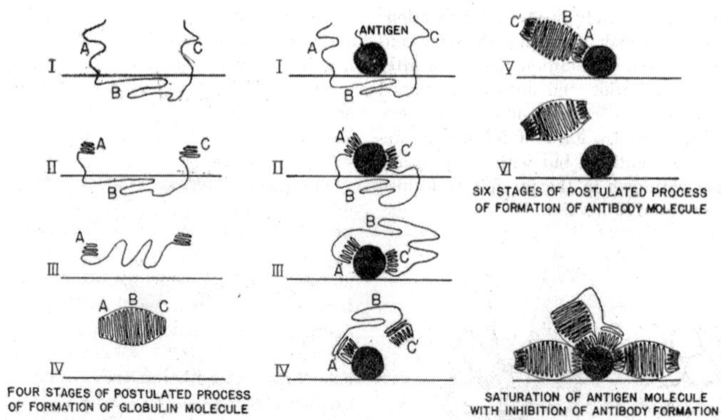

Fig. 1.—Diagrams representing four stages in the process of formation of a molecule of normal serum globulin (left side of figure) and six stages in the process of formation of an antibody molecule as the result of interaction of the globulin polypeptide chain with an antigen molecule. There is also shown (lower right) an antigen molecule surrounded by attached antibody molecules or parts of molecules and thus inhibited from further antibody formation.

Figure 7.1. Figure 1 on page 2645 of Pauling's 1940 *JACS* article. Original size (including caption): 15.6 × 9 cm (6 × 3.5 in). (Reprinted with permission from Linus Pauling, "A Theory of the Structure and Process of Formation of Antibodies," *Journal of the American Chemical Society* 62 (1940): 2643–2657. Copyright 1940 American Chemical Society)

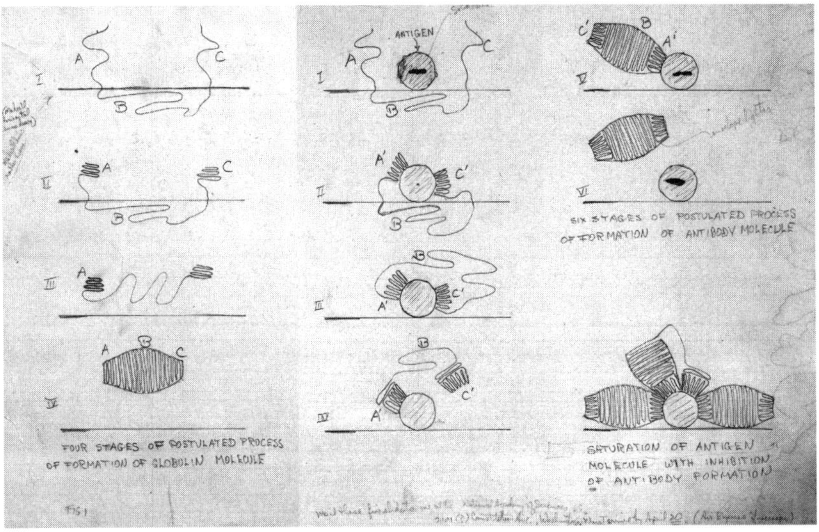

Figure 7.2. Hand-made sketch of the 1940 *JACS* diagrams. Original size: 56 × 35.5 cm (22 x 14 in). (Courtesy Ava Helen and Linus Pauling Papers, Oregon State University Special Collections, folder 1940a.2)

consultant artist named Roger Hayward (about whom, more below) to produce the required drawings. Pauling seems to have been very keen on delegating this task to Hayward. Since the artist was away on a trip, Pauling asked for a deadline extension on the grounds that: "Otherwise, I would have to be satisfied with much less interesting illustrations."[6]

The six figures featured in the 1948 *Endeavour* article, although obviously inspired by the 1940 *JACS* illustrations, are visually far more stunning than their predecessors (figures 7.3–7.6). Before examining them more closely, let us add that the *Endeavour* illustrations or variations thereon were re-used on several occasions, both for semi-popular and for professional, scientific audiences. These occasions include a 1949 *Scientific American* piece presenting an overview of Pauling's (and his colleague Beadle's) work (Gray 1949).[7] Since both Pauling and Hayward were unable to locate the original drawings, Hayward redrew them but this time in white on a black background (figure 7.7).[8] A set of glass slides using a similar white-on-black contrast was used at the First International Poliomyelitis Conference held in New York in 1948 (e.g., figure 7.8, where the antibody drawings look like bacteria under a microscope) and at several other conferences. In 1953, Pauling's longtime collaborator, David Pressman, republished the *Endeavour* drawings in a 1953 scientific review article on

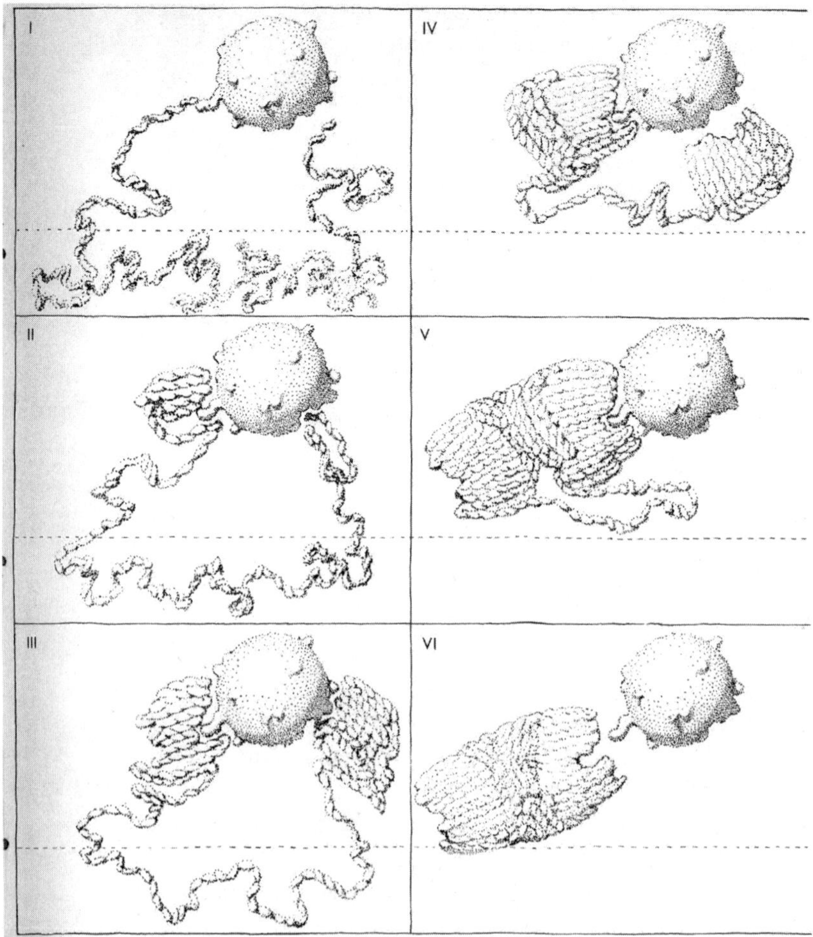

FIGURE 1 – *A postulated process of formation of antibody molecules. The polypeptide chain of the antibody precursor is folded, in the presence of an antigen molecule, into configurations complementary to the antigen. (See page 49.)*

45

Figure 7.3. Page 45 of Pauling's 1948 *Endeavour* article (Figure 1). Original size: 20.3 × 27.2 cm (8 × 11 in). (Reprinted with permission from Linus Pauling, "Antibodies and Specific Biological Forces," *Endeavour* 7, no. 26 (1948): 43–53, Copyright 1948 Elsevier Science.)

"Antibodies as Specific Chemical Reagents" (Pressman 1953).[9] They subsequently were reprinted in several medical textbooks in the 1960s and 1970s (see, for example, Humphrey and White 1973, 180–181; Carpenter 1965, 167–168; Weiser, Myrvik, and Pearsall 1969, 13). In short, the illustrations crossed the porous boundary between scientific and semi-popular representations many times.

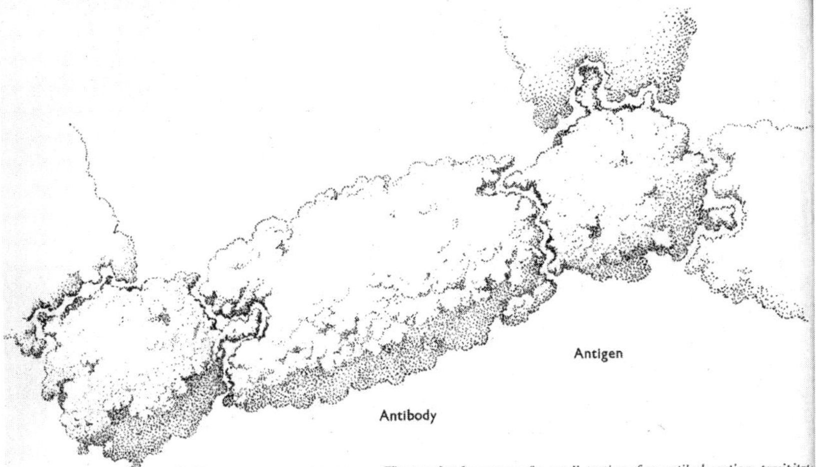

FIGURE 2 – *The postulated structure of a small portion of an antibody-antigen precipitate.*

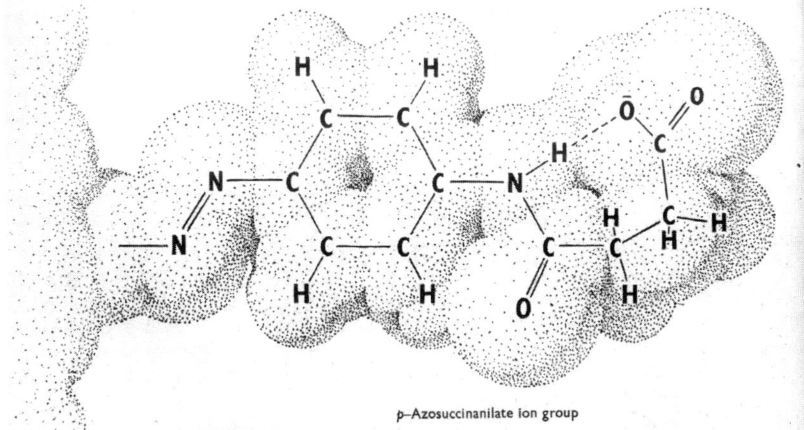

FIGURE 3 – *The structure of the haptenic group of an azoprotein, ovalbumin-*p*-azosuccinanilate ion.*

Figure 7.4. Page 46 of Pauling's 1948 *Endeavour* article (Figures 2 and 3). Original size: 20.3 × 27.2 cm (8 × 11 in). (Reprinted with permission from Linus Pauling, "Antibodies and Specific Biological Forces," *Endeavour* 7, no. 26 (1948): 43–53, Copyright 1948 Elsevier Science.)

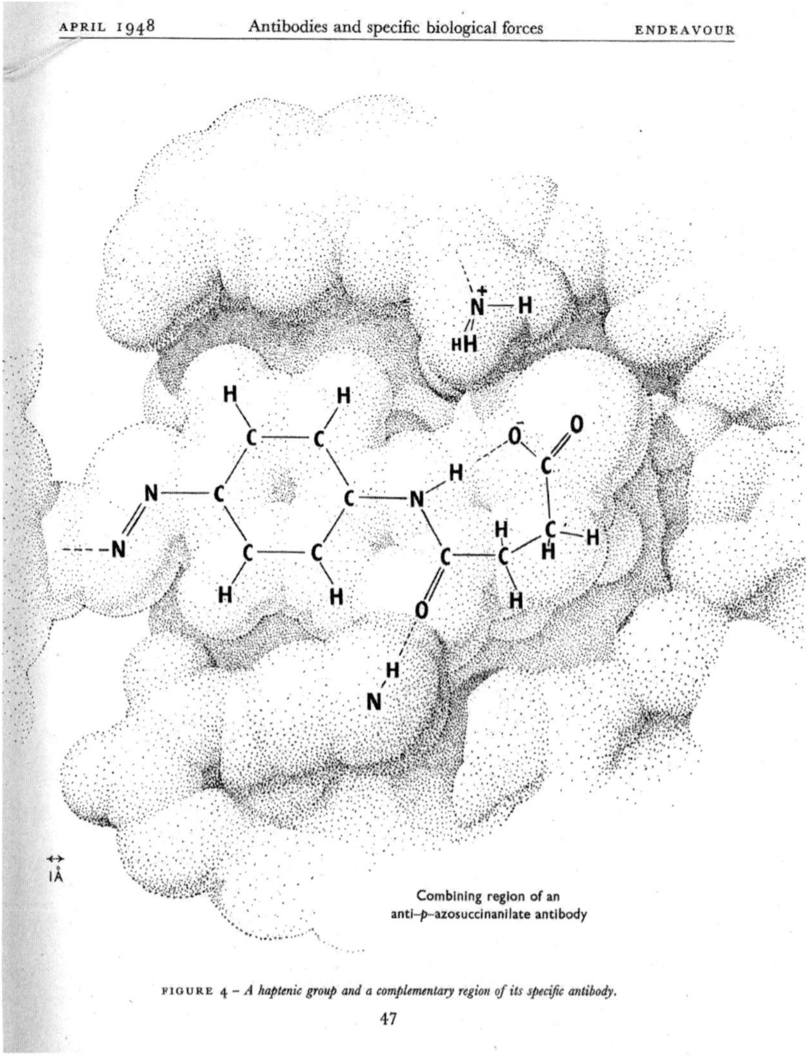

APRIL 1948 Antibodies and specific biological forces ENDEAVOUR

FIGURE 4 – *A haptenic group and a complementary region of its specific antibody.*

47

Figure 7.5. Page 47 of Pauling's 1948 *Endeavour* article (Figure 4). Original size: 20.3 × 27.2 cm (8 × 11 in). (Reprinted with permission from Linus Pauling, "Antibodies and Specific Biological Forces," *Endeavour* 7, no. 26 (1948): 43–53, Copyright 1948 Elsevier Science).

ENDEAVOUR Antibodies and specific biological forces APRIL 1948

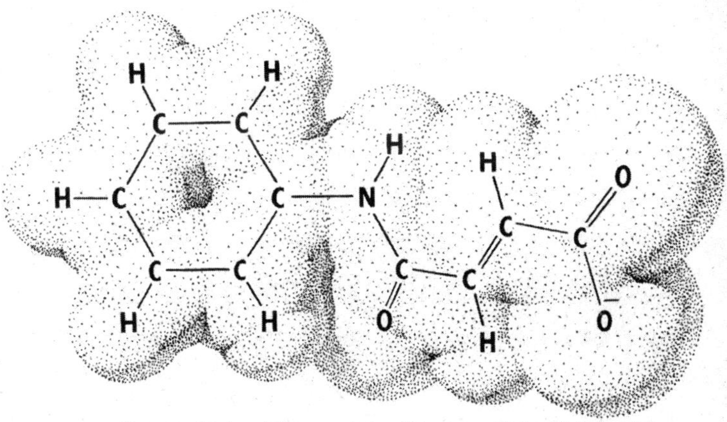

FIGURE 5 – *The fumaranilate ion, which has very small combining power with the antibody shown in figure 4.*

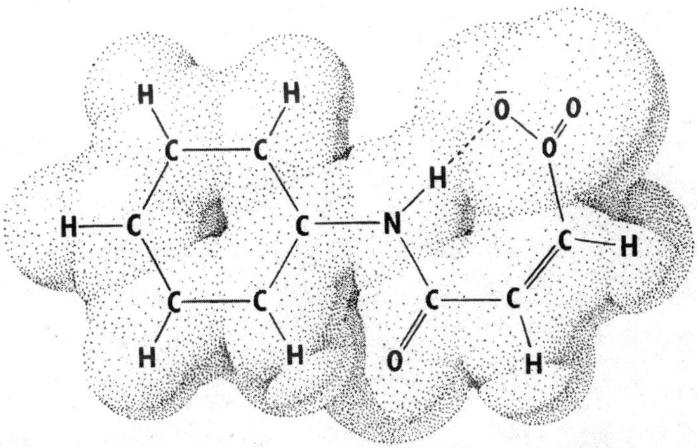

FIGURE 6 – *The maleanilate ion, which combines strongly with the antibody shown in figure 4.*

48

Figure 7.6. Page 48 of Pauling's 1948 *Endeavour* article (Figures 5 and 6). Original size: 20.3 × 27.2 cm (8 × 11 in). (Reprinted with permission from Linus Pauling, "Antibodies and Specific Biological Forces," *Endeavour* 7, no. 26 (1948): 43–53, Copyright 1948 Elsevier Science.)

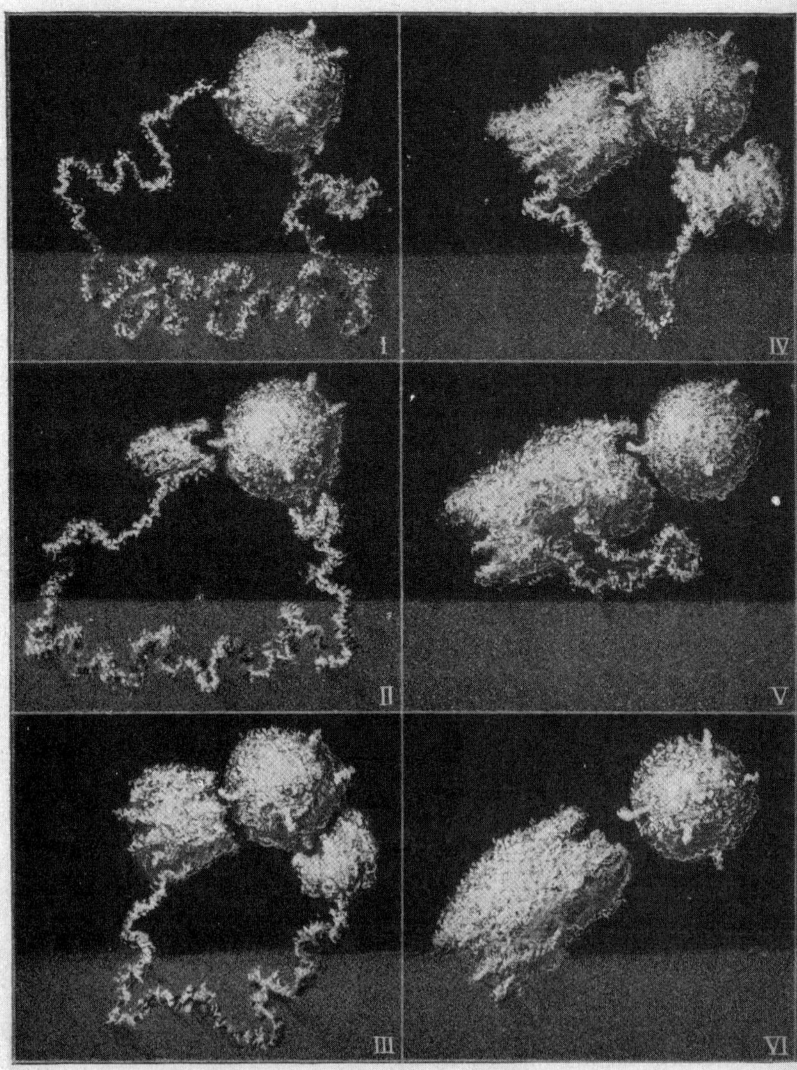

FORMATION OF ANTIBODY may proceed somewhat as shown in these drawings. Antigen molecule here is a roughly spherical structure of atoms. Precursor of antibody molecule is a long polypeptide chain. Two ends of the chain are folded to fit characteristic parts of the antigen (II, III, and IV). Two fitted ends are then joined by further folding (V). Final antibody (VI) is able to combine with two antigens to form clumps.

18

Figure 7.7. Page 18 of Gray's 1949 *Scientific American* article. Original size: 21 × 28.6 cm (8 × 11 in). (Reprinted from George W. Gray, "Pauling and Beadle," *Scientific American* 180 (May 1949): 16–21.)

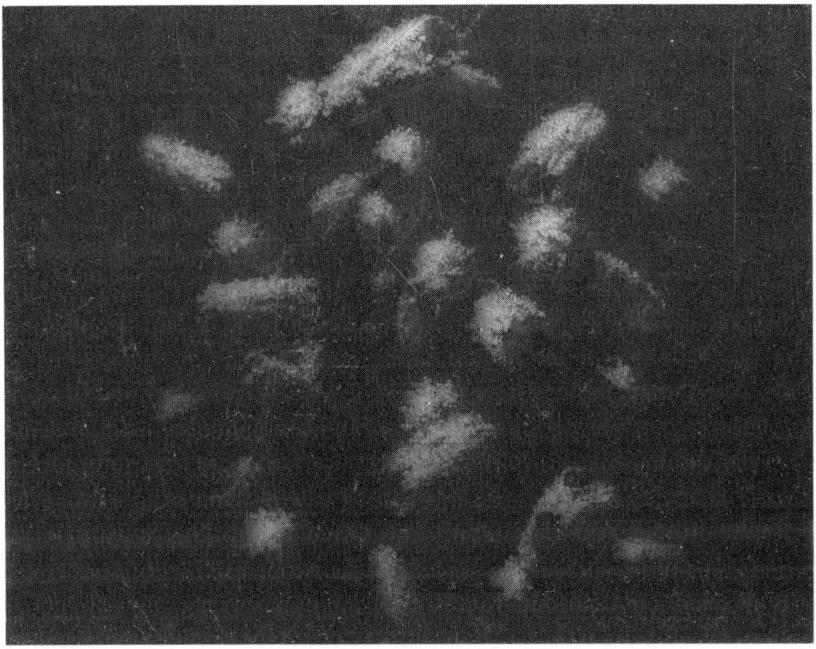

Figure 7.8. A 1948 glass slide ("Picture of Antibodies") used by Pauling during the First International Poliomyelitis Conference held in New York, 12–17 July 1948. (Courtesy Ava Helen and Linus Pauling Papers, Oregon State University Special Collections, LP Sci, Box 7.002, Folder 2.1)

2. The *Endeavour* Illustrations: Layout and Narrative Sequences

Before exploring the genealogy and argumentative role of the 1948 *Endeavour* illustrations, a short analytical description of their structure and layout is in order. We first observe that the six illustrations were reproduced in a special insert printed on glossy paper, unlike the rest of the article. Each illustration has its own short caption, so that they (and, indeed, the entire insert) could be consulted separately. This singular layout called readers' attention to the possibility of adopting a strategy similar to the one that offers itself to readers of illustrated magazines—leafing through the pages without really reading them (Jacobi 1999, esp. chap. 8). The relative autonomy of the visual component vis-à-vis the text of the article thus created the possibility of consulting the illustrations in two different ways: by browsing through the insert and/or by using the standard intratextual cross-reference system governing the text-figure relationship in scientific articles.

The differences among the six illustrations is fairly pronounced and is heightened by the page layout. The opening illustration (figure 7.3) is made up of a sequence of six smaller drawings located on the right-hand page. The central double page of the insert features the second and third illustrations on the same left-hand page (figure 7.4), followed by the fourth, showcase illustration (figure 7.5) standing on its own in the middle of the right-hand page. The final page juxtaposes the last two illustrations (figure 7.6).

By leafing through the insert, we can easily discern two intertwined narrative sequences. The first illustration (figure 7.3) functions as a sort of comic strip with its own narrative sequence detailing, step by step, how an amorphous, unspecific protein is turned into a specific, structured antibody in the presence of an antigen molecule. This first, self-contained sequence leads to a second sequence, a cinematic zooming-in linking three increasingly larger-scale close-ups spread out on the central double page (figure 7.4 and 7.5). It is as though the intention were to move from a quasi-biological form to a molecular, biochemical structure by increasing the resolution and the descriptive precision of the drawings.

The first narrative sequence (the comic strip: figure 7.3) is characterized by the presence of two characters: the antigen—a sphere spiked with little protuberances—and a polypeptide chain (an unfolded protein) whose loose configuration is represented as a long twisted filament one of whose two extremities almost touches one of the protuberances of the sphere. Very sober graphic means, namely patterns of more or less densely packed points, are used to produce three-dimensional effects.

From one image to the next, the shape and location of the antigen sphere remains stable, while the long polypeptide chain constantly moves and progressively changes its shape. The constant presence of a thin horizontal line marks the boundary, within the cell, beyond which the polypeptide chain, as if influenced by the force of attraction of the antigen sphere, will settle down into its new, well-ordered antibody form, as molded by the antigen. Each image in the sequence shows, in a fictional way, the progressive transformation of the chain into a shape somewhat reminiscent of a ball of wool, with lateral and central folds at right angles and the extremity neighboring the antigen bearing the telltale impression hollowed out by the antigen's protuberance.

The second narrative sequence (the cinematic zoom: figures 7.4 and 7.5) begins where the previous one ended: We have no difficulty recognizing the spherical antigen (with the same thinly dotted contour line and pattern of points to provide shadow and create the impression of volume) and the elongated shape of the antibody (albeit now with a lumpy look, having lost the strands of fabric that gave it the appearance of a ball of wool). As though to remove all possible ambiguity, the entities are identified by name in the drawing (antigen, antibody, antigen). On the left and right, the antibody is attached at two antigen spheres, which are themselves clipped by their appendices to other antibodies, of which the drawing shows only the beginnings. The presence of incomplete portions of additional antibody molecules on both sides of the figure—using a graphical technique called "paving," based on the repetition of a pattern—suggests an extension of the drawing beyond the confines of the page and thus the existence of a virtually continuous mesh. It therefore harks back to the graphic representation of an antigen-antibody lattice or framework that figured prominently in the experimental system of immunochemical investigations, as we will see below.

The third illustration (bottom of figure 7.4) shows a detail of the protuberance implicated in the antigen-antibody relation, at a much larger scale. The semiotic means have now become heterogeneous, for a chemical formula is superimposed on the electronic cloud represented by a pattern of points. In fact, this third illustration is simply a transition toward the theatrical high point of the second sequence, which is delivered by the showcase figure featured on its own on the third page (figure 7.5). Here, complementariness is made visible by displaying the biochemical framework of the antigen-antibody conjugate at the level of individual atoms and their links. The relations between antibody and antigen are marked by distance and contacts, with a denser, depth-conferring shading pattern for the space between the antigen and the antibody and chemical formulas (as well as an indication of the distance in angstroms) pointing to the existence of close range hydrogen bonds between them. As in the previous illustration, the showcase illustration is a sort of semiotic hybrid, the outcome of a translation that operates by focusing on a detail (a high resolution rendering of the complementary regions) and by superimposing a distinctive biochemical structure as represented by chemical conventions on the largely arbitrary shapes produced by Hayward's artistic skills.

The two final illustrations (figure 7.6), also semiotic hybrids, fall under a different modality: Their purpose is to present the reader with a comparison between two antigenic structures that, despite their resemblance, have very different antibody combining powers. Readers thus can witness the accuracy of Pauling's complementariness postulate: They can see how even small structural differences can account for large variations in experimental results.

So, to sum up, the *Endeavour* illustrations manage to construct a rich and complex account of the generation of antibodies by resorting to a few graphic means: Dotted lines and repetitive patterns create the effect of shadow and volume; the polysemic graphical dimension of the drawings is juxtaposed to the more restricted semiotic code of chemistry; and the main characters are given simple, contrasting forms (chain-like, elongated, spherical), while changes of scale and playing with grids and page borders (e.g., paving) confer an additional dynamic dimension to the illustrations. Of particular interest is the sudden intrusion, starting with the third illustration, of the chemical code that signals a switch in the interpretative discourse. The search for causality moves from a supra-molecular to a molecular focus and in so doing reveals, more than any other sign, Pauling's point of view. Yet, the dotted patterns used by Hayward are more than a decoration embellishing the chemical hard core, for, as we will see in the next section, there is nothing accidental about the particular shapes and the graphical means to which the Pauling-Hayward duo resorted.

3. From Ehrlich to Pauling: Visual Continuities

Mid-century immunologists leafing through Pauling's 1940 and 1948 papers must have experienced a sense of *déjà vu,* for it was not the first time that a highly speculative theory of antibody formation came fully armed with even more speculative drawings. Indeed, the success and, simultaneously, the controversy generated by the first major theory of antibody formation, Ehrlich's "side-chain theory" proposed approximately fifty years before, was in large part due to the illustrations that had accompanied it (Cambrosio, Jacobi, and Keating 1993, figure 7.9).

Before attending to Ehrlich's visual productions, we should note that major conceptual differences distinguished Pauling's template theory of antibody formation from its most famous forerunner. Ehrlich's theory—the first comprehensive account of the recently discovered substances responsible for the highly specific anti-microbial properties of blood serum—postulated that the physiological machinery of cells included "side-chains," later called receptors, that protruded from the cell surface and were responsible, *inter alia,* for anchoring nutrients. However, they could be blocked by various poisons, for instance by diphtheria and tetanus toxins, whose combining site, in this account, had a shape similar to the corresponding nutrient. Frustrated in its normal access to nutrients, cells reacted by overproducing receptors, which were then shed into the blood stream where, because of their capacity to combine with toxins, they acted as neutralizing substances, thereby performing the functions of antibodies.

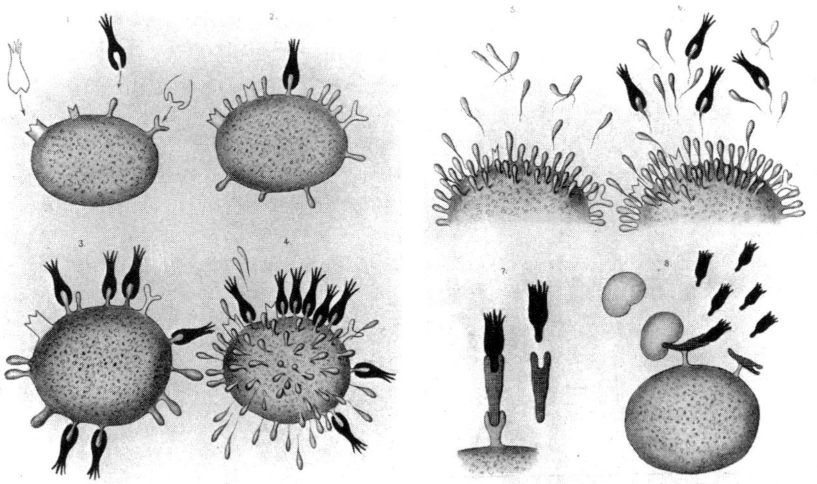

Figure 7.9. Ehrlich's 1900 representation of the process of antibody formation. Original size: 10.7 × 12.3 cm (first plate) and 10.7 × 12.9 (second plate) (each ~ 3 × 5 in). (Reprinted from Paul Ehrlich, "On Immunity with Special Reference to Cell Life," *Proceedings of the Royal Society of London*, 1900, 66: 424–448, unpaginated plates in front of pp. 438 and 439).

Ehrlich's theory shared with Pauling's subsequent account the hypothesis that antigens and antibodies had complementary shapes: Using a metaphor he borrowed from chemist Emil Fischer, Ehrlich spoke of antigens and antibodies combining as "lock and key." Yet, Ehrlich's theory differed from Pauling's on three counts. First, in Ehrlich's view, specific antibodies existed in the organism prior to their eventual encounter with the antigen against which they reacted: They were part of the normal machinery of the body. This aspect of Ehrlich's theory soon was criticized by a rising band of immunochemists on the empirical ground that since antibodies easily could be raised against newly synthesized chemicals, how, then, could organisms have developed pre-existing defense mechanisms against not-yet-existing substances, and how could they possibly harbor antibodies against a potentially unlimited number of antigens? Pauling's theory, as we have seen, eliminated this difficulty by postulating that non-specific protein molecules were molded into antibodies with a specific shape by whatever antigen they happened to encounter: The theory-laden "coin and die" metaphor thus replaced the similarly theory-laden, yet more generic "lock and key" metaphor.

A second, more subtle, yet essential difference between Ehrlich's and Pauling's theory concerned the nature of the "lock-and-key" bond between antibody and antigen. While Ehrlich conceived of it as a normal (*covalent*), chemical bond, Pauling, as already mentioned, opted for hydrogen bonds, thus emphasizing the role of the weak interatomic forces that, according to him, were responsible for the configuration of protein molecules and thus also for the complementary structure of the antigen-antibody combining sites.

A third difference implicated the broader framework of the two theories. Ehrlich was accused by other immunologists of having adopted a chemist's (as opposed to a biologist's) approach, and it is indeed the case that chemical thinking was a distinctive feature of his theorizing. Yet, as pointed out by Silverstein, Ehrlich never lost contact with medicine, always returning to the clinical value of his work (Silverstein 2002, 139–140) or, as we have described elsewhere, by constructing a new, hybrid representational space that borrowed elements from both biology and chemistry (Cambrosio, Jacobi, and Keating 1993). Not only had his theory been prompted by his work on anti-diphtheria therapy, but one of the key immunological phenomena he set out to explain was acquired immunity, that is, the protection offered to the organism by the presence of pre-formed antibodies following initial contact with, say, a microbial antigen. Pauling, on the contrary, adopted a decidedly chemical orientation. Following in the footsteps of other immunochemists such as Michael Heidelberger (1888–1991) and Karl Landsteiner (1868–1943), he championed a complete reduction of immune phenomena to chemistry, resorting to the extensive use of small, artificial antigens (*haptens*). Accordingly, the precipitation (or, to use the immunological term: precipitin) reaction (the combination in a test-tube of soluble antigen and antibody leading to the formation of an insoluble precipitate) became the model for his immunochemical investigations. As he put it explicitly:

> Immunological phenomena are many and varied, and I have ignored the complicating effects of complement, lipids, etc. We have pretty well concentrated on the precipitation reaction—although even I know that immunology is very broad.[10]

Yet, in spite of these substantial conceptual and experimental differences, visual renderings played a central role in the constitution of both Ehrlich's and Pauling's theories, and we can trace significant continuities between them. When he published the first, well-illustrated account of his theory, Pauling obviously was aware of the controversial drawings first introduced by Ehrlich in his 1900 Croonian Lecture in London (Ehrlich

1900). As noted, these drawings, notwithstanding the sometimes scathing comments they elicited, were immensely successful. Widely reproduced in immunology textbooks, they continued to be circulated, for instance in serology textbooks, long after the side-chain theory had been dismissed by immunochemists.

Ehrlich's 1900 drawings consisted of two plates with four diagrams each, numbered consecutively from 1 to 8 (figure 7.9). In the first picture, we see two food-stuffs and a toxin, homing in on their complementary side-chains sitting on a cell. In addition to geometrical shapes, arrows are used to emphasize complementariness. The toxin stands out because of its dark color. In the next picture, the toxin has united with a side-chain, creating an exact fit. In the third picture, five other toxins have joined the original toxin. In the fourth picture, the cell, reacting to the blockade caused by the anchoring of an increasing number of toxins, has begun secreting a large number of side-chains identical to those blocked by the toxin. The next two pictures are drawn as a sort of close-up. Picture 5 shows the excess of newly secreted side-chains being pushed off into the blood stream, and, finally, picture 6 shows the free side-chains uniting with, and thus inactivating, free toxins in the blood stream, protecting the cells from further toxic attack. Finally, pictures 7 and 8 show the purported mechanisms of immune reactions involving other substances, such as complement.

Ehrlich's iconography was extremely suggestive. In addition to a cyclic sequence and a layout verging on that of a comic strip, Ehrlich used injunctive marks (the arrows of picture 1), the black-white contrast, changes of scale, and a kind of chronic instability in the relations between the entities featured on the drawings. We are also struck by the figurative dimension of the pictures: Not only can we easily follow the adventures of the featured entities (all of which are identified clearly), we also can equate them with moving, active entities, in a word, living animals. The forms themselves suggest a liquid or submarine world, particularly in the case of picture 7, which resembles an underwater protozoan clinging to the seabed while its tentacles quiver with the ripples of the water. From a graphic point of view, this performativity is obtained through means at once restrained and limited in number: line drawings, simple shapes, and austere tone contrasts—from white to gray and from gray to black. We also note a certain Manichaeism at work in the black-white opposition (black being equated with poison, or evil).

If we now compare Ehrlich's and Pauling's diagrams, we can isolate five significant continuities. First, and most obviously, a comparison of

Ehrlich's first six pictures with the first of Pauling's 1948 figures (figure 7.3; see also the central and top right column of figure 7.1), shows that both authors organized their visual opening salvo as a time series, as a sort of comic strip or narrative. Second, in both cases the drawings lay emphasis on the complementary shapes of antigens and antibodies: The means used to this effect are not exactly the same—Ehrlich's simple, flat geometrical shapes were replaced confidently, almost fifty years later, by cloud-like atomic pictures, suggesting three-dimensional arrangements (more on this below)—but the reader's gaze is nonetheless steered in both cases toward the lock-and-key fit between the combining molecules.

A third continuity arises out of the sequential magnification mechanism embodied in both series of drawings: As the argument unfolds, the initial pictures are replaced by higher-resolution drawings.

The fourth element shared by the two sets of illustrations is, in fact, the outcome of the first three elements: Both series draw their strengths from a combination of static and dynamic dimensions. The fit between the corresponding antigens and antibodies shows the specificity of their union and corresponds to a static understanding of the immune reaction. The dynamic dimension is introduced by the detailed narrative of how an organism will encounter specific antigens and then react by producing antibodies that will protect it from further damage. In other words, the drawings' *tour de force*, first introduced by Ehrlich and re-utilized by Pauling, was to establish a visual domain that could capture, and could thus be used to account for, immune reactions in terms of both a sequence of events and of a particular kind of chemical interaction.

Finally, a fifth continuity between the two sets of illustrations concerns the two final drawings (figure 7.6 and the last two diagrams of figure 7.9). Unlike the preceding pictures in each series, they can be viewed separately, and they engage readers in a sort of "virtual witnessing." (Shapin and Schaffer 1985). Readers are encouraged to join the authors in verifying a hypothesis by inspecting the visual rendering of experimental results (in Ehrlich's case, experiments designed to ascertain the presence and interaction of different immune substances, and, in Pauling's case, experiments designed to demonstrate the difference that even small structural changes make in terms of complementariness) or, in other words, by comparing the organic structures that presumably account for the development of laboratory-based phenomena.

To sum up, Pauling clearly borrowed key elements of Ehrlich's representational order. Like Ehrlich, he adopted a hypothetical description, cast in visual terms, to report a series of controlled laboratory observations.

Judging from the success enjoyed by these proposals, both representations established a convincing framework generating credible accounts of experimental observations. By bestowing from the outset the form of a figurative sequence on his theoretical explanations, Pauling thus called upon previous representations (Ehrlich's diagrams, but also, more in general, the narrative, cyclical mechanism much beloved by biologists; see Ratcliff 1999) whose features he borrowed in order to prompt recognition of the elements and features of his argument.

If, as we have seen, Ehrlich's diagrams constitute a significant part of the visual background of his illustrations, Pauling borrowed additional elements from the immuno-chemical visual tradition. The latter's repertoire was somewhat limited, since in the intervening decades between the publication of Ehrlich's and Pauling's diagrams, immuno-chemists had used drawings sparingly. Immediately following Ehrlich's foray into this domain, and in the absence of dependable information about the chemical structure of antibodies, the overly liberal use of hypothetical representations of still largely unknown substances came under strident criticism as, at best, naïve and, at worst, dangerous. Even Pauling had to confess, in 1936, that "little reliable information" was available about the precise configuration of proteins (Mirsky and Pauling 1936, 442), adding, in 1946, that "we do not know in detail how the amino acids are combined to form proteins. We do not know, except very roughly, even the shapes of such important molecules as serum proteins, enzymes, genes, the substances which make up the protoplasm" (Pauling 1946, 1375).

When, in 1930, Felix Haurowitz (1896–1988) wrote an article in collaboration with Fritz Breinl (1888–1936) in which they proposed one of the very first template or instructionist theories of antibody formation—duly acknowledged in Pauling's 1940 article—the final publication in *Hoppe-Seyler's Zeitschrift für Physiologische Chemie* did not include any illustrations (Haurowitz and Breinl 1930). Haurowitz, however, apparently had felt the urge to concoct a visual representation of the postulated mechanism of antibody formation, for the archival collections at Indiana University (Bloomington) contain a 1930 ink-drawing (figure 7.10: see our translation of the original caption for explanations) accompanied by the following 1980 handwritten note by Haurowitz: "sent to the editors of Hoppe-Seyler Zeitschrift in 1930 (vol. 192, pp. 45–57). The editors (Prof. Knoop and Thomas) asked me to omit this diagram because it appeared to them too phantastic [sic] (Jan. 18, 1980)."[11] A quick comparison with Pauling's 1940 diagram shows that the Haurowitz and Breinl drawing comes nowhere near to the former in terms of sketching details

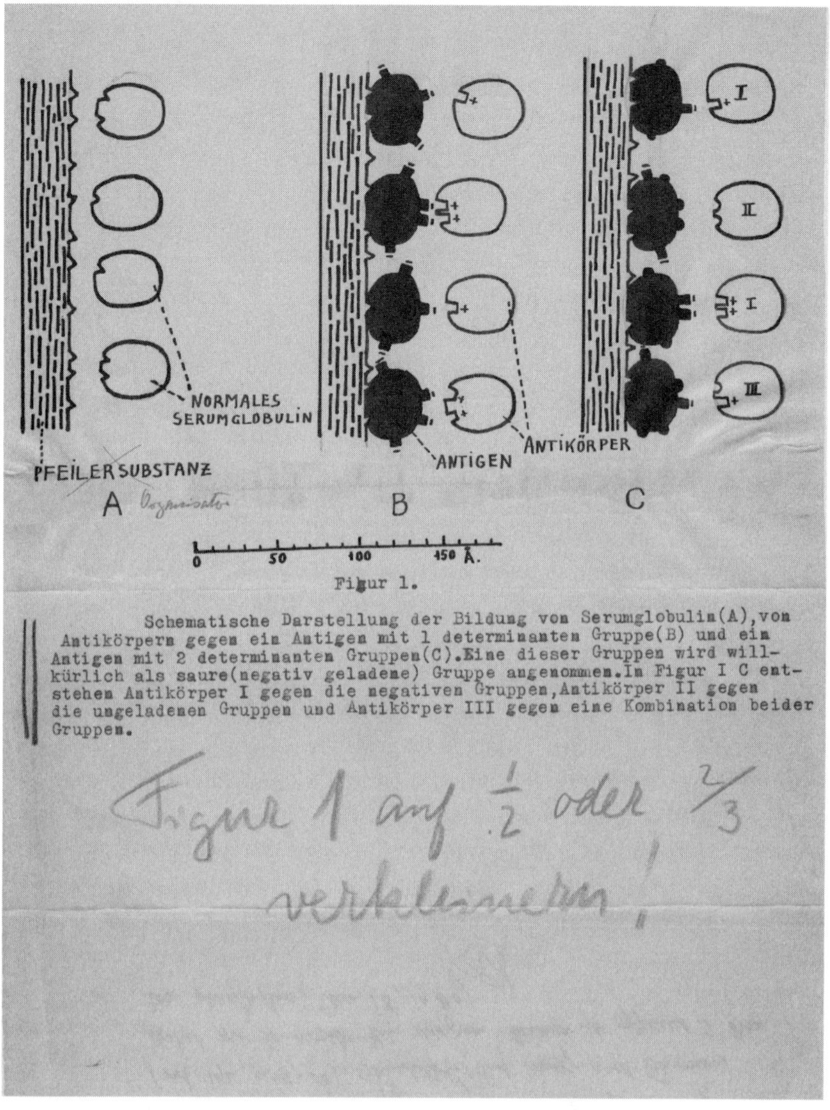

Figur 1.

Schematische Darstellung der Bildung von Serumglobulin(A),von Antikörpern gegen ein Antigen mit 1 determinanten Gruppe(B) und ein Antigen mit 2 determinanten Gruppen(C).Eine dieser Gruppen wird willkürlich als saure(negativ geladene) Gruppe angenommen.Im Figur I C entstehen Antikörper I gegen die negativen Gruppen,Antikörper II gegen die ungeladenen Gruppen und Antikörper III gegen eine Kombination beider Gruppen.

Figur 1 auf 1/2 oder 2/3 verkleinern!

Figure 7.10. Haurowitz's 1930 diagram of his template theory of antibody formation, refused by the editor of the journal in which the article appeared. Original size: 20.3 × 27.2 cm (8 × 11 in). The original German caption reads: "Schematic representation of the formation of a serum globulin (A), of antibodies against an antigen with a single determinant group (B) and an antigen with two determinant groups (C). We arbitrarily assumed that one of the groups is an acid (negatively charged) group. In Figure I C, antibodies I are produced against the negative groups, antibodies II against neutral [uncharged] groups and antibodies III against a combination of both groups" [our translation]. (Courtesy: Felix Haurowitz Manuscripts, Lilly Library, Indiana University (Bloomington), Box 2 Folder: Correspondence, Ho-Hz) and narrative structure. It is striking, however, that both diagrams are organized into three columns and that, in both cases, the left column shows the process by which a normal protein (globulin) is formed, whereas the central and right columns concern the formation of antibodies, leaving us to wonder whether Pauling somehow had been aware of the unpublished diagram.

Karl Landsteiner, another leading immuno-chemist, used only tables with + and o signs to denote presence or lack of immunological reactions. Similarly, his colleague Michael Heidelberger used mostly tables and chemical equations. His only foray into the visual domain came in a famous series of articles dissecting the model precipitin reaction (figure 7.11). The very simple diagram describing the formation of a precipitate consisted of letters (e.g., A for antibody, S for the antigen) surrounded by little dots, representing the arrangement of the molecules in a precipitate. John R. Marrack (1886–1976) developed a similar (visual) model of what became known as the lattice hypothesis of the precipitin reaction (figure 7.12). Marrack's hypothesis presupposed that antibodies had at least two combining sites that allowed them to form a lattice (or framework, as Pauling preferred to call it) with the antigen molecules. The lattice would precipitate out of solution and form a sediment at the bottom of the test tube.

Pauling was quite familiar with these diagrams. In an undated research note, he summarized Marrack and Heidelberger's approach with a quick sketch (top of figure 7.13).[12] His 1940 *JACS* article featured a reworking of the Marrack-Heidelberger diagram (figure 7.14) using, instead of letters and dots or circular shapes, the oval or cigar-like antibody shapes derived from the article's first composite illustration (figure 7.1). The second illustration

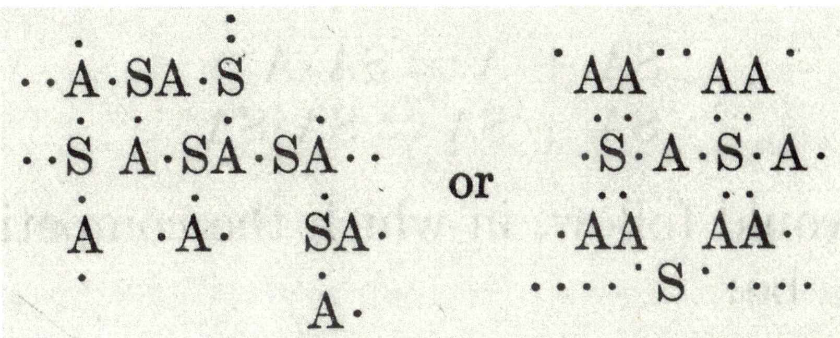

Figure 7.11. A diagrammatic rendering of the precipitin reaction by Michael Heidelberger. Original size: 6 × 2.5 cm (~ 2 × 1 in). (Reprinted with permission from the American Society for Microbiology, from Michael Heidelberger, "Quantitative Absolute Methods in the Study of Antigen-Antibody Reactions," *Bacteriological Reviews* 3 (1939): 49–95, p. 58)

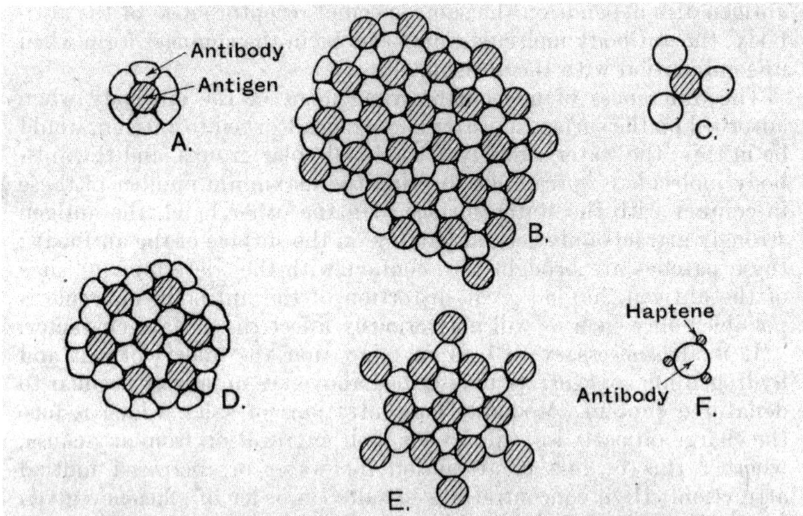

Fig. XXII. Diagrams of possible arrangements of antigen and antibody molecules in the complex. (A) simple unit; (B) complex structure at optimum proportions; (C) antigen excess; (D) antibody excess; (E) antigen excess; (F) antibody and haptene.

Figure 7.12. Visual representation of Marrack's lattice hypothesis. Original size (including caption): 11 × 8 cm (4 × 3 in). (Reprinted from John R. Marrack, *The Chemistry of Antigens and Antibodies* (London, 1934), p. 106)

of the 1948 article (top of figure 7.4), as we have seen, uses the paving technique to produce a sophisticated graphic reworking of a lattice.

Compared to the boldness of Pauling's immunological imagery, its (post-Ehrlich) immuno-chemical forerunners look very poor indeed. The strength of Pauling's illustrations derives from the fact that, even in the absence of detailed information about the shape of the antibody molecule itself, he did not refrain from translating physico-chemical assumptions about protein structure into visual shapes. The oval—elongated or cigar-like—shape of antibodies resulted from calculations based on hydrodynamic (ultracentrifuge) experiments published by Hans Neurath in 1939 (Oncley 1949; Neurath 1939). Figure 7.15 summarizes the limited information available in the 1940s, on that basis, about the likely shape of proteins (including γ-globulins, corresponding to the antibody fraction of serum proteins). Even more striking, the *Endeavour* drawings have a distinctive three-dimensional feeling that was to become somewhat of a trademark of Pauling's imagery, for Pauling's attempt to visualize the invisible world of organic and inorganic molecules went beyond flat illustrations. Max Perutz, the Nobel prize–winning chemist who determined the

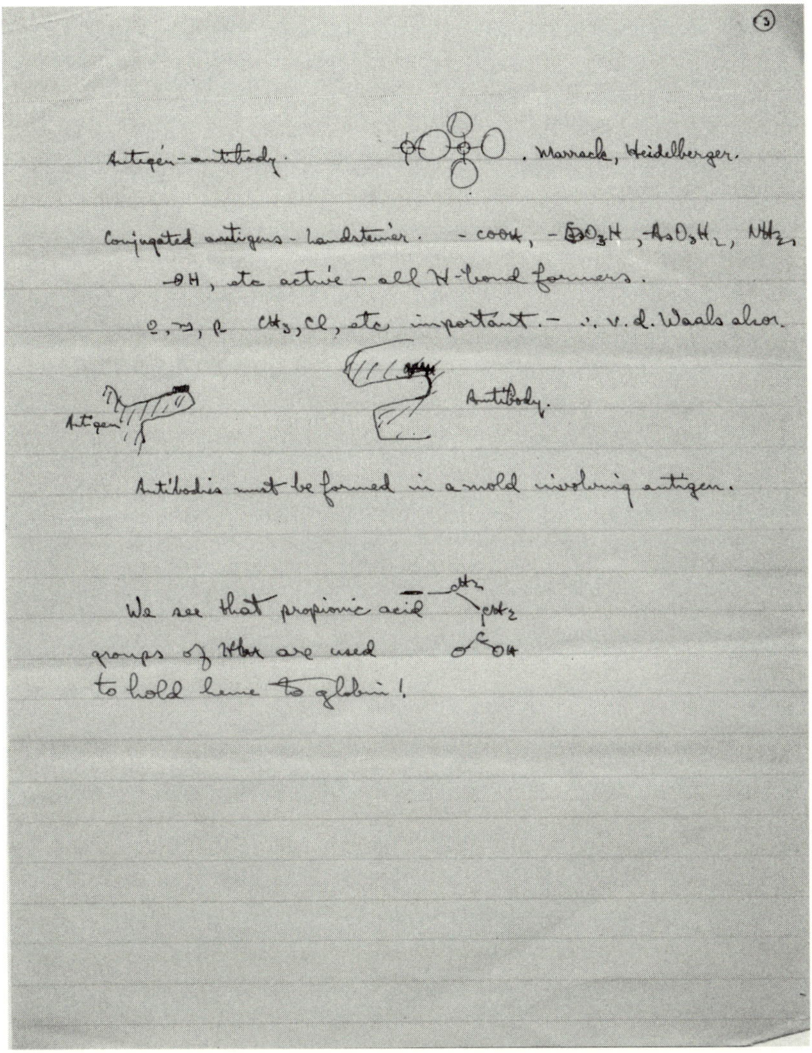

Antigen-antibody. . Marrack, Heidelberger.

Conjugated antigens - Landsteiner. - COOH, - SO$_3$H, -AsO$_3$H$_2$, NH$_2$,
-OH, etc active - all H-bond formers.
o, m, p CH$_3$, Cl, etc important. - ∴ v. d. Waals attr.

Antigen
Antibody.

Antibodies must be formed in a mold involving antigen.

We see that propionic acid groups of Hem are used to hold heme to globin!

Figure 7.13. Undated (circa late 1930s) research note by Linus Pauling showing, on the top, a sketch of Marrack and Heidelberger's lattice hypothesis and, below, a somewhat phallic rendering of the complementary shape of antigen and antibody. (Original size: 20.0 × 25.3 cm (8 × 10 in). (Courtesy Ava Helen and Linus Pauling Papers, Oregon State University Special Collections, LP SCI Box 6.008, folder 8.1)

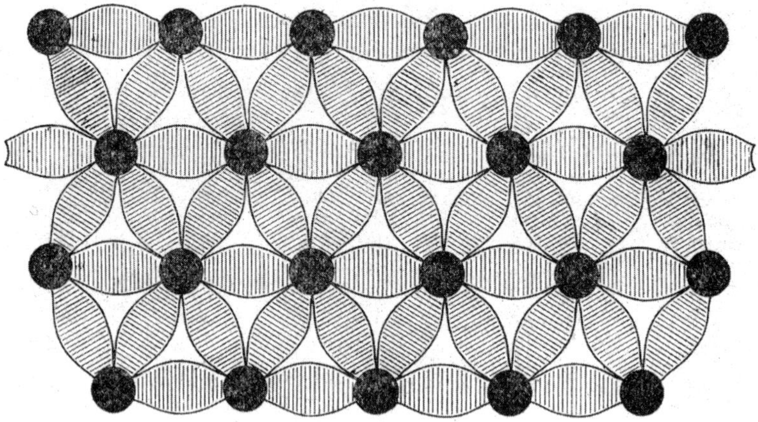

Fig. 3.—A portion of an ideal antibody–antigen framework. One plane of the structure corresponding to the value twelve for the valence of the antigen molecules is shown.

Figure 7.14. Pauling's rendering of the lattice hypothesis. Figure 3 from page 2647 of his 1940 *JACS* article. Original size (including caption): 8 × 6.5 cm (3 × 2.5 in). (Reprinted with permission from Linus Pauling, "A Theory of the Structure and Process of Formation of Antibodies," *Journal of the American Chemical Society* 62 [1940]: 2643–2657. Copyright 1940 American Chemical Society)

structure of hemoglobin, claimed that Pauling transformed "the chemical flatland . . . into a world of three-dimensional structures" (Perutz 1994).

Beginning in 1927, Pauling developed paper and wire, and subsequently wooden and plastic models of the molecules that he investigated. In the late 1940s, he and his co-worker Robert Corey at Caltech produced three-dimensional molecular models that displayed the space filled by the electrons orbiting about the nucleus of each atom. These models played a significant role in their work on the structure of proteins and became standard laboratory equipment for research on macromolecular structures. An improved version, known as CPK models (Corey-Pauling-Koltun) was later commercialized, following promotion and support by the National Science Foundation and the National Institutes of Health in the early 1960s (Francoeur 2001; 1997).

This is not the place to entertain a detailed discussion of the constitution, role, and distinctive properties of three-dimensional molecular models (on

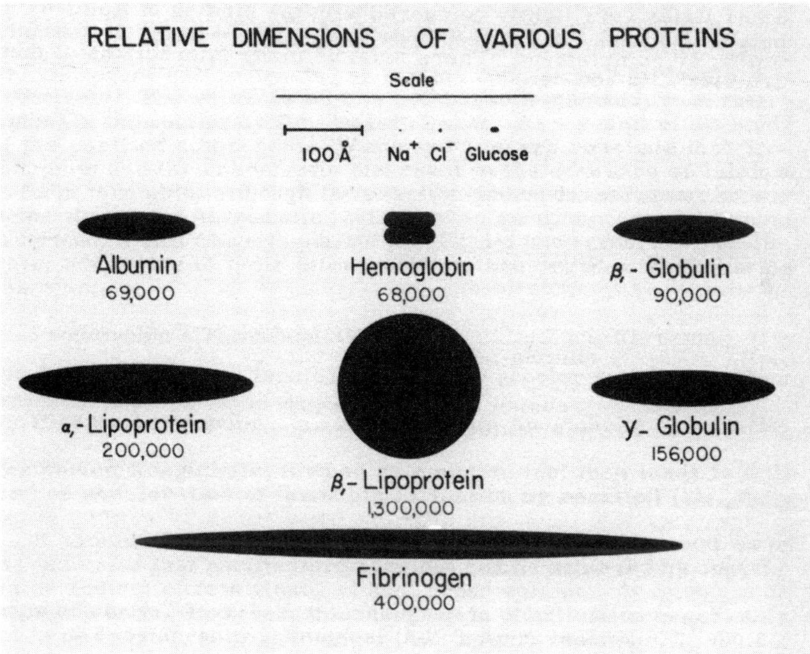

Figure 7.15. The shape of proteins as derived from hydrodynamic calculations. Original size: 12 × 9.5 cm (4.7 × 3.7 in). (Reprinted from J. L. Oncley, "Properties of Plasma Proteins with Special Reference to their Interactions with the Formed Elements of the Blood," in *The Preservation of the Formed Elements and of the Proteins of the Blood* (Boston, 1949), p. 79)

three-dimensional models in general, see de Chadarevian and Hopwood, forthcoming) except to note that the latter clearly have informed the *Endeavour* drawings. As we already have mentioned, Hayward resorted to dotting patterns to create three-dimensional effects and shapes that are closely reminiscent of the space-filling models that began circulating in the early 1930s and were perfected in the late 1940s by Pauling and Corey (Francoeur 2001, 100).[13] The implied or projected volumes promoted visual and semiotic uncertainty by urging the viewer to imagine what is on the other (non-existent) side of the image. The visual meaning of this kind of imagery therefore could be fully grasped only by referring to the three-dimensional molecular models whose production, as we just saw, was championed by Pauling. Beyond the specific argument concerning the antibodies and antigens depicted in the drawings, the latter thus contained a more general argument about the kind of representations best suited to account for biochemical phenomena.

Yet, the relation goes in both directions. Compare, for instance, figure 7.5 with figure 7.16, taken from a 1957 publication by Pauling's trusted collaborator David Pressman. The latter basically shows a three-dimensional interpretation of the former (which, in turn, uses graphic means to suggest three-dimensional arrangements), but notice the use of a transparent (plastic?) sheet wrapped around the model to somehow mimic the overall, cloud-like shape of the antigen protuberance (and/or the surrounding complementary antibody region). We find here another instance of a hybrid semiotic code, as in the *Endeavour* pictures, and, moreover, one that travels back and forth between (three-dimensional) models and (flat) drawings. The 1948 *Endeavour* drawings clearly were derived from the 1940 diagrams, the latter bearing no obvious traces of three-dimensional models. A few small sketches found in Pauling's hand-written research notes also show that flat drawings very much informed Pauling's account of antigen-antibody reactions. In a sketch dating from 1938 (figure 7.17), two years before the publication of the article on the theory of formation of antibodies, Pauling placed side-by-side a visual and a textual statement of his postulate according to which the antibody "is formed by being poured into a mold about two or more [antigens]."[14] Another, undated (circa late 1930s) research note (figure 7.13) shows a somewhat phallic rendering of the hypothesis that antibodies are formed "in a mold involving the antigen." As primitive or simplistic as they appear, these sketches are part of a visual continuum linking them to their more sophisticated counterparts in the published articles. Compare, for instance, the little sketch of figure 7.13 with Hayward's representation of the complementary shape of the antigen-antibody combining site (figure 7.5). In spite of an enormous difference in terms of pictorial rendering, the former functions as a rough sketch of the latter. We therefore would plead for some sort of relative independence of the two approaches (drawings and models). To use Foucault's terminology, they entertain a relation of similitude rather than of resemblance (Foucault 1983).

So, to summarize our argument so far, a genealogy of Pauling's illustrations highlights the role played by immunology's visual tradition in their production. This role is not limited to the aspects most germane to Pauling's own approach (namely, pre-existing visual renderings of the lattice hypothesis accounting for the key immuno-chemical experimental system, the precipitin reaction); rather, it also includes elements that are antithetical to it, as is the case with Ehrlich's theory. To understand Pauling's theory of antibody formation, one therefore must consider work antedating his 1940 with Delbrück paper on complementariness.

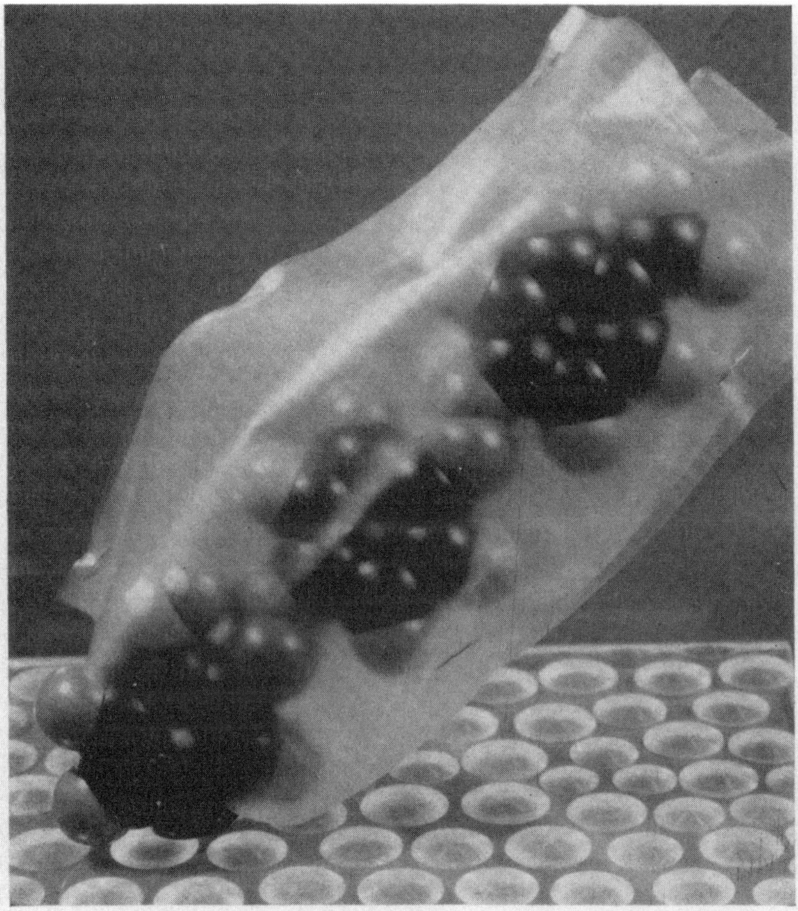

Fig. 1a

FIG. 1. Possible orientation of hapten with respect to antigen during antibody formation. a. Extended from surface of antigen. b. Lying flat on surface. c. Lying perpendicular to surface.

Figure 7.16. A CPK, three-dimensional rendering of an hapten protruding from an antigen to serve as a mold for the antibody. Notice the transparent (plastic?) sheet surrounding the CPK model. Compare with figure 7.5. Original size (excluding caption): 10.6 × 12 cm (4 × 4.7 in). (Reprinted with permission from David Pressman, "Molecular Complementariness in Antigen-Antibody Systems," in *Molecular Structure and Biological Specificity*, ed. Linus Pauling and Harvey A. Itano (Washington, D.C.: AIBS, 1957), 1–17; p. 2 Copyright, American Institute of Biological sciences)

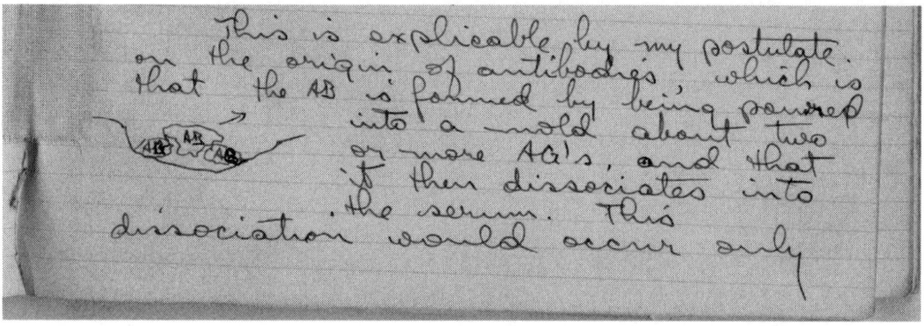
This is explicable by my postulate on the origin of antibodies, which is that the AB is formed by being poured into a mold about two or more AG's, and that it then dissociates into the serum. This dissociation would occur only

Figure 7.17. A 1938 sketch by Pauling showing an antibody molded by two antigens. Original size: 18.8 × 7 cm (7 × 3 in). (Courtesy Ava Helen and Linus Pauling Papers, Oregon State University Special Collections, LP SCI Box 6.008, folder 8.3.)

Not simply a forerunner of the continuing saga of molecular biology, Pauling also appears as a transition figure, firmly anchored in early-twentieth-century representational spaces. Pauling's illustrations, of course, bear the telltale mark of his innovative structural-chemical imagery, grounded in the developing material culture of molecular biology. But the latter, decisive as it was in adding credibility to Pauling's visual speculations on antibody formation mechanisms, was grafted onto a pre-existing body of images that can be traced to a distinctive disciplinary culture.

4. Visual Arguments: Their Role and Dynamics

In the preceding section, we outlined the genealogy of Pauling's immunological imagery. It is now time to look at its role within the discursive economy of Pauling's immunological work. Insofar as our starting point consists of images embedded in texts, rather than taken in isolation, we immediately confront the thorny issue of the relation between text and image. The latter focuses on questions such as: Is the illustration an afterthought (so to speak) to the conceptual and linguistic informational core? Is the text the unique source of the image? In seeking the meaning of the image, is it sufficient to pursue the exegesis of the document in which it is inscribed? Or does the operation work in both directions, with the image clarifying (from the Latin *illustrare*) the text (Schapiro 1996)? Our take on this issue is that the text-image relation is, to use an old-fashioned word, clearly dialectic. Textual references enable readers to make sense of the imagery, but the imagery, which has a distinctive genealogy, as we have seen, in return casts a different light on the text, guiding interpretation and forcing readers to reconsider it (Damisch 2000).

From this point of view, images as well as words can be considered part of the overall argumentative strategy of a given text. We use the term "argumentation" as defined by authors such as Ducrot and Grize; that is, as a constitutive and permanent, albeit context-dependent, element of language-as-a-performance (as opposed to language-as-a-faculty) (Ducrot 1980; Anscombre and Ducrot 1983; Grize 1982). One finds arguments whenever a speaker attempts to convert another speaker to his/her own perception or description of a given situation ("It is cold" or "Your home is far away"). We argue that images can be analyzed as visual arguments and that one can distinguish between different categories of visual arguments (see also Miller 1998, 30).

Let us begin our own argument about the argumentative role of images by noting that, as already hinted, Pauling himself took images very seriously, devoting much time and attention to their production and design. Of particular interest to us, in this context, is the relation between Pauling and Hayward, the artist who drew the illustrations for *Endeavour.* Hayward began working with Pauling in 1933, helping to illustrate his lectures, and went on to become Pauling's major collaborator in this domain, rising to the status of co-author of a widely cited book, *The Architecture of Molecules* (Pauling and Hayward 1964).[15]

Their professional relationship was clearly hierarchical—Pauling defined what could be represented and, to a large extent, how—but, as the *Endeavour* illustrations make clear, the enterprise also involved creative interaction. We unfortunately were unable to find archival evidence directly connected to the production of these illustrations, but an example of the two men's *modus operandi* is provided by a 1951 request by Pauling to Hayward to prepare a series of drawings representing the structure of the Na_2Cd_{11} molecule. An excerpt from Pauling's letter shows the extent to which the latter provided Hayward with detailed instructions:

> I would like to have drawings made of the structure of Na_2Cd_{11}, closely similar to those that you have already made for me in pastel. . . . The drawings to be made now should be line drawings from which cuts can be made—that is, black and white line drawings, with stippling or some other type of shading of the atoms. . . . The five drawings needed are, as before, a central atom surrounded by the skeleton of an icosahedron. Then the complex of 13 atoms surrounded by the skeleton of a pentagonal dodecahedron. Then two views of the rhombic triacontahedron, one a skeleton

> drawing, as before, and the other a drawing showing 32 atoms at the corners of the triacontahedron. In this case, however, 12 of the 32 atoms are to be represented as somewhat larger than the others. . . . I think that the first four drawings might be put on a single sheet, about 11 inches square, which could be reduced to 4 1/2 × 4 1/2. The second drawing could then be reduced to full page width, and the balls would come out about the same size as in the first drawing.[16]

Having received a draft version of the drawings, Pauling replied with a detailed list of comments and corrections, of which the following is a short excerpt:

> The Figure D does not show a sufficient difference in size between the small atoms and the large atoms. The large atoms should be about 20 percent greater in diameter than the small atoms. The two figures B and C do not look right to me. The left side seems to be too narrow. . . . As for E, I think that it would be worthwhile to draw it in a different way. I suggest that the edges of the cubic unit of structure be outlined, and that a triacontahedron resembling C be circumscribed about each of the corners of the unit cube, in such a way that rhombic faces are shared.[17]

Most interestingly for us, Pauling justified his attention to minute details as follows: "This [Na_2Cd_{11}] is a pretty complicated structure, and the success of our paper will be determined in considerable part by the clarity of the drawings."[18]

Clearly, as shown by a comparison between figure 7.1, most likely sketched by Pauling himself, and the corresponding version by Hayward (figure 7.3), the added value brought by the latter's artistic and graphic skills was far from negligible. Yet, the care and attention with which Pauling oversaw Hayward's work, as exemplified by the above exchange, shows that Hayward's artistic zeal was not left unchecked. In other words, no detail of the *Endeavour* illustrations was left to chance, and this applied both to visual elements that remained purposely vague and to elements that, to the contrary, were carefully motivated by immuno-chemical considerations.

As an example of the first, let us mention the "protuberances and hollows" on the surface of the antigen sphere of figure 7.3. Their function is to provide the mold around which the antibody shapes itself. The picture, however, does not provide graphical information concerning the extent to which the "protuberances" are intended to be similar or identical (compare, for instance, with Ehrlich's drawings in figure 7.9, where there are striking geometrical differences between the various combining sites). This lack of differentiation among protuberances is hardly an oversight; rather, it expresses a major problem with Pauling's theory. As we have seen, immuno-chemists had set up the precipitin reaction as a model reaction. In order to form a lattice and a precipitate with the antigen, antibodies had to be at least bivalent, that is, they had to have at least two combining sites directed against the same structure (or, in other words, carrying the same specificity). If differently shaped protuberances had been drawn on the surface of the antigen sphere, nothing could prevent the two ends of the same polypeptide chain from being molded by different structures and thus from acquiring a different specificity (Silverstein 1989, 71). Pauling's pictures neither promoted nor denied this possibility; they were silent, or ambiguous, in this matter. In choosing to show certain things, Pauling also chose to eliminate or hide others.

As for elements that were carefully motivated, let us mention the fact that in the initial 1948 illustration (figure 7.3) the filamentous protein almost touches the antigen during the antibody formation process. In the subsequent drawings (e.g., figure 7.5) the antigen and the antibody are intertwined closely and the illustrations specify not only the underlying biochemical structure but also the short distances (in angstroms) separating atoms and molecules. As we have seen, Pauling had estimated the forces at work in the antibody formation process (and, more generally, in the stabilization of complementary structures) and had concluded that they were such that they involved short-range interatomic forces such as hydrogen bonds.

This was not, however, the only possibility. Eminent physicists such as the Nobel prize–winner Peter Debye (1884–1966) felt that organic structures similar to antibodies could be held together by longer-range forces.[19] At a conference held in the late 1940s focusing on work on the formation of protein structures in thin films carried out by the protein chemist Alexander Rothen and the immunochemist Herman Eisen at the Rockefeller Institute, Pauling came under direct attack. As Pressman reported back to Pauling:

> The final speaker was Dr. Langmuir. He was very eager to have people work on this effect of Rothen claiming that we have great need for a concept of long range forces. Some people (he said) "like Pauling" think that

> these molecules, antigens and antibodies, have to come right close to each other and there must be a good fit of the antigen and antibody, but that may not necessarily be so.

When Pressman protested, he apparently generated some consternation: "many people are shocked that there is any question of Rothen's results. They just swallowed long-range forces directly since the concept is so convenient."[20] What thus at first may appear as anodyne graphical details of the *Endeavour* drawings, such as the closeness of both ends of the polypeptide chain to the antigen sphere, turn out to embody decisive theoretical claims, of which the beholder of the drawing may not actually be aware.

Turning now to the explicit relations between text and images, we should note first that the figures occur in a text already informed by a vivid, metaphorical tone. As we have seen, Pauling's crucial postulate, for which antibodies were but an example, albeit a prototypical one, concerned the existence of a "detailed 'complementariness' in configuration extending over a considerable surface of the antigen molecule and the corresponding combining region of the antibody molecule" (Pauling 1948, 44). Although modern readers may not be immediately aware of it, terms such as "surface" and "region" that instantiate the idea of complementariness are in fact metaphors. Not only are these terms transferred from geography or geometry to the invisible world of chemical compounds, but, as pointed out by Laszlo, we now "tend to forget how relatively recent, in biochemistry, are structural considerations," and overlook the fact that the notion of molecules having a "shape," and thus also of consisting of "regions" and "surfaces," was long held to be a fiction (Laszlo 1986, 39–40).[21]

More directly, Pauling also engaged in a rough metaphorical comparison:

> The phenomenon [of antibody formation] *is the same* as a production of a coin by a die, or in general of a replica by the process of pressing a plastic material against a mould and permitting it to harden. The polypeptide chain, with its power of assuming alternative configurations, is the plastic material, and the surface of the antigen serves as a die or mould. The process of hardening is the result of the operation of the weak forces between different portions of the polypeptide chain that find themselves in juxtaposition. (Pauling 1948, 49; emphasis added)

Not content with mobilizing this mechanistic image, he turned the comparison into a full-fledged analogy, replacing the words "coin" and "die" by the corresponding immunological entities ("The polypeptide chain . . .

is the plastic material, and the surface of the antigen serves as a die or mould").[22]

Metaphors such as coin-and-die partake, of course, of textual imagery. The latter, however, does not necessarily overlap with visual imagery. In the present case, the visual was not a literal rendering of the textual: the illustrations featured no coins or dies. Rather, the two kinds of imagery intersected, the coin-and-die working as a first approximation, soon relayed by a visual portrait of the chemical structures allegedly involved in the process. Highly schematic in the case of the 1940 *JACS* version, the structures became increasingly *realistic* in the 1948 version. In the latter, the textual-visual intersection was marked by the sudden appearance of the subject and object pronouns "we" and "us" ("Let *us* describe the immediate precursor of a molecule of normal γ-globulin . . . *We* assume that this polypeptide chain can become either a molecule of normal γ-globulin or . . ."); Pauling 1948, 44; emphasis added), signaling the beginning of intra-textual references to each of the figures (e.g., "In Figure 1 the antigen molecule is represented as a roughly spherical aggregate of atoms, with a surface structure shown by protuberances and hollows") (Pauling 1948, 49). Henceforth, deictic marks—"is represented," "shown by," "might represent," "is shown as," "being indicated," "are seen"—became permanent features of the text. The author—describing and commenting on what was now made visible on the page—invited readers to join in witnessing the development of the process.

We are now in a position to ask in what respect Pauling's illustrations amount to an argumentative device. Based on what we have said so far, we can recognize a series of visual-argumentative strategies that can be collected, with some overlap, under two major headings: *figuration* and *demonstration*.

Beginning with *figuration,* we have just indicated that the *Endeavour* illustrations acted by instituting another kind of reader-text relation. Once rendered into a series of drawings, the mechanism of antibody formation as postulated by Pauling lost its abstract and hypothetical qualities and was turned into a concrete mechanism. Immuno-chemical entities were equated with specific forms—with a stable shape and a frozen, simplistic visual identity—and readers were able to follow their transactions and interactions without difficulty. We call *figuration* the process by which drawings concretize notions and proposals by impressing a graphic form on them, thus offering a material expression of a hypothetical mechanism for which they simultaneously create a contextual framework.[23] Figuration opened the way to two kinds of effects that were impossible to achieve through the textual means of chained sentences: Readers could view at a glance the different stages of a complex mechanism and, at the same time, focus on its dynamic aspects. As the readers' eyes moved from one window to another, from one column to another, and finally from one page to the next, they could observe the successive stages of the postulated mechanism of antibody formation. The strip-like cartoon, depicting a sequence of invisible events, tracking spatial or temporal relations within it, lent a narrative tension to the account.

Figuration unveiled and exposed phenomena, by pointing, as we saw, to unknown, hidden, or invisible elements, by emphasizing details and making particular features stand out against an acknowledged background, while concealing others. Seduction was also part of the process, as various aesthetic elements and three-dimensional effects were added to the drawings. Even today, the *Endeavour* illustrations strike us as handsome and attractive: Such is the quality and finesse of the contours, the volumetric rendering of forms, the patterns of points, the precision of the chosen shapes, that one is almost tempted to approach these visuals as autonomous productions. Hayward's professional treatment of the imagery unquestionably lent credibility to the drawings and the account they proposed. The imagery was coherent and homogeneous: Readers were confronted with a unique landscape in which the immuno-chemical entities moved and acted continuously, as though the author wished to usher beholders into his own visual world in order to share his vision.

As we saw, figuration operates by referring, at least in part, to the visual conventions of a given research domain, that is, by conforming to the legitimate, and thus authoritative models shared by researchers in the field (Rowley-Jolivet 2000). In the life sciences, these included cyclical representations that were used to represent visually all kinds of complex developmental processes (Ratcliff 1999). In the case of 1940s immunology, their legitimacy was reinforced by the scientific popularity of Ehrlich's drawings. Pauling also borrowed and re-specified the visual models of the lattice hypothesis used by Marrack and Heidelberger.

It is instructive to compare Ehrlich with Pauling on this account. Ehrlich engaged in what Umberto Eco calls "invention as code-making," creating a new coding convention (his representation of antibodies, which became a regular feature of serological manuals in the subsequent decades) while simultaneously establishing the existence and properties of the new entities (antibodies and related substances) that he represented through that convention (Eco 1979, 250). Pauling, as we saw, built on this work, but while Ehrlich, operating before the blossoming of structural chemistry, was prevented from mobilizing its resources, Pauling was able to use chemical formulas in his drawings. However, he also included additional iconographic elements reminiscent of Ehrlich's vaguely biological shapes. Pauling's purpose obviously was to establish an explanatory correspondence between supra-molecular, biological shapes and molecular structures (the atoms and the forces they carry). From this point of view, the third illustration in the *Endeavour* series (bottom of figure 7.4) occupied a liminal position: Once the first two illustrations had construed a biological reality made up of simple forms, it displayed a chemical formula and replaced the fluffy appearance of the previous drawings with the round shapes of the atoms and their orbitals. The fourth, "flagship" illustration then became the culmination of a narrative/argumentative crescendo, combining and condensing all the previous information: the antigenic protuberance, the corresponding invagination of the antibody forming a gigantic nebulous pattern, the chemical notations accounting for the specific match between antigen and antibody. The *Endeavour* illustrations thus conflated different degree of resolution (along the macro-micro continuum) and the two ontological registers (conventional-realistic) of bio-chemical representation, while simultaneously unveiling Pauling's candidate cause of immunological reactivity: molecular complementariness.

Let us now turn to the second heading, *de-monstration* (from the Latin *monstrare:* to show). This process builds on figuration: Once the immunochemical entities and processes had been made concrete and visible, they were used as part of a kind of reasoning that partakes more of the show-and-tell exercise than of logical proof. By juxtaposing different forms and shapes, by building visual metaphors, by extrapolating from a limited number of results, Pauling's illustrations engaged in analogical reasoning. The form of the antigen, the fact that it had a single large protuberance, the length of the polypeptide chain, its contortions, the horizontal dashed line that materialized the frontier within which the antigen acted on the antibody precursor, the exact shape and position of the antibody ball were all arbitrary design decisions that nevertheless constrained what followed and, as such, were part of Pauling's analogical account.

A particularly remarkable version of this visual mechanism is to be found in the last two stereo-chemical drawings of the *Endeavour* article, through which Pauling de-monstrated how related molecules may not have the same immunological effects. The solution to the visual enigma posed by the last drawings was predicated upon the descriptive narrative tension provided by the previous drawings, with its progressive zooming-in on the relevant details. Yet, as we saw, these final drawings also may be viewed separately, as an invitation to share in the verification of an experimental hypothesis. This de-monstrative dynamics took advantage of the fact that the drawings allowed for the combination of heterogeneous semiotic means (formulas and polysemic shapes) within the same illustrations, while simultaneously channeling vision in a certain direction, namely toward the complementary features that, according to Pauling, held the secret of life.[24]

5. Concluding Remarks

Studies of visuals in scientific documents have shown convincingly that this kind of representation tends to re-present scientific objects as if experimental work consisted simply in uncovering what was, in principle, already there for everybody to witness, rather than actively participating in the production of the object being re-presented (Lynch and Woolgar 1990). Our case study of Pauling's antibody illustrations, while agreeing with this analysis, moves beyond it by approaching scientific illustrations as visual arguments, and showing, in this respect, that re-presentation involves two main processes: figuration and de-monstration. These two processes might be equated to the two periods that make up the argumentative process, according to specialists of communicative action, namely the constitution of a common frame of reference followed by attempts to impose one's point of view (Breton 1996). Our analysis of visual-argumentative structures shows, however, that in the case of visuals these two stages of argumentation are superimposed and intertwined: drawings and figures, in a kind of synergy designed to amplify their efficiency, simultaneously summon up and conflate these two stages.

To reach this conclusion, we have not confined ourselves to a rhetorical or semiotic analysis of text-image relations. Rather, we have engaged in a genealogy of Pauling's imagery, describing mechanisms (e.g., visual quotations from authoritative diagrams) that anchor visual argumentation in distinctive disciplinary traditions and that also leave space for innovation. If this is so, and if images, as we claim, do not merely illustrate theoretical views, but are means of arriving at and embodying such views, then a history of science limited to theoretical and textual pronouncements misses part of the action. To return to our opening paragraphs, while it is certainly correct to argue that information has replaced complementariness as the secret of life, it is also essential to understand, in historical perspective, the visual argumentative structure that brought complementariness to the forefront, before its eventual demise.

Notes

1. Pauling never wrote the book mentioned in the epigraph. Pauling used the term "complementariness," rather than the usual "complementarity," to avoid confusion with the physicists' notion of "complementarity."
2. In his book, Elkins summarizes and re-specifies work by authors such as Galison and Daston, Knorr-Cetina, Latour, Lynch, and ourselves. See his book for detailed references. See also Seeing Science (1992).
3. A polypeptide chain consists of a chemically linked sequence of amino acids.
4. Ava Helen and Linus Pauling Papers, Oregon State University Special Collections (henceforth AH&LPP), folder 1940a.2.
5. Holmyard to Pauling, 10 June 1947; AH&LPP, Correspondence/*Endeavour,* folder 1947a.4a.
6. Pauling to Holmyard, 9 September 1947; Holmyard to Pauling, 15 September 1947; AH&LPP, Correspondence/*Endeavour,* folder 1947a.4.
7. Gray was the scientific reporter of the Rockefeller Foundation.
8. Flanagan to Pauling, 29 March 1949; Pauling to Flanagan, 18 April 1949; AH&LPP, Correspondence/*Endeavour,* folder 1947a.4a.
9. See also Pressman to Pauling, 9 January 1952; Pauling to Pressman, 14 January 1952; AH&LPP (David Pressman correspondence) LP Correspondence, Box 250, folder 7.
10. Linus Pauling, "Chemical Studies of the Structure of Antibodies" (handwritten outline of a conference given on 22 April 1943, at the Chemical Society of Washington and Washington Academy of Sciences); AH&LPP, folder 1943s.2.
11. Felix Haurowitz Manuscripts, Lilly Library, Indiana University, Bloomington, Box 2, Folder: Correspondence, Ho-Hz. Haurowitz subsequently routinely used illustrations in his immunology papers.
12. Notes and Reprints: Early notes on proteins, 1935–1951; AH&LPP, LP SCI Box 6.008, folder 8.1.
13. Hayward also collaborated in the development of Pauling's three-dimensional molecular models.
14. Research Notebook: "Proteins," Spring 1938; AH&LPP, LP SCI Box 6.008, folder 8.3. Notice that in this early version the antibody is at the center, being molded by two antigens.
15. Hayward graduated in architecture from the Massachusetts Institute of Technology in 1922 and moved to California, where he worked for an architectural firm (his designs included the Los Angeles Stock Exchange and the Doheny Library at the University of Southern California) while pursuing artistic activities such as landscape painting.
16. Pauling to Hayward, 19 July 1951; AH&LPP (Roger Hayward Correspondence), LP Correspondence, Box 120, folder 9.
17. Pauling to Hayward, 24 July 1951; AH&LPP (Roger Hayward Correspondence), LP Correspondence, Box 120, folder 9.
18. Pauling to Hayward, 24 July 1951; AH&LPP (Roger Hayward Correspondence) LP Correspondence, Box 120, folder 9.
19. Pressman to Pauling, 28 May 1948, p. 2; AH&LPP (David Pressman correspondence), LP Correspondence, Box 250, folder 7.
20. Pressman to Pauling, 28 May 1948, p. 2; AH&LPP (David Pressman correspondence), LP Correspondence, Box 250, folder 7.

21. On the popularization of atomic models in the 1950s, see de Chadarevian (2002, 151–159).
22. In addition to the coin-and-die metaphor, Pauling resorted to an analogy internal to chemistry, namely crystallization processes: see Laszlo (1986, 81–84).
23. Some semioticians call this process *figurativization;* see Nöth (1995).
24. This, of course, still applies to contemporary representations of proteins; see Richardson et al. (1992).

References

Anscombre, Jean-Claude, and Oswald Ducrot. 1983. *L'argumentation dans la langue.* Bruxelles: P. Mardaga.

Breton, Philippe. 1996. *L'argumentation dans la communication.* Paris: La Découverte.

Cambrosio, Alberto, Daniel Jacobi, and Peter Keating. 1993. Ehrlich's "beautiful pictures" and the controversial beginnings of immunological imagery. *Isis* 84: 662–699.

Carpenter, Philip L. 1965. *Immunology and serology.* 2nd ed. Philadelphia: Saunders.

Damisch, Hubert. 2000. La peinture prise au mot. Preface to Meyer Schapiro, *Les mots et les images* (French edition of *Words, script and pictures: Semiotics of visual language*). Paris: Macula.

de Chadarevian, Soraya. 2002. *Designs for life: Molecular biology after World War II.* Cambridge: Cambridge University Press.

de Chadarevian, Soraya, and Nick Hopwood, eds. Forthcoming. *Displaying the third dimension: Models in the sciences, technology and medicine.* Stanford: Stanford University Press.

Ducrot, Oswald. 1980. *Les échelles argumentatives.* Paris: Éditions de Minuit.

Eco, Umberto. 1979. *A theory of semiotics.* Bloomington: Indiana University Press.

Ehrlich, Paul. 1900. On immunity with special reference to cell life. *Proceedings of the Royal Society of London* 66: 424–448.

Elkins, James. 1999. *The domain of images.* Ithaca: Cornell University Press.

Foucault, Michel. 1983. *This is not a pipe.* Berkeley: University of California Press.

Francoeur, Eric. 1997. The forgotten tool: The use and development of molecular models. *Social Studies of Science* 27: 7–40.

———. 2001. Molecular models and the articulation of structural constraints in chemistry. In *Tools and modes of representation in the laboratory sciences,* edited by Ursula Klein. Amsterdam: Kluwer.

Gray, George W. 1949. Pauling and Beadle. *Scientific American* 180 (May): 16–21.

Grize, Jean-Blaise.1982. *De la logique à l'argumentation.* Genève: Librairie Droz.

Haurowitz, Felix, and Fritz Breinl. 1930. Untersuchungen des Präzipitates aus Hämoglobin und Anti-Hämoglobin-Serum und Bemerkungen über die Natur der Antikörper. *Hoppe-Seyler's Zeitschrift für Physiologische Chemie* 192: 45–57.

Humphrey, J. H., and R. G. White. 1973. *Immunology for students of medicine.* 3rd ed, 3rd pr. Oxford: Blackwell.

Jacobi, Daniel. 1999. *La communication scientifique: Discours, figures, modèle.* Grenoble: Presses de l'Université de Grenoble.

Kay, Lily E. 1993. *The molecular vision of life: Caltech, The Rockefeller Foundation and the rise of the new biology.* New York: Oxford University Press.

———. 2000. *Who wrote the book of life? A history of the genetic code*. Stanford: Stanford University Press.

Laszlo, Pierre. 1986. *Molecular correlates of biological concepts*. Vol. 34A of *Comprehensive biochemistry,* edited by Albert Neuberger and Laurens L. M. Van Deenen. Amsterdam: Elsevier.

Lynch, Michael, and Steve Woolgar, eds. 1990. *Representation in scientific practice*. Cambridge, Mass.: MIT Press.

Miller, Thomas.1998. Visual persuasion: A comparison of visuals in academic texts and the popular press. *English for Specific Purposes* 17: 29–46.

Mirsky, A. E., and Linus Pauling. 1936. On the structure of native, denaturated, and coagulated proteins. *Proceedings of the National Academy of Sciences of the United States of America* 22: 439–447.

Neurath, Hans. 1939. The apparent shape of protein molecules. *Journal of the American Chemical Society* 61: 1841–1844.

Nöth, Winfried. 1995. *Handbook of semiotics*. Bloomington: Indiana University Press.

Oncley, J. L. 1949. Properties of plasma proteins with special reference to their interactions with the formed elements of the blood. In *The preservation of the formed elements and of the proteins of the blood*. Boston: Harvard Medical School.

Pauling, Linus. 1940. A theory of the structure and process of formation of antibodies. *Journal of the American Chemical Society* 62: 2643–2657.

———. 1946. Molecular architecture and biological reactions. *Chemical and Engineering News* 24: 1375–1377.

———. 1948. Antibodies and specific biological forces. *Endeavour* 7 (26): 43–53.

———. 1984. Interview by John L. Greenberg. Palo Alto, California, May 10. Oral History Project, California Institute of Technology Archives, 19: http://resolver.caltech.edu/CaltechOH:OH_Pauling_L. (retrieved April 2003).

Pauling, Linus, and Max Delbrück. 1940. The nature of the intermolecular forces operative in biological processes. *Science* 92: 77–79.

Pauling, Linus, and Roger Hayward. 1964. *The architecture of molecules*. San Francisco: W. H. Freeman.

Perutz, M. F. 1994. Linus Pauling (1901–1994). *Nature Structural Biology* 1: 667–671.

Pressman, David. 1953. Antibodies as specific chemical reagents. *Advances in Biological and Medical Physics* 3: 99–152.

Ratcliff, Marc J. 1999. Temporality, sequential iconography and linearity in figures: The impact of the discovery of division in infusoria. *History and Philosophy of the Life Sciences* 21: 255–292.

Richardson, Jane S., D. C. Richardson, N. B. Tweedy, K. M. Gernert, T. P. Quinn, M. H. Hecht, B. W. Erickson, Y. Yan, R. D. McClain, M. E. Donlan, et al. 1992. Looking at proteins: Representations, folding, packing, and design. Biophysical Society National Lecture. *Biophysical Journal* 63: 1185–1209.

Rowley-Jolivet, Elizabeth. 2000. Image as text: Aspects of the shared visual language of scientific conference participants. *ASp* 27/30: 133–154.

Schapiro, Meyer. 1996. *Words, script and pictures: Semiotics of visual language*. New York: G. Braziller.

Seeing science. 1992. Special Issue. *Representations* 40 (Fall).

Shapin, Steven, and Simon Schaffer. 1985. *Leviathan and the air pump: Hobbes, Boyle, and the experimental life*. Princeton: Princeton University Press.

Silverstein, Arthur M. 1989. *A history of immunology*. San Diego: Academic Press.
———. 2002. *Paul Ehrlich's receptor immunology: The magnificent obsession.* San Diego: Academic Press.
Weiser, Russel S., Quentin N. Myrvik, and Nancy N. Pearsall. 1969. *Fundamentals of immunology for students of medicine and related sciences*. Philadelphia: Lea & Febiger.

8

Discipline and the Material Form of Images

An Analysis of Scientific Visibility

Michael Lynch

A characteristic feature of scientific activity is the production of visual displays of objects, processes, relationships, and theoretical constructs. Scientific publications often include illustrative photographs, diagrams, graphs, and other data displays. Visual displays are not only valuable as illustrations in scientific texts; they are irreplaceable as documents that enable objects of study to be perceived and analyzed initially. Such displays systematically transform specimen materials into observable and mathematically analyzable data.

Objects and relationships that initially were invisible become visible and palpable as a result of highly technical skills and complex instruments. What laboratory scientists perceive and work upon is thus artificial in the extent to which its appearance depends upon such technologies (Garfinkel, Lynch, and Livingston 1981; Knorr-Cetina 1981; Latour and Woolgar 1979).[1] Artificial features of visual displays sometimes are blamed

This chapter initially was published as an article in *Social Studies of Science* 15 (1985): 37–66. I am grateful to Sage Publications, Ltd., for permission to reprint the article. I am grateful to Harold Garfinkel for his pervasive influence on this work. I would also like to thank John O'Neill, Ken Morrison, David Weinstein, Melvin Poliner, Gus Brannigan, Richard Hadden, Renee Anspach, Richard Hilbert, Steven Shapin and four anonymous reviewers from this journal for comments and criticism during the preparation and drafting of the paper. Bruno Latour, John Law, Elibu Gerson, and Susan Leigh Star sent me their published and unpublished writings on related topics while I was working on this manuscript. The members of the neuroscience's lab discussed in this essay patiently instructed me about their technical practices, put up with my strange and naïve questions, and tolerated my surveillance of their laboratory activities.

for illusions, misrepresentations, or distortions, but in fields such as electron microscopy, the artificial appearance of a specimen is what enables it to be observed and analyzed in the first place (Toner and Carr 1968).[2]

In this chapter I analyze two cases of scientific illustration and discuss conventions used by laboratory researchers to make their objects of study visible and analyzable. The conventions produce features of visual displays that reflect the disciplinary organization of scientific labor as much as they do the organization of natural objects and relationships.

My resources for making the analysis are diverse. They include previous studies on the details of laboratory practice and scientific literatures.[3] The discussion also is informed in a more general way by extant sociological and historical research on methods of order production in the human and natural science disciplines.[4]

Another source of information about visual displays are the conversations by laboratory practitioners while constructing and analyzing displays.[5] Such conversations demonstrate that what can be seen and said of such displays varies with the situation—with the purposes of a project, the composition of a work group, and the informal methods used in a laboratory and in the field at large to establish what counts as an adequate display of an object.

What I want to address, however, is not how any particular conversation refers to a visual display or influences the further course of its production: I want to be able to characterize general features of the production process as a whole. The visual documents and conversations that appear at various stages of a project are part of a larger and more inclusive set of operations involving complex instruments, simple hand tools, manual tasks, specimens and their residues, and literary and mathematical facilities such as scales, graphs, tables, labels, indices, and so forth. The process of working with such materials cannot be characterized adequately by focusing only on verbal references to an object, literary inscriptions (Latour and Woolgar 1979; Latour 1986), the relationship between written accounts and visible displays, or embodied actions at the lab bench (Lynch, Livingston, and Garfinkel 1983; Schrecker 1980; see also Amerine, and Bilmes 1990). What I will present instead is an account of how a specimen is transformed into visible and analyzable data through the various stages of a project. I will describe the rendering practices that utilize a variety of modalities of action to transform the documentary form of a scientific object from one stage of a project to another.[6]

The documents reproduced in figures 8.1 through 8.5 will be used as an occasion to discuss practices of constructing scientific data. Although I believe that such documents act as an organizational focus of the various tasks that make up a project, I will not be discussing semiotic elements that can be inferred from the pictures by themselves.[7] The pictures will be discussed inasmuch as they are featured in a labor process. Some of the features of the displays that reflect that process became obvious only after I had observed laboratory work in situ.

The first of the two cases discussed in this paper (figure 8.1) is taken from a popular field guide of reptiles and amphibians, and was chosen because of its utter "proto-scientific" simplicity. Specific themes will be taken from the analysis of that case and then applied to the second instance. The second case is a series of documents used during a neuroscience research project (figures 8.2–8.5). The laboratory that conducted the project was the subject of an ethnography I performed several years ago. During that study, I examined the overt features of the lab's graphic displays, observed the series of procedures that were used to construct them, tape-recorded several discussions while lab members were assembling and interpreting the displays, and read several drafts of a research publication that resulted from the project (Lynch 1985; Lynch 1982).

I believe that the properties of graphic representation that the two cases illustrate are very common in science. This does not mean that they are universal in science, or that they are essential to science as a demarcated domain of activity. Instead, they are more like typical *instruments* of graphic and pictorial representation that enjoy widespread use in many scientific fields, although science surely does not hold a monopoly on their use.

1. A Proto-Scientific Example

The diagram in figure 8.1 is a rudimentary display of *scientific* activity. It is selected for analysis here as a case that should be comprehended readily, as its reading requires no extensive training in any of the specialized natural sciences. The simplicity and ordinariness of the display make it useful for pedagogical purposes, and, I will argue, they allow us to identify elements that are common to more complex and esoteric representational procedures in science.

This figure is part of a set of instructions for doing a field study of reptiles. The instructions do not describe an experiment; they outline certain preparations for systematic observation and experimentation, and are in that sense "proto-scientific."[8] No explicit hypotheses are formulated or tested; instead, a set of elementary operations is performed that clears the way for observation and experimentation.

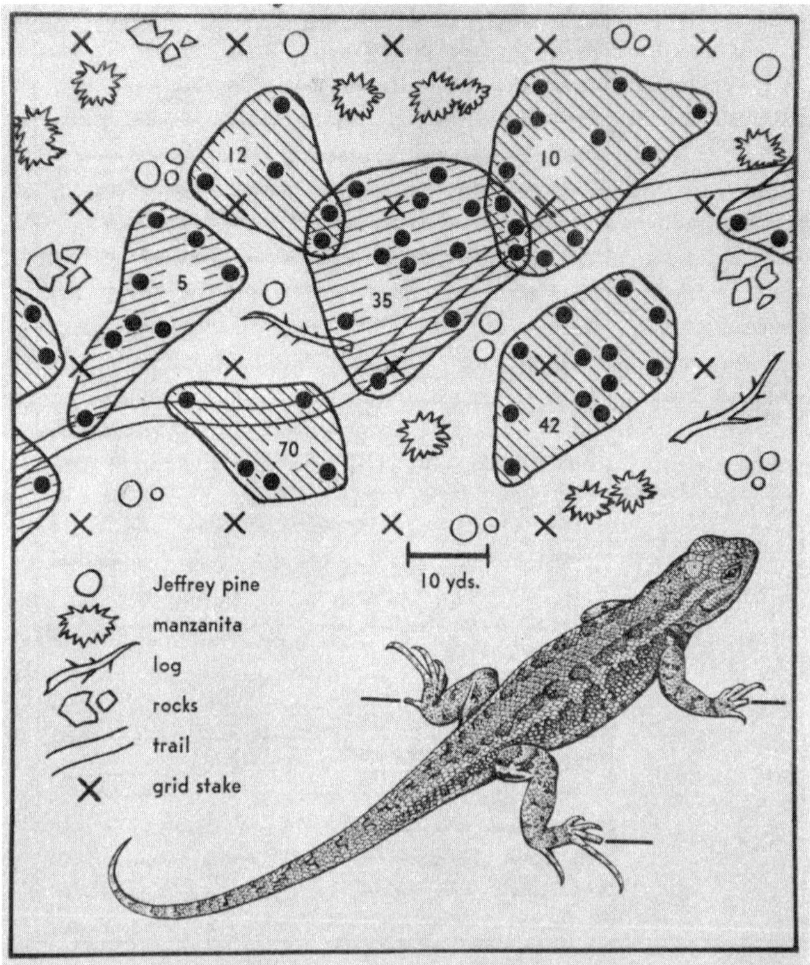

Figure 8.1. Map of study area, showing home ranges of marked lizards.
The figure illustrates the kind of results that can be obtained. The home ranges of six Sagebrush lizards are shown. The lizards were given identification numbers and were permanently marked by removal of two or more toes in combinations. Recaptures (black dots) were plotted in reference to numbered stakes (X) arranged in a grid at 20-yard intervals, and were located by pacing the distance to the two nearest stakes. Toes removed from the lizard shown are right fore 4, right hind 3, left hind 5, expressed as RF4-RH3-LH5 (Figure and legend from *A Field Guide to Western Reptiles and Amphibians,* 3/e by Robert C. Stebbins. Copyright © 2003 by Robert C. Stebbins. Reprinted by permission of Houghton Mifflin Company. All rights reserved.)

The diagram is organized as a map, with certain identified landmarks, a set of coordinates, and a distribution of territories. The accompanying caption provides instructions for constructing the map: (1) marking lizards; (2) constituting a grid; and (3) placing a series of observations within the coordinates of the grid. The instructions can be summarized as a set of procedures for turning a *natural* setting into a geometricized workplace.[9] A landscape is transformed into the graphic grounds of careful observation and measurement, and the figures of the lizard territories are constituted on the ground of that transformation. The three themes, *marking, practically constituting graphic space,* and *rendering observations as points* identify a program of widespread use in science—one that will be examined in reference to other examples later in this essay. First, however, I will supply preliminary characterizations of each of the three themes in reference to the proto-experiment described in figure 8.1.

Marking: Counting on the Fingers of a Lizard

The text in figure 8.1 matter-of-factly describes a procedure for marking and coding lizards by amputating a unique combination of toes on each captured individual. A literary record of the individual's mark is produced as a code for limb and amputated toe (for example, *RF4-RH3-LH5,* as a notation for *right front fourth* digit, etc.). Each captured animal thus is given a unique *mathematical* identity for any subsequent observations, through the construction of a *mark* as a naturally visible code of the animal's place in a numerical order.

The naturalistically visible lizard is no longer just a lizard, as it becomes the bearer of a numerical code. Marking preserves a class of lizards, equivalent in all respects except for the unique identity of each mark within the set of marks.

Constituting Graphic Space

The caption identifies each x in the illustration as a "numbered stake arranged in a grid at 20-yard intervals." The arrangement of stakes produces the coordinates of a graph through a concrete construction on the ground surface, analogous to a surveyor's rendering of a landscape into grids or plots to set the stage for residential development. *Mathematical* space comes to dwell within the *natural* terrain. The two surfaces coincide at the visible points marked by the stakes.

The placement of stakes performs a double role. On the one hand, it is clear that the stakes mark off equivalent intervals in a two-dimensional grid, and that the placement attempts to approximate a geometric overlay on the uneven surfaces of the terrain. On the other hand, the selection of interval size appears to be *phenomenally* and *practically* oriented. Twenty-yard intervals are selected. The selection is not arbitrary, since it already includes reference to the sorts of size that lizards' territories might show. Three-inch intervals are not recommended, nor are 100-yard intervals. The interval is selected to be on a scale appropriate to the texture of territories it resolves. Too large an interval does not readily resolve the lizards' movements, while too small an interval creates a tedious and disruptive army of stakes. In other words, theoretical assumptions about lizard behavior already are built non-discursively into the graphic device's scalar properties.

The ultimate reconciliation of the grid and the surface it marks off occurs on the surface of the map. The two-dimensional paper brings together the identified landscape and the mathematical grid. Graphic formats come to embody properties analogous to the Kantian categories of time, space, and causality. The map transfers the categories from an ideal transcendental context to a concrete manipulatory region.[10] No longer are they transcendental in an absolute way; they transcend the field of operations that they prepare for, and yet they are themselves constructed in human activity. I suppose what this procedure amounts to is "standing Kant on his hands." Although we are analyzing the chart on the page, its ordered properties do not originate there. A "proto-map" is built into the natural landscape itself. The stakes driven into the surface of the ground mark off the outlines of a scale and the geometrical form of a graph.[11]

Normalizing Observations as Equivalent Dots

Individual specimens are identified graphically by their place on the grid. Each sighting of a lizard typically is governed by the constraints of place-on-the-grid. The temporal collection of sightings is spatialized and geometrized in the synthesis of animal territories. (The sequential order of sightings is not notated, presumably since it reflects the vicissitudes of observing and capturing specimens and not systematic patterns of the lizard movement.)

Note that each observation of a marked individual is rendered equivalent to all others through the use of the device of the *dot*. The only material difference between one dot and another on the chart is its locale.

Locales are reckoned in terms of the grid of stakes, and all other circumstantial features of observation *drop out.*

2. The Docile Object: Merging Graphic Properties with the Visible Residues of a Specimen

The problem of visibility in science is more than a matter of providing illustrations for publication. Published and unpublished data displays, in the form of graphs, photographs, charts, and diagrams, constitute the material form of scientific phenomena. By "material" I mean sensible, analyzable, measurable, examinable, manipulatable, and *intelligible.* Although the procedures for making the object scientifically knowable implicate an independent object, they simultaneously achieve a graphic rendering of the object's materiality.

The question is not, "How do objective properties correspond to graphic devices for isolating and ordering those properties?"; it is, "How do graphic properties merge with and come to embody the *natural object?*" The latter question inquires about how science initially determines what is natural on the basis of what its graphic devices disclose.

From our brief examination of the "proto-scientific" illustration, we have introduced three themes: marking, constituting graphic space, and normalizing observations. A docile object is the product of those rendering procedures.[12] It is an object that *behaves* in accordance with a program of normalization. This does not mean that it fails to resist, or that its recalcitrance does not serve to adumbrate its objective news for science. It is to say that, when an object becomes observable, measurable, and quantifiable, it already has become civilized; the disciplinary organization of civilization extends its subjection to the object in the very way it makes it knowable.[13] The docile object provides the material template that variously supports or frustrates the operations performed upon it. Its properties become observable-reportable in reference to the practices for revealing them.[14] If the object was not compliant to such a program, its attributed properties would be incompletely or *unscientifically* observable.

3. Constituting the Visible Anatomy of Axon Sprouting

The themes of marking, constituting graphic space, and normalizing observations now be will explicated in more detail in reference to a more complex case—a project for anatomically demonstrating a regenerative process in the hippocampus. The report of the project was published in a neurosciences journal and was one of several similar studies by members of a small university research laboratory in California. I was able to observe this project in situ, and to tape-record several conversational interactions among laboratory scientists as they performed various tasks during the project.

The focal point of the project was the construction and analysis of electron-micrographic *montages* of the rat hippocampus (see figure 8.2). The montages consisted of several electron micrographs that were placed together to show a continuous view of tissues, similar to the way an overlapping array of aerial photographs is used to display a more extensive terrain. Each montage consisted of a *map* of two adjacent regions of axon *terminals* (the end *boutons* of axons that form synapses with neuronal dendrites). The two fields of axons were said to arise from neurons in distal areas of the brain: One field arose from an area of the cortex, and the other from the hippocampus on the opposite side of the corpus callosum. The fields were said to be *layered* in a direction roughly parallel to another layer of *granule cell* bodies. The neurons in the granule cell layer sent tree-like branches of dendrites coursing across the two fields of axons, synapsing with terminals in both layers.

The montages were used as part of an experiment for assessing the anatomical changes occurring in the adjacent axon layers after the destruction of the area of the brain from which one of the axon layers originated. A lesion was performed on anaesthetized animals (Sprague-Dawley rats—a strain of rat bred for uniformity of brain dimension and behavioral docility), and the animals were revived and then killed at a variable number of days following the lesion. The brains were removed and preserved, and the hippocampus was dissected. Electron micrographs were taken at standardized locales in the dentate gyrus, and were pasted carefully together to produce the montages. The montages then were analyzed, and cross-sections of *intact* (apparently *normal*) and *degenerating* terminals were marked with color-coded outlines drawn with felt-tipped pens. The edges of the montages were used as scales for constructing a grid of sectors, and counts of intact terminals were taken in each sector. These then were plotted on a graph (figure 8.3), where one axis of the graph was a scale of distance in microns from the granule cell layer and the other axis displayed density of intact and degenerating terminals. The broken and solid lines indicate, respectively, *mean terminal counts* of intact terminals at four days and at eleven days post-lesion. The graph can be read as evidence for a change in the distribution of intact terminals between four and eleven days after the lesion.

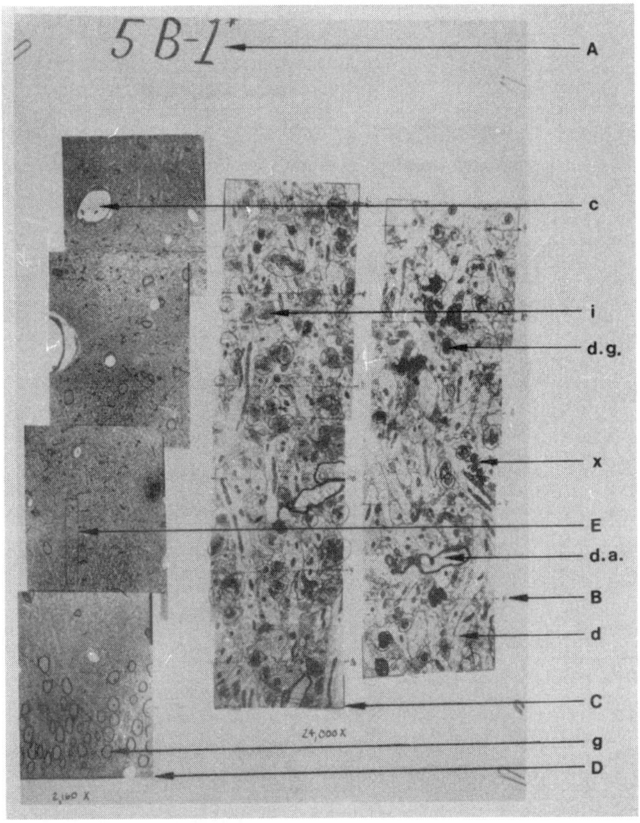

Figure 8.2. Electron micrographic montages.

Key to figure 8.2

A = index identifying montage (5 = number of days post-lesion before specimen's sacrifice; B = index for particular section of tissue used for montage; 1 = identification of particular animal used as specimen)

B = scale of micron equivalents drawn along margin of column. Numbers on the scale represent vertical distance from the granule cell bodies.

C = high-power electron-micrographic montage (24,000X)

D = low-power electron-micrographic montage (2,160X)

E = hand-drawn rectangle marking placement of high-power montage in the field of low-power montage (see close-up in figure 8.5)

d.a. = marked degenerating myelinated axon. Dark outline and parallel slash marks within the borders of axon were drawn by hand to mark it.

d.g. = marked degenerating axon terminal. Black pen was used to trace outlines of degenerating terminals on a clear plastic overlay. Circle around the terminal marks it as a countable instance.

i = marked intact axon terminal. Red pen was used to trace intact terminals. Circle around the terminal marks it as a countable instance.

x = marked degenerating axon, with an X mark notating that terminal is a doubtful instance. Often an X was used to mark a case where the terminal candidate was not associated visibly with a post-synaptic density (a conspicuous feature of a synapse). In other cases, an X noted that the analyst was uncertain about whether the instance was intact or degenerating.

c = capillary (in cross-section)

d = dendrite (presumably arising from granule cell layer)

g = granule cell bodies (nucleii outlined with marking pen)

The low-power montage was used as a locational reference for the high-power montage. Analysis of axon terminal density and distribution was done on the high-power montage by counting numbers of marked and countable intact terminals for each 5-micron sector of the montage. A statistical display of the outcome of the analysis is shown in figure 8.3.

(Photographs in this figure are taken with permission from unpublished data displays in an electron-microscope study of axon sprouting.)

Together with other evidence from the lab's researches, these data were used to make a case for *axon sprouting,* a compensatory mechanism in the face of irreversible damage to nerve cells. Axons from the undamaged field *sprouted* new branches and formed new terminals partially to reoccupy synaptic sites vacated by the destroyed terminals in the other cell layer. Continued research on the topic over the past several years has substantiated these anatomical findings, and, further, has suggested a biochemical basis for the regenerative phenomenon.

The graph in figure 8.3 is the literary end-point of a series of renderings that transformed cellular arrangements into graphic space. This transformation is expressed nicely in the figure, where the horizontal axis of the graph is extended to the left, across the page, to intersect with the schematic camera lucida drawing of a granule cell. Figure 8.4 further schematizes the naturalistic origins of the graph, tracing them back successively to a geometric slab of hippocampal tissue, a cross-section of the hippocam-

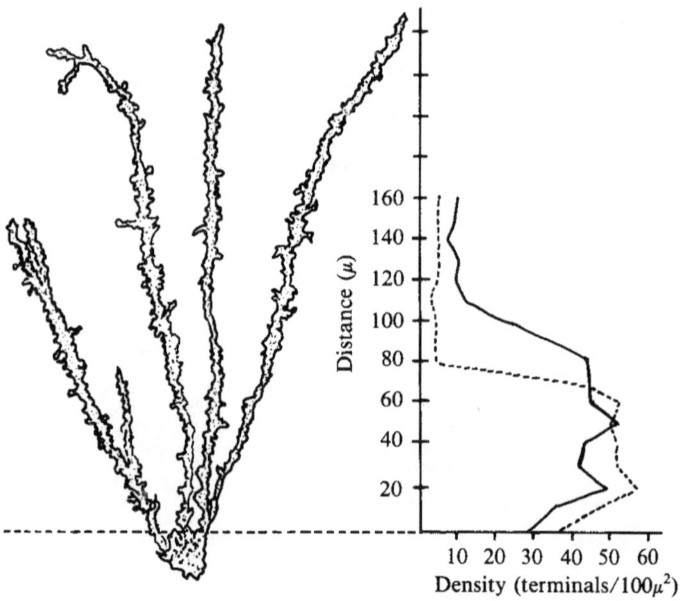

Figure 8.3. A camera lucida drawing of a Golgi [stain]-impregnated granule cell on a common vertical axis with a graph depicting density (terminals/100μ^2) of the molecular layer as a function of distance (microns) from the top of the granule cell layer. The plots are mean terminal counts of the 4-day (dotted line) and 11-day (solid line) post-lesion groups. (Figure and legend were taken with permission from a draft of a paper written in 1975 by members of the neurosciences laboratory discussed here. The specific names of the researchers and title of their publication is not mentioned in order to preserve anonymity.)

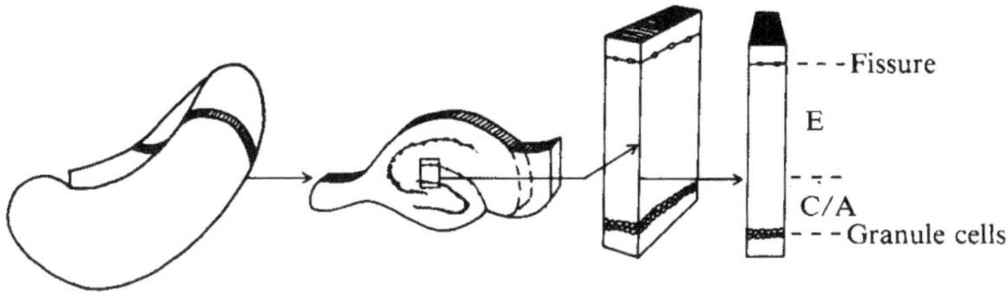

Figure 8.4. A schematic illustration of the position and orientation of the region sampled by electron-microscopic montage. At left is the hippocampal formation ipsilateral to the entorhinal cortex lesion. A 125-μ coronal section from the dorsal portion of the formation is cut, and from that section is removed a block of tissue oriented perpendicular to the plane of the granule cell extending from below the granule cell layer to above the hippocampal fissure. Ultra-thin sections are cut from the medial edge of this block. The zones of afferent occupation in the molecular layer are indicated E = entorhinal, C = commissural, A = associational. (Figure and legend taken with permission from a draft of a paper written in 1975 by members of the neuroscience's laboratory discussed here. The specific names of the researchers and title of their publication is not mentioned here in order to preserve anonymity.)

pus, and the hippocampus as a whole (the banana-shaped item on the far left of the figure).

Given the naturalistic linkages between the graph and the hippocampal material, we now can examine in detail how the material was successively *geometricized, chronologized,* and *mathematized* such that its scientific visibility and demonstrability were achieved. The three themes of *marking, constituting graphic space,* and *normalizing observations* now will be reintroduced in the discussion of this particular instance.

Marking

Marking was achieved at numerous junctures in the course of the project, and in widely different ways. As mentioned above, color-coded marks were hand drawn upon the surfaces of the montages to distinguish *intact* from *degenerating* terminals. Among the several tasks achieved by such marks was that of indexing.

Indexing Early in the project, while the animals were still alive, a varying number of ink slashes were drawn on the tails of each individual rat to index the animal's unique numerical identity in the corpus of specimens. The fragmentary remains of the animal were also indexed with marks and numerals to constitute a coherent *biography*.[15]

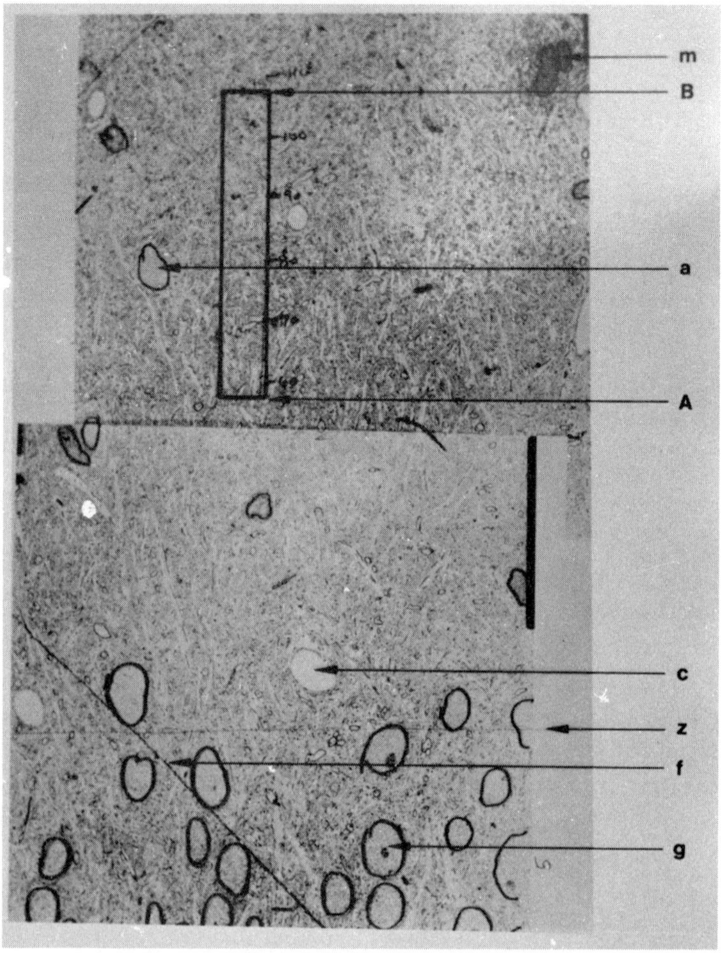

Figure 8.5. Enlargement of low-power montage from figure 8.2, showing placement of high-power montage in the field above the granule cell bodies.

Key to figure 8.5

A = *lower* edge of rectangle showing placement of high-power montage in the field of the low-power montage. The scale of micron equivalents begins at about 60μ above the granule cell bodies.

B = *upper* edge of rectangle, approximately 110μ above granule cell layer. Note, microns were computed in reference to the magnificational power of the photographs and a metric scale was constructed to mark *micron equivalents* for each montage.

a = astrocyte cell nucleus marked with blue line drawn around border. [Note: These were not discussed in the present paper.]

m = two *marked* microglia cell nucleii. [Note: These also were not discussed here.]

g = *marked* granule cell nucleus in granule cell layer (nucleolus is dark spot in center of nucleus)

c = capillary cross-section

f = artifactual line presumably due to the folding of the thin tissue section during preparatory procedures.

z = line drawn to interpolate *upper edge* of the granule cell layer for purposes of measuring distances from the cell layer. The line acts as the *zero* point on the vertical scale of microns.

(Photographs in this figure are taken with permission from unpublished data displays in an electron-microscope study of axon sprouting.)

Marking was constituted not only through a relation between a practitioner's hand and a specimen. It already was built into the operation of the procedures that made *tissue structure* visible in the first place.

Labelling The term "label" was used as a synonym for the various stains applied to tissues to make them visible for light and electron microscopy. Specimen materials were immersed in solutions of metallic compounds in order to distinguish protein structures by selectively covering them with dense precipitates. The structure of deposits composed a visible architecture of light or electron-absorbent material that enhanced the visibility of membranal structures. Cellular structure was thus dependent upon the devices of *labelling* for its accountability.

Although writing in the familiar mode was not involved, the term "label" was apt. The stain, like a name, defined, pointed to, and consolidated the visibility of its object. Furthermore, labelling was performed under a programmatic *Cartesian* separation between the object-labelled and the artifice of labelling. The artifice of labelling was distinguished, in principle, from what the label revealed. Like language, however, labelling was shadowed by a devil's playground full of possible mistakes, errors, oversights, and fraudulent uses. Staining artefacts could obscure the object or create fictional resemblances; it could constitute ambiguities in the form of the object versus the form of the label. Daily lab activities largely were concerned with the practical maintenance of that Cartesian division. Questions on the essential artificiality of anything that becomes accountable within such a program are entertained in the neurosciences, though only as a matter of "philosophical" interest.

Upgrading Visibility As in the case of labelling, varieties of practices were used for *upgrading visibility:* to enhance and select thematic constituents for appropriation within the continuing development of the project.[16] These practices might be summarized as an externalized and instrumental labor of selective perception. For instance, marking terminals on the surface of the montages was not merely a matter of indexing. Each time a terminal was outlined, and thus coded, its visibility against the background of ultra-structural materials was highlighted and disambiguated. The number of marked terminals did not correspond to the number of *actual* terminals in the field, nor could they. Instead, each time a terminal was marked, it was counted as "good enough" to stand as a defensible case. Some possible cases were not marked because they were ignored. Others were marked as indistinct, within the plane of the photograph, on whether they were intact rather than degenerating or axon terminals rather than something else (see figure 8.2). After they were marked, the traced instances stood out as a corpus of clear cases, easily distinguished and readily counted.

In other areas of the lab's work not directly affiliated to the particular project, an amplification and clarification of literary analyzability was produced instrumentally. Complicated electro-physiological equipment was used to transfer mechanically, within pre-coded ranges of frequency, the electrochemical events of a region of brain cells to the graphic surface of a moving scroll of paper. Such electrophysiological monitors worked much like a seismograph on a different terrain to enhance and analyze faint and otherwise undetectable movements.

The camera lucida drawing in figure 8.3 is a further instance of an upgraded specimen. The drawing is interesting for the way it straddles the vague line between drawing and photography—between the realm of *manual* transfers, and the more respected realm of *instrumental* transfers of imagery.[17] The camera lucida projects a light microscopic field onto a sheet of paper. The projected conformations of a prepared slide of tissues then can be traced by hand on the surface of the paper. Unlike a photograph, the drawing can superimpose, in one continuous line, courses of dendrites and axons that snake in and out of the focal field of the microscope; a series of adjustments of focus can be collapsed into one schematic trace. Outlines can be clarified and continued across gaps and blurriness. *Inessential* constituents can be omitted or glossed, as in the undifferentiated texture of dark shading in the interior of the granule cell in figure 8.3. The drawing therefore retains its visible relation to the projected slide while offering the advantages of a schematic record that has been clarified in accordance with the analytic aims of the project.

Constituting Graphic Space

Practices for constituting graphic space can be described as prelinguistic modes of order production. This does not mean that language and verbal conceptualization are not involved. Instead, it means that material media, in addition to the coded sounds or marks of verbal and written language, already define to an indefinite degree what becomes *knowable* or *reportable* in linguistic or conceptual terms. Intelligibility is built into the visible form of materials in the way they are brought under scrutiny.

In the case described here, all stages of the production of the *sense data* for axon sprouting involved the location and use of empirical approximations of the perfected forms of geometry.[18] These practical geometries were appropriated both from *endogenous* conformations of the cellular material examined and *exogenous* conformations of the instrumental fields and literary formats used in rendering the materials examinable. The sequential stages of the project exhibited a kind of *bootstrapping operation*, which upgraded the visible form of *endogenous* geometries to bring them into alignment with angular and dimensional properties of *exogenous* graphic formats. There was thus a reflexive relationship between the regular features of the observing and measuring devices and the regularities observed in the tissue. The consequence of the operation was to identify "natural" residues of the specimen in and as the instruments and scales used scientifically to display and assess their properties. I will try to substantiate and clarify these assertions in the following discussion of *exposed geometries, upgrading geometricity*, and *utilizing formats to compose scales.*

Exposed Geometries The dentate gyrus is characterized anatomically as a region of the brain's hippocampus where the various neuronal constituents differentiate in a distinctly stratified fashion.[19] Granule cells, pyramidal cells, and axons synapsing with the dendrites from those cells are sorted into distinct layers in cross-sectional representations of the dentate gyrus. In contrast, the cellular constituents of the cerebral cortex are less easily distinguished into homologous anatomical regions.

The distinctive organization of the dentate gyrus was appropriated as a *natural laboratory* for experimentation. The layering and internal coherence of the anatomical structures provided a happy coincidence of anatomy and classification. Categorical groups of neurons were isolated readily in a visible and manipulatory space. The grammar of experimentation—of selection, control, and comparison—was operated in the hippocampus with an ease that could not be approached elsewhere in the brain. The project discussed here, along with the bulk of the particular laboratory's many other projects, utilized the hippocampus as a particular region for exploring generic brain structures and processes. The studies of axon sprouting, for instance, concerned a recovery process in the entire brain, with the hippocampus as a proximal "window."

The exposed layering of the cells in the dentate gyrus also made possible its use as a naturally occurring graphic template. Figure 8.3 provides a genealogy of the scale for distance along the vertical axis of the graph by tracing the origin of the scale to a line that intersects the cell body of the granule cell schematized in the camera lucida drawing (on the left-hand side of the figure). Figure 8.4, in turn, traces the linear conformation of the granule cell layer from a *slab* of tissues extracted cross-sectionally from a section of the hippocampus. The tissue slices exposed the cell layer and excised it in such a way as to place the lower frame of the rectangular slice in a parallel alignment with a relatively straight segment of the curving granule cell layer (second from left drawing in figure 8.4). The slab was cut so that its edges *located* a parallel alignment of pre-delineated anatomical structures.[20]

Upgrading Geometricity A progressive series of operations ever more closely arranged the specimen materials into the form of a representational two-dimensional graph. The anatomical layout of the slab displayed the dendrites of the granule cells as vertical components and the axons synapsing with those dendrites as horizontal components. This, of course, relied upon the relatively parallel orientation of each dendrite and each axon with the other members of its class. It also used the roughly graphic anatomy of axons and dendrites, the fields of which are aligned at approximately a 90 degree angle to each other. The horizontal and vertical axes of the graphic arrangement of the cellular constituents were exposed in turn through the orientation of the cross-sectional slices constituting the slab. Each linear slice instrumentally constituted planes and lines of orientation referenced, with a superior "exactness," to the exposed linearities of the anatomical divisions.

The uniformity and cross-sectional orientations of slabs and sections brought together the visible properties of tissue and the graphic properties of a page of text.[21] The thin sections were constructed as if they were pages of brain tissue to be read for their graphic imprint. This coincidence was no accident, of course. Tissue choppers and microtomes were used, respectively, to slice thick and thin sections in preparation for microscopic viewing. Instrumentally designed planes of operation for the movement of the blade through the tissue assured the precise conformation, for all practical purposes, of a two-dimensional rectangular strip of tissue. Electron micrographs further accentuated the two-dimensionality and rectilinear form of the data. Under high-power magnification, a thin section appeared as a plane extending beyond the horizons of the field of the microscope. Photographs were aimed within that field carefully to frame selected features within a vertical column of prints identified with an up-down orientation to a reader.

The uniformity of the tissue grain was accentuated not only through the imposition of instrumental planes of operation and manually inscribed lines. The selection of sites for shooting the series of micrographs expropriated locales where the cellular layering was especially *straight* and *parallel*. The visible micro-anatomies showed considerable variance from the practical ideal of linearity, and occasional moments of alignment were framed selectively for use as templates of graphic formatting. (By "graphic formatting," I mean practices that compose a display by placing data within the parameters of a graph.)

Utilizing Formats to Compose Scales The series of practices for upgrading natural formations show progressive degrees in the extent to which visible materials are *mathematized*. Proto-mathematical properties of objects are selectively exposed and are placed carefully within lines and planes of orientation and operation that construct the material form of visible displays. Those lines and planes then provide linear edges for the location of scales, flat surfaces for the superimposition of grids, and discrete sites for use as points within the grids.

Montages were produced in reference to the cell layer. A regional view of tissue under low-power electron microscopic magnification was selected as an appropriate field for a series of higher-power photographs (see figures 8.2 and 8.5).

The *lower-most* of the photographs was aligned at a standardized point *above* a line, drawn across the surface of the low-power field (figure 8.5), representing the mean upper edge of the granule cell bodies. The drawn line constituted a graphic axis from which a perpendicular column of photographs (figure 8.2) was reckoned. The perpendicular line simultaneously became a *scale* whose standardized units, in microns, were calculated in reference to the power of magnification. Each photograph was shot in the series, moving *upward from the cell layer,* so that its lower border overlapped with the upper border of the photograph below it and its left and right borders aligned with the borders of the others.

Intracellular *landmarks* were used as mnemonics for reckoning the placement of each photograph in reference to the others in the series, and for pasting together the photographs, after they were all developed, to reconstitute the continuous terrain. The resulting strip exhibited roughly aligned right and left edges, which then were used as a scale for assessing positional relationships of axon terminals in the montages. The strip was designed to enclose the key area of overlap between the two adjacent axon regions *above* the granule cell layer, and to allow the observation and measurement of the extent of overlap between the *intact* axon terminals in one of the two regions and the *degenerating* terminals in the other (see the graph in figure 8.3 for a schematic account of the analytic results).

Once the scale was put in place for ordering the tissual plane in reference to measurable dimensions and locales, a measured distance from the *starting point* (the line across the granule cell bodies) was used to standardize the locale of each montage extracted from each distinct specimen. This standardization was further assured by the selection of a domestic breed of rats whose brains were said to exhibit a high degree of dimensional conformity from one individual to another.

Within the linear frame of the montages, the visibly intact and degenerating axon terminals were categorically identified and counted as graphically significant points in the two-dimensional spatial grid of the photographs. Marking the terminals transformed them into usable points, equivalent (except for their spatial locale) to other such points identified with the same color-coded mark. These points, although they facilitated the categorical transformation of terminals into *numbers,* were simultaneously inspectable as photographic images of singular terminals. One could still see the interior of the terminal outlined with the color-coded mark; each instance was a pivotal entity—a meeting place of a graphic-mathematical point and the direct image of an empirical residue extracted from the remains of an animal.[22] Note that in figure 8.3, the lines on the graph lose such concrete reference to the visible residues of the *animal,* and reside unambiguously in a statistical space.

Normalizing Observations

The marked *points* on the surface of the montage and the broken and solid lines in figure 8.3 are graphic devices for aggregating observational cases as categorically uniform objects. *Intact* axon terminals, outlined in black, are aggregated within each sector of a grid of distances. (The vertical dimension of the montage becomes *distance from the granule cell bodies,* and the horizontal dimension becomes a swath of standardized width in which distributions of terminals are sampled.) Each marked point on the montage represents a single observation. The montage presents each observation as a spatial point within its grid of measured distances. Without contradicting or impugning such a reading, we also can see how each marked terminal is the product of a practical course of actions that prepared the specific terrain for visible assessment and measurement before including the specific case into the corpus of acceptable observations. The graphic surface of the montage confines and "simplifies" (Star 1983) the singularity of each such observational procedure to a two-dimensional spatial position.

The graph in figure 8.3 preserves the aggregate properties of the marked cases, while dropping any direct reference to the singularity of micrographed terminals. Unlike the montage, it no longer displays the image of each terminal within the crude outlines of the montage's graphic *points*. Instead, it provides an average density of terminals in each sector of distance for two temporal classes of *animals* (4-day and 11-day). The graph aggregates and homogenizes the collection of different marked points and specimen animals into a unitary spatial display.

The graphic display normalizes the properties of each animal and each counted terminal. The specimen *animal* becomes both more than, and less than, a laboratory rat. It becomes more than a nervously staring creature living out its life in a wire cage, since the fine structures of its nervous system revealed through dissection of the animal, histological treatment of its tissues, and microscopic magnification are not at all apparent from the outset. The structures revealed by rendering the animal's remains are virtually abstracted while retaining their attachment to the concrete residues of the animal's body. It becomes less than the ordinary animal since the original animal literally is thrown away in favor of the residues retained for inspection.[23] In the end, the animal is *sacrificed* along with a cohort of others for the sake of an aggregate line on the graph. Its practical history drops off: the graphic line is indifferent to the fact that a particular animal may have struggled fiercely before going under anesthesia, that a crisis occurred on whether the perfusing fluid would *take* during the preservation of the brain during dissection, or that staining of thin sections of the animal's neural tissues created an unusual grainy texture that almost led to its being discarded from the corpus of analyzed specimens. The lines on the graph no longer represent rats in their ordinary, familiar mode; they represent measurements performed on methodically processed extracts of the animals' dissected brains.

The spatial coherence of the graph stands on behalf of a discipline of laboratory activities, and is tied intimately to the lab's structure of activities. The graphic aggregation of singular observations represents a claim about the standardization and efficacy of the rendering procedures applied to each animal. The solid and broken lines represent the lab's work as a production of essentially equivalent actions applied to each case, in each cohort, and by each separate lab technician involved in the production of the project. They further imply that such equivalency is explainable and accountable within the lab's program; that members have resources for detecting instances of operational deviancy, and for keeping such deviancy in check so far as it penetrates the realm of the scientific data produced by members' activities. Such a program includes matters of temporal scheduling, division of labor, monitoring, and evaluation of technical activities.

The chart can be read as a disciplinary specific organizational scheme of activities—an idealized account of the lab's work. It bears both overt and hidden impressions of the labor that constructed it, and can be read by practitioners to evaluate whether the constructive practices were performed well, mistakes were made, or improvements should be devised. Furthermore, features of a bivariate curve, micrograph, or diagram analytically reveal the organization of the work that produced them, since, for instance, temporal scales calendrically organize the scheduling of lab activities by defining beginning and end-points for each replication of an observation. A graphic display articulates the requirements of a scheme of action (the selection and imposition of objectives, means and constraints, and so on), while being no different from a specific representation of natural phenomena.

If, in the end, a line on the graph represents a cohort of animals, it acts as a claim about the unremarkable character of the singular histories of each specimen, and of the practical actions and numerous assessments on the adequacy of the actions that accompanied and guided that history. The line stands on behalf of that history, reduces and objectifies it to a bivariate trajectory, and embeds it within an emptiness of graphic space. This emptiness is infused with moral significance inasmuch as it involves the tacit claim of scientific integrity, with motives assumed to be beyond reproach, and is offered with an unstated presumption that, if anything significant should have been said about the operational history of the graphic line, it will have been stated. Nothing intrinsic to the practices of reducing and objectifying cases to graphic lines guarantees or detracts from the foundations of that emptiness.[24]

4. Methodology as an Externalized Retina

The practices described above make up a "social" program of perception in science (Douglas 1982).[25] It is not a psychological program in the familiar sense, nor a social-psychological program (if we understand social psychology to treat the interaction between society and mentalistic perception). Graphic formats, instrumental fields, and preparatory techniques in histology penetrate both the field of what is visible and the means for perceiving it. It is as though they operate as elements of an externalized retina, activating the perceptible and schematically processing it. Analogous to the specialized retinal cells that are said to constitute our visual world by detecting lines, edges and coded ranges of color, the externalized retina acts flexibly to manipulate graphic and instrumental lines, edges, and codes in order to constitute the sensible, palpable, tangible, and appreciable properties of data. Although a psychological subject is necessary to process what an instrument shows, the instrument does not merely extend the sensitivities of sensory perception. An instrument, and the accompanying project of "domesticating" (Goody 1977) or routinizing space and time in accordance with the instrument's use, prepares the way for perception by precoding, geometrizing, and normalizing the properties of what comes to be perceived. An active reconstruction of the world is achieved.

5. Conclusion

The descriptions of scientific activity in this paper justify the conclusion that science is a constructive activity (Knorr-Cetina 1981). However, if we review just how the cases described here are constructive, we see that scientific data are artificial in only a certain sense. The data were neither wholly *out there* in the animals' anatomy, nor wholly constructed out of thin air.

In the case of electron micrographs, features of the tissue were exposed and upgraded through a series of operations that enabled them to be perceived and analyzed. Laboratory practices were produced carefully to retain a continuity in analyzable form from one rendering to another. Such practices failed when, for instance, a high-powered photograph could not be located within the frame of a lower-power photograph, or when a blurry or grainy texture of a tissue section made it difficult to mark the outlines of specific neuronal organelles reliably.

The resultant data were therefore neither wholly constructed, nor simply a "mirror of nature" (Rorty 1979) arising from an encounter between a rational mind and an inherently orderly nature.[26] Instead, the representational adequacy of the data depended upon a tenuous coherence of actions established in the social environs of the laboratory.

Notes

1. A variety of metaphors make the point that laboratory workers do not relate directly to the *natural objects* their accounts describe, but instead work with highly processed materials. Harold Garfinkel, Michael Lynch and Eric Livingston (1981, 137) use the terms "cultural object" and "potter's object" to describe the character of the object

scientists actually work upon. Karin Knorr-Cetina (1981), describes laboratory work as a manufacturing process, and Bruno Latour and Steve Woolgar (1979) insist that "literary inscriptions" provide the materials for scientific investigations.

2. Peter Toner and Katharine Carr (1968, 91) define a "histological artifact" as "any appearance or image in the specimen not present in that form in the living state but introduced by tissue processing or some other factor." They go on to say that all light and electron-microscopic observations are observations of artefact, but because some systematic artefacts are reproducible and consistent, they are useful and meaningful in the context of cell biology.
3. A sampling of *laboratory studies,* as they have come to be called, are: Garfinkel, Lynch, and Livingstone (1981); Knorr-Cetina (1981); Latour and Woolgar (1979); Law and Williams (1982); Lynch (1985); McKegney (1979); Mulkay and Gilbert (1982); Pinch (1985); Star (1983); Traweek (1980); Zenzen and Restivo (1982); and a collection of studies in Collins (1981). Studies of scientific texts that were particularly influential for the approach taken in this study are: Morrison (1989); and O'Neill (1981).
4. Historical studies of the development of visual conventions in scientific research include: Edgerton (1976); Ivins (1973); Rudwick (1976); and Tilling (1975). For a discussion of the role of visualization in the rhetoric of early scientific writings, see Shapin (1984). Also pertinent are Goody's investigations of literary phenomena such as lists and tables and their constitutive relationships (1977). Less specific, although no less inspiring for this study, are Michel Foucault's discussions of perceptual reasoning in the human and medical sciences (for instance, 1973).
5. Studies of laboratory conversation and their relation to scientific objects include Garfinkel, Lynch, and Livingstone (1981); Latour and Woolgar (1979); Lynch (1982; 1985); Williams and Law (1980); and Woolgar (1990).
6. "Rendering" is a central theme in Harold Garfinkel's recent researches. It describes the transformation of lived-activity into documentary phenomena. See note 12 below for further elaboration. An analogy is made here to Merleau-Ponty's description of the embodied "modalities" of perception (1962). The analogy relies upon personal communications with Melvin Pollner and Christopher Pack at UCLA, who suggested that phenomena such as typing paper, typewriter keyboards or, in this case, visual fields and displays in science could be viewed as modalities of perception in the way Merleau-Ponty describes.
7. Latour (1986), in his theory of literary inscription, argues that literary artefacts link the transient and situated world of the laboratory to the extensive and relatively permanent and extended world of the scientific literature by reducing the various activities of scientific research under the heading of literary inscription. The theory also provides for the application of the tools of literary criticism for the analysis of scientific work. Latour's theory borrows from historical arguments on the development of print technology and the growth of science: see Eisenstein (1979). The present paper does not rely upon any claim about the necessity of print or any other specific kind of media for the accomplishment of scientific work, as it describes the rendering practices that use and rely upon various media. Although the present study is not a semiotic analysis of illustrations in scientific texts, studies of illustrations in scientific texts by Bastide and Jacoby independently bear out some of the points made here: see Bastide (1990), and Jacobi (1983).

8. Rudwick (1976, 178) describes how visual materials from mining and surveying, "which were not even 'proto-geological' in their intentions," provided precedents for geologists' specialized diagrams. My use of the term "proto-scientific" is slightly different. I am speaking of contemporary methods of preparing a field for scientific operations.
9. One of the identifying traits of *constructed* form is the extent to which the simple limiting forms of geometry (the point, line, circle, square, sine wave, and the like) are ever more closely approximated. Jacques Monod (1971) argues that the unusual geometricity of the crystal and the bilateral symmetry of life forms were at one time assumed to be evidence of a deity's intelligence. Mathematical forms are, of course, assumed by most scientists to be intrinsic limiting elements of the structure of scientific objects, despite the otherwise "messy" appearance of empirical instances. At the same time, the ever more exact approximation of these forms identifies the civilized landscape of the "modern" world, visible in the utilitarian design of architecture and equipment.
10. George Herbert Mead (1980, 140–160) has discussed the reference of manipulatory activity to scientific objects.
11. Edmund Husserl (1970, 23–59), argues that Galileo set the stage for modern consciousness by imaginatively displacing the geometric limit forms used for the mundane work of surveying into an eternal and inherently mathematical realm of nature; see also Heidegger (1967). Although Husserl's grand thesis may lay too much responsibility on Galileo, there is some circumscribed historical support for the thesis that mathematical formalism has mundane practical origins. Rudwick (1976, 169), in a study of the historical development of geology, states that "a fully structural approach to the interpretation of the complex phenomena of geology was most readily attained within a social context of practical mining and mineral surveying." Tilling (1975, 195) in a study of the historical origins of graphs, describes how early graph-like devices came into use as instrumental recording devices without, at first, being recognized for their analytic value as graphs. Although I am not making an historical argument in the present case, there seems to be an analogous, locally ordered, temporal development in the ordinary work of *surveying* and rendering a specimen to prepare it for *mathematization*.
12. The notion of the "docile object" borrows themes from Foucault's discussion of "docile bodies": see Foucault (1979, 135–169); and from Garfinkel's idea of the "docile record." The latter idea is applied to documentary records in the form of writing, tape recordings, photographs, charts, maps, and so on. Garfinkel describes such records as "renderings" of the work that produced them. They are significant more for what they hide about their practical genealogies than for what they reveal about them. The implication for ethnomethodological studies is to go beyond an analysis of residue documents to investigate the embodied real-time production of such "docile records": see Garfinkel, Lynch, and Livingstone (1981,136).
13. Foucault's writings (1979) disclose how certain modern forms of knowledge of the body became possible in the nineteenth century through the institution of programs for regimenting the body's movements, subjecting it to detailed code and constant surveillance, and evaluating it through the use of bureaucratic records. Goody (1977) more specifically relates the growth of modern forms of knowledge and social organization to the development and widespread use of literary facilities such as tables, written instructions, and lists. Where Foucault identifies modern civilization with a new

method of translating power into action, Goody locates it as an outgrowth of literary developments.

14. The hyphenated phrase, "observable-reportable," is adopted from Garfinkel's usage: "Ethnomethodological studies analyze everyday activities as members' methods for making those same activities visibly-rational-and-reportable-for-all-practical-purposes, i.e., *accountable,* as organizations of commonplace everyday activities" (Garfinkel, 1967, "Preface," vii).
15. At one time a lab assistant remarked casually to one of his colleagues about a particular animal, "Old number 65 was a good animal all the way down the line." I understand this to mean that the animal was docile when alive, and afterward its corpse yielded "good" results at each stage of its processing.
16. The theme of upgrading is introduced in Garfinkel, Lynch, and Livingstone (1981,154). The theme was suggested by Eric Livingston (personal communication).
17. I do not mean to claim that automated images and readings are intrinsically less suspect than their manual counterparts. However, they do tend to be treated that way under many, although not all, occasions of scientific research.
18. The mention of "sense data" is playful. Clearly, there is an attempt here to resort to *sense data* as an empirically solid ground of perception and knowledge. From my point of view, the *sense data* in scientific investigation are docile objects—their features exhibit designs that are a consequence of human activity, while at the same time those features generate what counts as scientific knowledge.
19. The stratified structure of the dentate gyrus was first detailed at the turn of the century in the anatomical studies of Santiago Ramon Y Cajal (1968). Rudwick's study of geology (1976) offers an interesting parallel. Rudwick points out how early geological drawings and maps utilized exposed cliff faces, mines, or other outcroppings as observable bases for constructing formalizations of the underlying strata.
20. "Located" does not necessarily mean "located what already existed," but means something more like "produced alignment with." To use a homely analogy, this sort of "locating" is like "locating a place on line" in a queue. The activity of "locating" reflexively constitutes the place it finds. This example is from Harold Garfinkel's lectures on queues (1979–1981).
21. Morrison (1989) and O'Neill (1981) develop an analysis that includes the concrete features of the literary page as constitutive elements of a text's sensible reading.
22. Different degrees of evidentiary value are assigned to renderings depending on how the literary representation *packages* objective residues. A rendering that utilizes residues, extracts or other material relics of the object generally is evaluated more highly than one that mediates the representation of the object with a human act of verbal description or drawing. This, of course, is not peculiar to scientific evidence. The reverence and controversy surrounding the Shroud of Turin revolves around the claim that the *picture* is constituted through *pigments* impressed into the fabric and exuded from Christ's wounded face. A sketch, though perhaps of evidentiary value, would not be taken as a *relic.*
23. James L. Wilkins, University of Toronto, is fond of saying that statistical analysis in sociology "throws out the baby and leaves the bath water for analysis." My similar remarks here do not imply that the residues of an animal left for inspection have nothing to do with the original animals, instead, they have plenty to do with them within certain methodological restrictions.

24. This is to say that the adumbrated moral significances of the graph can be interpreted as products of Mertonian norms of science: see Merton (1942, 115–126). However, the visible features of the graphic display do not by themselves compel such an interpretation. In the present case, finding "normality" ultimately is dependent upon a good faith reading.
25. Douglas's "grid-group" scheme is, of course, not used in the present paper.
26. The phrase "mirror of nature" is used in a critique of correspondence theories of knowledge: see Rorty (1979).

References

Amerine, Ronald, and Jack Bilmes. 1990. Following instructions. In *Representation in scientific practice,* edited by Michael Lynch and Steve Woolgar. Cambridge: The MIT Press.

Bastide, Francois. 1990. Iconographie des textes scientifiques: Principes d'analyse. Translated by Greg Myers. In *Representation in scientific practice,* edited by Michael Lynch and Steve Woolgar. Cambridge: The MIT Press.

Collins, Harry, ed. 1981. Knowledge and controversy: Studies in modern natural science. *Social Studies of Science* 11 (1): 3–158.

Douglas, Mary, ed. 1982. *Essays in the sociology of perception.* London: Routledge & Kegan Paul.

Edgerton, Samuel. 1976. *The Renaissance discovery of linear perspective.* New York: Harper and Row.

Eisenstein, Elisabeth. 1979. *The printing press as an agent of change.* Cambridge: Cambridge University Press.

Foucault, Michel. 1973. *The birth of the clinic: An archeology of medical perception,* translated by Alan Sheridan. New York: Random House.

Foucault, Michel. 1979. *Discipline and punish: The birth of the prison,* translated by Alan Sheridan. New York: Random House.

Garfinkel, Harold. 1967. *Studies in ethnomethodology.* Englewood Cliffs, N.J.: Prentice-Hall.

———. 1979–1981. Lectures on queues. Department of Sociology, UCLA.

Garfinkel, Harold, Michael Lynch, and Eric Livingston. 1981. The work of a discovering science construed with materials from the optically discovered pulsar. *Philosophy of the Social Sciences* 11: 137.

Goody, Jack. 1977. *Ten modes of reasoning and practice: The domestication of the savage mind.* Cambridge: Cambridge University Press.

Heidegger, Martin. 1967. *What is a thing?* translated by W. B. Barton and Vera Deutsch. Chicago: Henry Regnery Co.

Husserl, Edmund. 1970. *The crisis of European sciences and transcendental phenomenology,* translated by David Carr. Evanston, Ill.: Northwestern University Press.

Ivins, William. 1973. *On the rationalization of sight.* New York: Plenum Press.

Jacobi, Daniel. 1983. Vulgarisation et illustration dans les sciences de la vie. Paper presented to a conference on Visualization and Cognition, December, at Ecole Superieure des Mines, Paris.

Knorr-Cetina, Karin. 1981. *The manufacture of knowledge: An essay on the constructivist and contextual nature of science.* Oxford: Pergamon Press.

Latour, Bruno. 1986. Visualization and cognition: Thinking with eyes and hands. *Knowledge and Society: Studies in the Sociology of Culture Past and Present* 6: 1–40.

Latour, Bruno, and Steve Woolgar. 1979. *Laboratory life: The social construction of scientific facts*. London and Beverly Hills, Calif.: Sage.

Law, John, and Rob Williams. 1982. Putting facts together: A study of scientific persuasion. *Social Studies of Science* 12: 535–558.

Lynch, Michael. 1982. Technical work and critical inquiry: Investigations in a scientific laboratory. *Social Studies of Science* 12: 499–533.

———. 1985. *Art and artifact in laboratory science: A study of shopwork and shop talk in a research laboratory.* London: Routledge & Kegan Paul.

Lynch, Michael, Eric Livingston, and Harold Garfinkel. 1983. Temporal order in laboratory work. In *Science observed,* edited by Karin Knorr-Cetina and Michael Mulkay. London and Beverly Hills, Calif.: Sage.

McKegney, Doug. 1979. Inquiry into inquiry: Local action and public discourse in wildlife ecology. Master's Thesis, Simon Fraser University.

Mead, George H. 1980. *The philosophy of the present*. Chicago: The University of Chicago Press.

Merleau-Ponty, Maurice. 1962. *Phenomenology of perception*. New York: Humanities Press.

Merton, Robert. 1942. Science and technology in a democratic order. *Journal of Legal and Political Science* 1: 115–126.

Monod, Jacques. 1971. *Chance and necessity,* translated by Austryn Wainhouse. New York: Alfred A. Knopf.

Morrison, Ken. 1989. Some researchable recurrences in disciplinary-specific inquiry. In *The Interactional Order: New directions in the study of social order,* edited by D. T. Helm, W. T. Anderson, A. J. Mehan and A. W Rawls. New York: Irvington.

Mulkay, Michael, and Nigel Gilbert. 1982. Accounting for error: How scientists construct their social world when they account for correct and incorrect belief. *Sociology* 16: 165–183.

O'Neill, John. 1981. The literary production of natural and social science inquiry. *Canadian Journal of Sociology* 6: 105–120.

Pinch, Trevor. 1985. Towards an analysis of scientific observation: The externality and evidential significance of observational reports in physics. *Social Studies of Science* 15: 3–36.

Ramon Y Cajal, Santiago. 1968. *The structure of Ammons horn,* translated by Lisbeth M. Kraft. Springfield, Ill.: C. C. Thomas.

Rorty, Richard. 1979. *Philosophy and the mirror of nature*. Princeton: Princeton University Press.

Rudwick, Martin. 1976. The emergence of a visual language for geological science: 1760–1840. *History of Science* 14: 149–195.

Schrecker, Friedrich. 1980. Doing a chemical experiment: The practices of chemistry students in a student laboratory in quantitative analysis. Unpublished paper. Department of Sociology, UCLA.

Shapin, Steven. 1984. Pump and circumstance: Robert Boyle's literary technology. *Social Studies of Science* 14: 481–519.

Star, Susan L. 1983. Simplification in scientific work: An example from neurosciences research. *Social Studies of Science* 13: 205–228.

Stebbins, Robert C. 1966. *A field guide to reptiles and amphibians*. Boston, Mass.: Houghton Mifflin.
Tilling, Laura. 1975. Early experimental graphs. *British Journal for the History of Science* 8: 193–213.
Toner, Peter, and Katharine Carr. 1968. *Cell structure: An introduction to biological electron microscopy*. Edinburgh and London: E. S. Livingstone.
Traweek, Sharon. 1980. Culture and the organization of scientific research in Japan and the United States. *Journal of Asian Affairs* 5: 135–148.
Williams, Rob, and John Law. 1980. Beyond the bounds of credibility. *Fundamental Scientae* 1: 295–315.
Woolgar, Steve. 1990. Time and documents in researcher interaction: Some ways of making out what is happening in experimental science. In *Representation in scientific practice,* edited by Michael Lynch and Steve Woolgar. Cambridge: The MIT Press.
Zenzen, Michael, and Sal Restivo. 1982. The mysterious morphology of immiscible liquids: A study of scientific practice. *Social Science Information* 21: 447–473.

9

Edward Tufte and the Promise of a Visual Social Science

John Grady

Trained as a statistician and as a political scientist, Professor Edward Tufte of Yale University is an unlikely candidate to become the most influential figure in contemporary graphic design. Yet he has achieved an almost guru-like status among designers working in advertising, public relations, and a wide array of occupations and industries concerned with the analysis and display of data and information. His renown has been due to the success of three astonishingly beautiful and fascinating books, which he has published under his own imprint (Graphics Press): *The Visual Display of Quantitative Information* (1983); *Envisioning Information* (1990), and *Visual Explanations* (1997a). Tufte has also been tireless in running extremely popular one-day workshops that include the purchase of his books as part of the price of admission.

Interestingly enough, it appears that Tufte's training as a quantitative social scientist has played an important role in shaping his message. Graphic designers have too often seen themselves—and, of course, have acted—as hucksters who excite the public's baser instincts to entice them to embrace products and ideas they otherwise might find unnecessary. Tufte, however, proposes a more elevated purpose for their craft, something far closer to the traditional social scientific quest for a truthful and usable knowledge:

> What is to be sought in designs for the display of information is the clear portrayal of complexity. Not the complication of the simple; rather the task of the designer is to give visual access to the subtle and the difficult—that is, the revelation of complexity. (Tufte 1983)

Tufte challenges designers to accept their craft as a calling to minister to the intellectual need for reliable information that is required of citizens in a democracy.

If Tufte is a prophet, however, his influence in his own land has been limited. Few social scientists cite him and even fewer appear to have been influenced by him to develop an interest in the visual display of data. While this is generally true, it is especially so of sociology, which has produced no equivalent to Steven Kosslyn in psychology. Kosslyn, a professor at Harvard, is an internationally renowned authority on visual cognition, and his book, *The Elements of Graph Design* (1994), is the clearest and most useful introduction to the subject. Recently, however, several applied social scientists have produced very valuable guides to graphing quantitative data. Nevertheless, neither Gary Henry's *Graphing Data: Techniques for Display and Analysis* (1995) nor Anders Wallgren et al.'s *Graphing Statistics and Data: Creating Better Charts* (1996) have received much attention from sociologists.

This indifference to how evidence is, and should be, displayed visually is puzzling. Among the social sciences, sociology in recent years has distinguished itself by the attention it has paid to the tools of its craft. Sociologists have participated fully in the discussion and debates about what is entailed in qualitative research (Denzin and Lincoln 1994). They also have concerned themselves with other issues that tend to the nuts and bolts of the field (see, for example, Becker 1998; Lieberson 1985; Ragin 1994; Ragin and Becker 1992). In addition, the American Sociological Association has made a substantial investment in creating resources for teaching various aspects of the discipline. Why then are the contributions of Tufte and those who have been influenced by him neglected?

I suspect that the answer to that question is really quite simple. Tufte's books are beautifully written but are not organized around a clear line of analysis. In many ways, they consist of epigrams that emerge in page after page of encounters with a dazzling array of graphic representations. Tufte suggests that visual displays are important not just because they confirm or disconfirm testable theories, but because they also generate knowledge that would be unavailable in any other way. Furthermore, he argues that accessing the visual with the tools he describes and the craft he models nurtures an aesthetic sensibility that the mind desperately needs in order to think clearly and well. I am certain that many social scientists feel this stricture demeans the work that they currently do and burdens their already information-overloaded lives with demands that would not be easy to integrate into everyday professional obligations.

There are, of course, other intellectual currents in the social sciences that one would think might be stimulated by Tufte's work. Critical theory, post-modernism, feminism, cultural studies, and even much of visual sociology, exhibit a more flexible approach to theory and have responded positively to the analytic use of visual representations of one kind or another. Nevertheless, none of these groups has shown much interest in Tufte, probably because he is an old school social scientist who is less concerned with interpreting meaning than he is in deciphering data.

The social sciences and the growing interest in image-based research (Prosser 1998), however, would be enriched immeasurably by a sustained encounter with his work. Moreover, the goal of creating a more visual social science ultimately may depend on a thorough integration of Tufte's work with contemporary currents in visual sociology.

Put simply, Tufte has four major contributions to make to contemporary social science. First, Tufte clearly establishes that we should display data visually not so much to illustrate or embellish an argument but rather because graphics are an integral step in the analysis of data, and can shape an argument as much as be shaped by it. Such an emphasis on the power of visual displays could revitalize teaching and research in the social sciences by encouraging the integration of qualitative and quantitative modes of analysis. For example, a growing number of the most important and reliable databases—such as the U.S. Census and the General Social Survey—are on-line and can be accessed easily with on-site and very user-friendly software. It is now relatively simple for students to carry out increasingly imaginative secondary data analyses. In addition, teachers easily could integrate exercises with these real-life data sets into the classroom, allowing students to question and evaluate the often out-dated information that so litters contemporary textbooks (Davis 1987). Interpreting visual displays of data downloaded from the Web then would become an integral part of classroom instruction and foster graphic competence, if not sophistication, through practice (Henry 1995, 145).

Second, Tufte establishes that a wide array of graphic displays can be used to make different kinds of points or arguments more effectively, clearly, and in less space than is needed in either exclusively verbal or written communication. Furthermore, very different types of displays can be combined to form complex narratives, where the displays themselves form the spine of the argument. This suggests that social scientists may have limited the spread of their findings and their influence by restricting their means of communication to the traditionally formatted journal article and book. Reporting in a more visual social science format might take a variety of forms and utilize a wide array of media. In addition, reporting skills not only could include command over many existing graphic conventions—maps, charts of various sorts, and photographs—but also the capacity to improvise displays to suit the data and reporting media being used.

Third, the social sciences are generalizing disciplines. The models they construct, the theories they develop, the research methods they pursue, all aim to discover what is common in as many cases as possible. Life, however, is lived in the particular: moments, situations, events, and crises. Social scientists unfortunately often have a trained incapacity to deal with these particularities for reasons that Kai Erikson discusses in his award-winning study of the Buffalo Creek disaster:

> [T]he traditional methods of sociology do not really equip one to study discrete moments in the flow of human experience . . .
>
> I think an argument can be made that sociologists should do this kind of work more often anyway. The search for generalization has become so intense in our professional ranks that most of the important events of our day have passed without comment in the sociological literature. The aim of any science, I agree, must be to move from particular observations to general findings, but there are times when the need for generalizations must yield to the urgency of passing events, times, when the event must tell its own story. (Erikson 1976, 12)

Unlike conventional social science, Tufte's work has, if anything, pursued a trajectory that identifies the most successful graphs as those that use universal design principles to address a particular situation or need. Figure 9.1, for example, is a case in point. It is Charles Minard's Chart of Napoleon's march on and from Moscow, which Tufte has referred to as the greatest chart ever devised, and which he hands out in poster form to everyone who attends one of his workshops. We will see below that Tufte's penchant for displays of this sort—that illuminate the dynamics of particular situations and events—makes his work particularly useful for applied or policy research.

Finally, and most importantly, Tufte's oeuvre is permeated by an ethos that makes analytic work an aesthetic pleasure. In his view, sound analysis requires not only a consistent aesthetic but also that the task itself be art. In so doing, Tufte challenges the various dualisms that see art as completely distinct from work, science, and other spheres of human purpose. His work, therefore, models an understanding about the proper

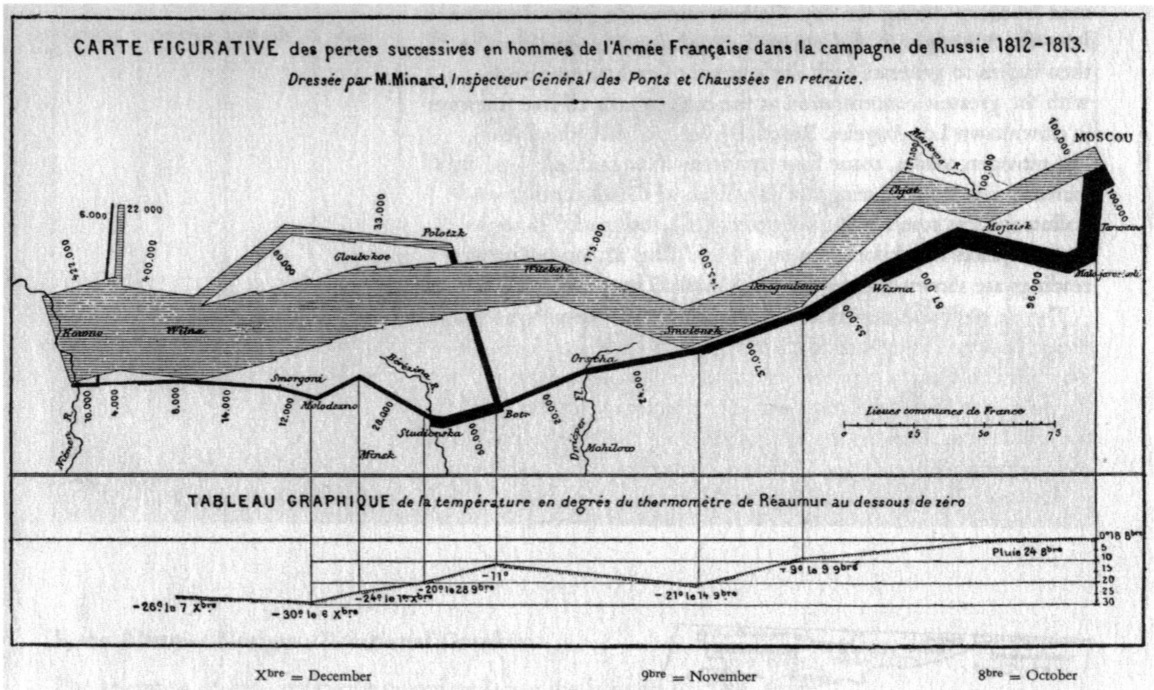

Figure 9.1. *Carte figurative* of Napoleon's march to Moscow. "[T]his combination of data map and time-series . . . portrays the devastating losses suffered in Napoleon's Russian campaign of 1812. Beginning at the left on the Polish-Russian border near the Niemen River, the thick band shows the size of the army (422,000 men) as it invaded Russia in June 1812. The width of the band indicates the size of the army at each place on the map. In September, the army reached Moscow, which was by then sacked and deserted, with 100,000 men. The path of Napoleon's retreat from Moscow is depicted by the darker, lower band, which is linked to a temperature scale and dates at the bottom of the chart. It was a bitterly cold winter, and many froze on the march out of Russia . . . and the army finally struggled back into Poland with only 10,000 men remaining . . . Six variables are plotted: the size of the army, its location on a two-dimensional surface, direction of the army's movement, and temperature on various dates during the retreat from Moscow. It may well be the best statistical graph ever drawn" (Tufte 1983, 40). Charles Minard, originally published in 1869, reproduced in Edward Tufte, *The Visual Display of Quantitative*

relationship between art and daily life that John Dewey captures beautifully in *Art and Experience:*

> In order to understand the esthetic in its ultimate and approved forms, one must begin with it in the raw; in the events and scenes that hold the attentive eye and ear of man, arousing his interest and affording him enjoyment as he looks and listens: the sights that hold the crowd—the fire-engine rushing by; the machines excavating enormous holes in the earth; the human-fly climbing the steeple-side; the men perched high in air on girders, throwing and catching red-hot bolts. The sources of art in human experience will be learned by him who sees how the tense grace of the ball-player infects the onlooking crowd; who notes the delight of the housewife in tending her plants, and the intent interest of her goodman in tending the patch of green in front of the house; the zest of the spectator in poking the wood burning on the hearth and in watching the darting flames and crumbling coals. These people, if questioned as to the reason for their actions, would doubtless return reasonable answers. The man who poked the sticks of burning wood would say he did it to make the fire burn better; but he is none the less fascinated by the colorful drama of change enacted before his eyes and imaginatively partakes in it. He does not remain a cold spectator. What Coleridge said of the reader of poetry is true in its way of all who are happily absorbed in their activities of mind and body: "The reader should be carried forward, not merely or chiefly by the mechanical impulse of curiosity, not by a restless desire to arrive at the final solution, but by the pleasurable activity of the journey itself."
>
> The intelligent mechanic engaged in his job, interested in doing well and finding satisfaction in his handiwork, caring for his materials and tools with genuine affection, is artistically engaged. The difference between such a worker and the inept and careless bungler is as great in the shop as it is in the studio. Oftentimes the product may not appeal to the esthetic sense of those who use the product. The fault, however, is oftentimes not so much with the worker as with the condition of the market for which his product is designed. Were conditions and opportunities different, things as significant to the eye as those produced by earlier craftsmen would be made. (Dewey 1934, 4–5)

From this standpoint, then, aesthetics may be defined as the impulse to find and build meaning in that which pleases us in our being in the world, while art is the construction of objects and events that so please us. Aesthetics may be the vehicle that energizes our engagement with, and prepares our mind for, life's practical challenges. It should not surprise us, therefore, that Donald Norman's criteria for designing information appliances apply just as strongly to the visual displays that so concern Tufte. Norman tells us that they should be "simple," "versatile," and "pleasurable"; and about the latter criterion, or axiom, Norman has this to say:

> The third axiom, *pleasurability,* might seem out of place. After all, what has pleasure got to do with appliances? These are serious matters, serious devices for serious uses. Not games, not amusement or entertainment.
>
> Good tools are always pleasurable ones, ones that the owners take pride in owning, in caring for, and in using. In the good old days of mechanical devices, a craftsperson's tools had these properties. They were crafted with care, owned and used with pride. Often, the tools were passed down from generation to generation. Each new tool benefited from a tradition of experience with previous ones so, through the years, there was steady improvement, each new generation ever-better fitting the requirements of the task and the needs of the owner. Woodworking tools, gardening tools. Tools for painting and drawing, tools for baking and cooking. Gloves and bats for baseball, fishing rods, boats, cars. A pleasurable tool is one that is made with quality, that fits the mind and body, that makes it fun to do a task. Tools that help. Tools that do not get in the way. Appliances should add pleasure to our lives. They can remove the drudgery of task and make them enjoyable, fun even. (Norman 1998, 67–68)

Shouldn't the same apply to the tools we use in social scientific work and communication? Putting the fun back in the social sciences might then be one of Tufte's most significant contributions.

That Tufte has made these four contributions to building a more visual and dynamic social science is not readily apparent in his work. For example, he does not directly address the question of the current state of the social sciences nor its relationship to his work. Additionally, although reading Tufte is always a delight, the way he organizes his arguments doesn't make that task any easier for social scientists. The next section of this paper will discuss Tufte's idiosyncratic writing style in hopes of making the road less bumpy for the reader. This section will be followed by one that summarizes and explicates the major points of his theory of graphic design. Then I will turn to discussing the particular challenges that Tufte's most recent book, *Visual Explanations,* poses for visual social scientists. Finally, in a concluding section, I will concern myself with some of the implications that Tufte's work has for the future of a more visual social science.

1. Reading Tufte

In 1974, Edward Tufte prefaced *Data Analysis for Politics and Policy,* by saying:

> This book demonstrates some statistical techniques useful in the study of politics and policy. My aim is to present fundamental material not found in statistics books, and, in particular, to show techniques of quantitative analysis in action on problems of politics and public policy. . . .
>
> The approach centers on fitting equations to data. More fundamental, however, is the illustration and development of good statistical thinking—a sense of judgment about what we can and can't learn about the world by looking at quantitative data. (1974, ix)

Much of the book is concerned with various examples of two-variable linear and multiple regression analyses and the like. Nevertheless, in a section

DATA SET 1		DATA SET 2	
X	Y	X	Y
10.0	8.04	10.0	9.14
8.0	6.95	8.0	8.14
13.0	7.58	13.0	8.74
9.0	8.81	9.0	8.77
11.0	8.33	11.0	9.26
14.0	9.96	14.0	8.10
6.0	7.24	6.0	6.13
4.0	4.26	4.0	3.10
12.0	10.84	12.0	9.13
7.0	4.82	7.0	7.26
5.0	5.68	5.0	4.74
DATA SET 3		**DATA SET 4**	
X	Y	X	Y
10.0	7.46	8.0	6.58
8.0	6.77	8.0	5.76
13.0	12.74	8.0	7.71
9.0	7.11	8.0	8.84
11.0	7.81	8.0	8.47
14.0	8.84	8.0	7.04
6.0	6.08	8.0	5.25
4.0	5.39	19.0	12.50
12.0	8.15	8.0	5.56
7.0	6.42	8.0	7.91
5.0	5.73	8.0	6.89

Table 9.1. Four data sets. Reproduced in Edward Tufte, *Data Analysis for Politics and Policy* (Englewood Cliffs, N.J.: Prentice-Hall, 1974), 132.

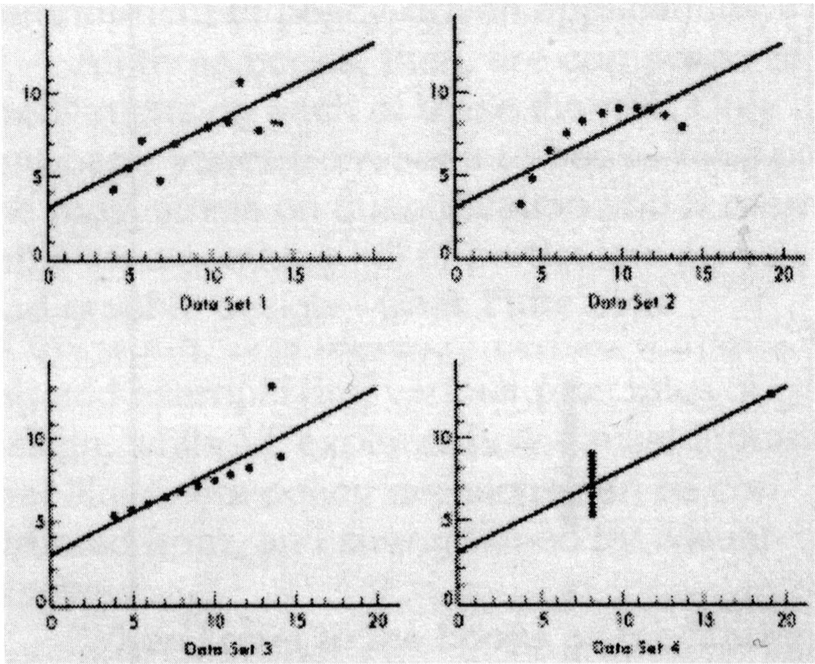

Figure 9.2. Figure 3-29: Scatterplots from table 9.1, published in Edward Tufte, *Data Analysis for Politics and Policy* (Chesire, Conn.: Graphics Press, 1974), 133.

entitled "Regressions Aren't Enough—Looking at the Scatterplot," Tufte cites an article by F. J. Anscome that shows how the four very distinct data sets in table 9.1 can yield exactly the same result when they are fitted to a linear regression model. Visually displaying the data with a scatterplot, however, permits the author to represent the fitted line but also the actual data itself as well as the patterns exhibited in figure 9.2. Anscome comments:

> A computer should make both calculations and graphs. Both sorts of output should be studied; each will contribute to understanding.
>
> Graphs can have various purposes, such as: (i) to help us perceive and appreciate some broad features of the data. (ii) To let us look behind those broad features and see what else is there. Most kinds of statistical calculation rest on assumptions about the behavior of the data. Those assumptions may be false, and then the calculations may be misleading. We ought always to try to check whether the assumptions are reasonably correct; and if they are wrong we ought to be able to perceive in what ways they are wrong. Graphs are very valuable for these purposes. (1973, 17)

I		II		III		IV	
X	Y	X	Y	X	Y	X	Y
10.0	8.04	10.0	9.14	10.0	7.46	8.0	6.58
8.0	6.95	8.0	8.14	8.0	6.77	8.0	5.76
13.0	7.58	13.0	8.74	13.0	12.74	8.0	7.71
9.0	8.81	9.0	8.77	9.0	7.11	8.0	8.84
11.0	8.33	11.0	9.26	11.0	7.81	8.0	8.47
14.0	9.96	14.0	8.10	14.0	8.84	8.0	7.04
6.0	7.24	6.0	6.13	6.0	6.08	8.0	5.25
4.0	4.26	4.0	3.10	4.0	5.39	19.0	12.50
12.0	10.84	12.0	9.13	12.0	8.15	8.0	5.56
7.0	4.82	7.0	7.26	7.0	6.42	8.0	7.91
5.0	5.68	5.0	4.74	5.0	5.73	8.0	6.89

N = 11
mean of X's = 9.0
mean of Y's = 7.5
equation of regression line: Y = 3+0.5X
standard error of estimate of slope = 0.118
t = 4.24
sum of squares $X - \overline{X}$ = 110.0
regression sum of squares = 27.50
residual sum of squares of Y = 13.75
correlation coefficient = .82
r^2 = .67

Table 9.2. Four data sets. Reproduced in Tufte, *VDQI*, 13.

Tufte concludes this book with a muted paean to what can be accomplished by applying new developments in computer software to huge data sets consisting of multiple variables. He then adds:

> The view is: If you're going to search for a model, why not search thoroughly?
>
> Of course, we would trade all those searches in for one good idea. And that idea might come from looking at the data. (1974, 163)

Nine years later comes *The Visual Display of Quantitative Information*, which opens with Anscome's table and figure, except that now the displays are more elegant (see table 9.2 and figure 9.3).

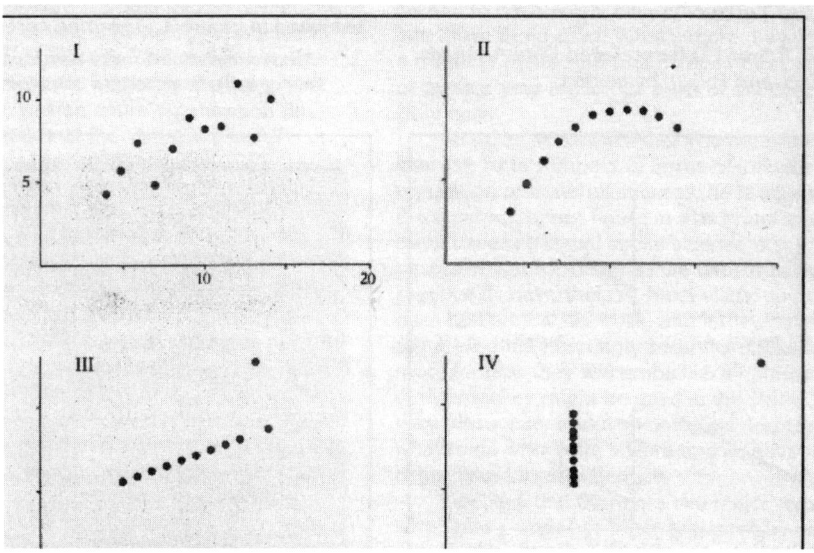

Figure 9.3. Scatterplots from table 9.2, published in Tufte, *VDQI*, 14.

In this short account of Tufte's earlier writing, we find the basis for all the elements of his subsequent work, in particular *The Visual Display of Quantitative Information* (hereafter *VDQI*), *Envisioning Information (EI)*, and *Visual Explanations (VE)*. As unique as his contribution may be, however, Tufte stands on the shoulders of giants. The first and most monumental of these figures is, as he often reminds us, William Playfair (1786, 1805, and 1821) who, in the late eighteenth and early nineteenth centuries, invented in one fell swoop most of the conventions of the graphic display of statistical information. These include "the bar chart and histogram, the surface chart, and the circle diagram, or 'pie-chart'" (Wainer 1997, 5–6). Closer to the present are the works of sociologist Calvin Schmid (1954) and, more significantly, statisticians John Tukey (1977) and William Cleveland (1985 and 1993).

As a guide to reading Tufte, the following distillation of his intellectual orientation may prove useful. It is characterized by:

- A primary focus on data that can be quantified;

- An abiding concern with the ways in which cognition and aesthetics are entwined. For Tufte, graphical excellence, or skill in the application of principles of visual legibility to the problem at hand, is a necessary component of clear thinking. His work thus translates into a unique Minimalist manifesto, where all design elements are disciplined to the task of rendering content unadorned[1]; and,

- An unwavering, passionate commitment to policy-driven applications.

All three books, then, are composed of meditations on each of these themes. Only the emphasis varies from book to book. *VDQI* puts the most stress on quantification and is memorable for a scathing bill of particulars on how bad graphic design—what Tufte calls "chartjunk"—distorts an accurate understanding of the world. *EI* is more concerned with articulating and exemplifying various principles of good design (see figure 9.4), while *VE* explores how causal arguments that illuminate policy decisions can be constructed from, and strengthened by, visual displays.

When I refer to the books as *meditations,* I mean that quite literally. None of the books is built around a sustained argument, although *EI* comes closest to having a traditional analytic structure. Rather, each book consists of a collection of topics or concerns. Often of chapter length, many of these short pieces stand alone as reflections and could be cut and

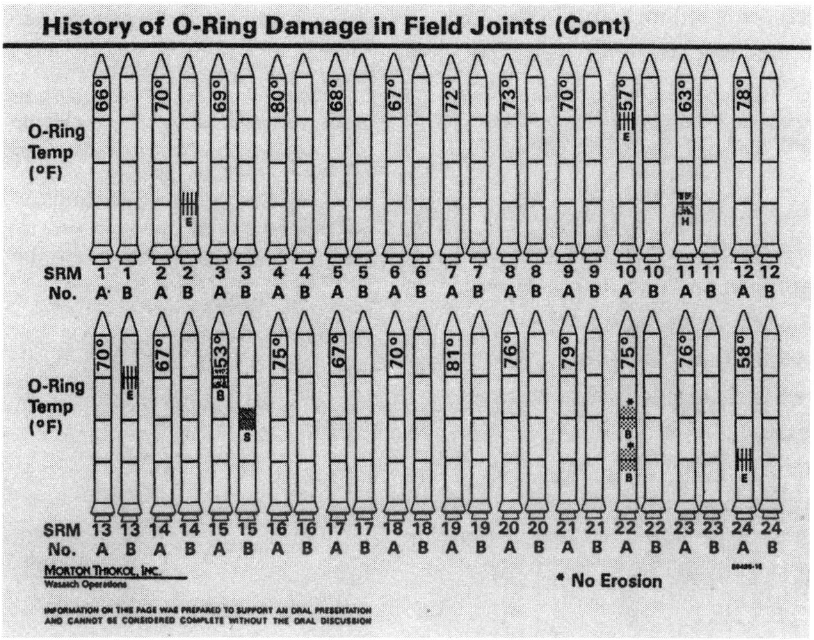

Figure 9.4. History of O-ring damage in field joints, reproduced in Edward Tufte, *Visual Explanations* (Chesire, Conn.: Graphic Press, 1997), 47. This diagram was prepared by engineers who had been very concerned that the O-rings might fail in the launch of the *Challenger* shuttle. They were not able to convince NASA and tragedy ensued. At the commission hearings held after the fact they were still unable to make their point clear as Tufte explains: "Shown . . . [above] are the O-ring experiences of all 24 previous shuttle launches, with 48 little rockets representing the 24 flight-pairs: Rockets marked with the damage code show the seven flights with O-ring problems. Launch temperature is given for each pair or rockets . . . [T]his display contains all the information necessary to diagnose the relationship between temperature and damage, if we could only see it. The poor design makes it impossible to learn what was going on." (Tufte 1997, 47)

pasted into any of the other books. In addition, each chapter often consists of loosely integrated discussions of the merit of particular displays. As the books are extraordinarily thoughtfully designed, these discussions tend to begin and end as the page turns and reveals yet another interesting display. This punctuation, if you will, makes the reading of the books appear more seamless than they actually are.

A topic invariably is explored with a narrative that follows what is depicted on the page. Chapters conclude in a philosophic mood, usually with an aphorism or two that are meant to encapsulate all that has come before. Each book also has an epilogue that performs the same function for the book as a whole. Tufte writes well and as a rhetorician he is generally quite successful, although the reader may be perplexed by how he concludes things and wonder: "Well, that certainly seems right; but surely there was more in what I just read."

In some cases, the fit between a chapter and its epilogue, or the chapter as a whole and the rest of the book, is a little too loose. It is apparent that the way Tufte organizes his argument provides him with handy containers for assembling many disparate things that interest him, appear to be relevant, and, in any event, are full of nice pictures. The most egregious example is "Explaining Magic: Pictorial Instructions and Disinformation Design," which is found in *VE*.

The idea upon which the chapter is premised is plausible enough. We can learn how to present information honestly by inverting the methods of those who's intent is dishonest: magicians and conjurers. The chapter establishes that magicians employ principles of stagecraft that we should studiously avoid in our own work. Tufte's inversion of these principles of disinformation into positive injunctions is intended, therefore, to provide a succinct guide to good pedagogy.

Unfortunately, much of the chapter is composed of illustrations that either demonstrate how to do magic tricks, or show failed nineteenth-century attempts to diagram famous tricks. These images, however, are more suited to another topic entirely, and that is the use of pictures to illustrate or deconstruct a process or event. Nevertheless, Tufte does not focus specifically on this topic anywhere in his oeuvre, although it is a concern that surfaces with increasing urgency in each new book and is an issue that I will address later in this paper. In any event, the upshot of Tufte's ways of organizing his material is to place these various materials in an inappropriate place simply because they all concern magic and touch on some of his abiding concerns.

The meditative quality of Tufte's work is reinforced further by his tendency to recycle materials from one book to another. Very often the same image will appear in a slightly different context in a later book. Quotes are used several times and stories about images with strong narrative elements are revisited and expanded. It is hard to know what to make of a melange that includes: Minard's *Carte figurative* (figure 9.1); Marey's diagrammatic timetable of the train schedule for Paris to Lyon in the 1880s (figure 9.5); and Dr. John Snow's map of cholera deaths in the area of the Broad Street Pump (figure 9.8). I suspect, however, that they function much like sacred icons or relics do in the practice of many traditional religions, as ways of focusing the entirety of our consciousness on a matter that the writer believes must be accorded substantial existential significance.

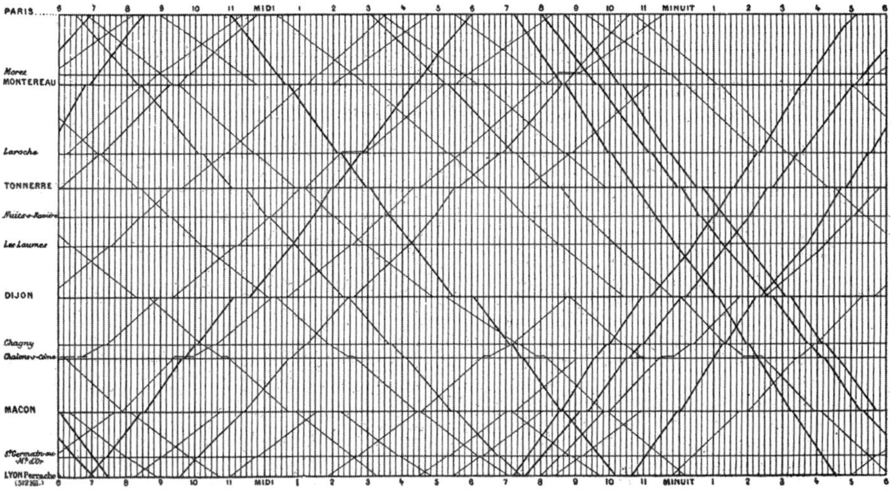

Figure 9.5. Timetable for train schedule from Paris to Lyon. Reproduced in Tufte, *VDQI*, 31.

These images that are recycled through his books share the following characteristics: graphic elegance; information that could easily be muddled; and that address matters of consequence. Moreover, they also represent instances of communion between producers and consumers of images where that which is communicated creates a moment of moral equality and participatory democracy.

The most important of these recurring images is Marey's timetable (see figure 9.5). Howard Wainer explains the graph and hints at why Tufte might find it so compelling:

> In it the time of day is depicted along the horizontal axis and the stations are spaced proportionally along the vertical axis. Trains from Paris to Lyons are naturally shown as lines descending from the left to the right. Trains in the opposite direction ascend from the left to the right. *The metaphor is complete; at a glance it shows both the general and the particular.* There are thirty-eight trains and thirteen stations. Train departures from Paris are at their most frequent in the early evening. Steeper slopes denote faster trains. We can see that the twenty minutes of extra sleep obtained by preferring the 6:50 train from Paris over the one at 6:30 is paid for by increasing the length of the journey to Lyons by five hours. (1997, 75; italics mine)

For Tufte, displaying both the general and particular and the macro and the micro in one image isn't just about simplifying things. Rather, the purpose of the diagram is to empower viewers who can grasp quickly how to read the chart and negotiate the fit between their wants and the exigencies of the particular system the chart represents.

Tufte appears to be calling for our special attention in one particular text that appears in both *VDQI* and *VE*. It is a quote from the American Minimalist painter, Ad Reinhardt:

> Painting is special, separate, a matter of meditation and contemplation, for me, no physical action or social sport. As much consciousness as possible. Clarity, completeness, quintessence, quiet. No noise, no schmutz, no schmerz, No fauve schwarmerei. Perfection, passiveness, consonance, consumateness. No palpitatings, no gesticulations, no grotesquerie. Spirituality, serenity, Absoluteness, coherence. No automatism, no accident, no anxiety, no catharsis, No chance. Detachment, disinterestedness, thoughtfulness, transcendence. No humbugging, no button-holing, no exploitation, no mixing things up. (Ad Reinhardt, statement of the catalog of the exhibition "The New Decade: 35 American Painters and Sculptors," Whitney Museum of American Art, New York, 1955, quoted in Tufte 1983, 122, and Tufte 1997a, 72)

Imagine a public reading of this passage rising to a crescendo and you have the core of Tufte's moral vision. It crests with an imperative that is something like a mantra for reading his books: *No exploitation! No mixing things up*! For Tufte, the social scientist who reasons with numbers is working with materials that few find easy to understand and that many are tempted to exploit. Because a situation like this can lead to much mischief and because visual displays very often help clarify what numbers are all about, graphical elegance has to be seen as more than a matter of taste. It is, in fact, one of those sacred obligations that builds trust in a secular civilization that would collapse without it. For Tufte, aesthetics is linked inextricably to moral and political issues as well as cognitive ones, and this is the master theme that flows through his books.

2. Tufte's Argument

The overriding purpose of Tufte's work, therefore, is to enlist principles of graphic design to display data so that this information will inform thoughtful policy analysis adequately. More specifically, he believes that good graphics should reveal correlations not otherwise evident and so facilitate causal analysis and decision-making about issues with serious consequences. Many of his meditations explicate two major *design strategies.* The first is the principle of the *smallest effective difference,* while the second is any convention that permits the viewer to visualize variation and change by comparing two or more sets of data displayed on a similar template. Moreover, an important goal of Tufte is to encourage others to design what he calls *visual narratives,* which are displays that elegantly and economically apply his two design principles and tell complex and compelling stories.

The Smallest Effective Difference

Tufte tells us that:

> Relevant to nearly every display of data, the smallest effective difference is the Occam's razor ("what can be done with fewer is done in vain with more") of information design. And often the happy consequence of an economy of means is a graceful richness of information, for *small* differences allow *more* differences. (1997a, 73)

In *VDQI,* Tufte provides the reader with a formula for determining how closely a graphic has approached this goal, the "data/ink ratio." Put simply, this ratio is produced by measuring the percent of actual ink that cannot be erased without a loss of data-information and subtracting it from 100 percent or 1. The closer the ratio comes to 1 the better (Tufte 1983, 93). Notice, for example, the difference between figure 9.2 and figure 9.3 and how Tufte has applied the principle meticulously to his own earlier representation of Anscome's chart.

In the art of visual display, the devil is always in the details, and so Tufte's argument for what he calls "data/ink maximization" (1983, 123) is a kind of fugue that reminds us to consider when lines need to be drawn, and when not; when shading should be used, and when not; what kinds of charts are useful for certain kinds of information, and which should be avoided for other—and in certain instances (like the pie-chart) for just about all—cases; and, finally, how charts should, and should not, be labelled. Tufte's most sustained, and detailed, focus on these matters is in *VDQI,* although Wainer's section on "graphical forms" in *Visual Revelations* is probably the leanest introduction to the issues involved (Wainer 1997, 85–124).

One of the recurring problems faced by anyone attempting to display complex information is how to avoid cluttering up even those charts that have maximized their data/ink ratio. To this end, Tufte instructs us that: "Among the most powerful devices for reducing noise and enriching the content of displays is the technique of layering and separation, visually stratifying various aspects of the data" (1990, 53).

A design can be layered by the use of different line width and color, so that one type of information about something can be rendered with one, while another can be represented simultaneously with another line width or color. Even arguments that depend upon many variables can be layered with virtual or actual transparent overlays of different information placed upon the same template.

Maps are one of the most ancient and successful examples of such a practice. Small roads may be drawn in blue and large roads in red. Different hues and shades represent variation in terrain. Topographical features may utilize light gray or tan contour lines to mark elevation, while bodies of water and mountains are rendered in shades of blue and steel gray. Once we learn the legend, most of us can read well-designed maps easily. If we want to know where the turn-off from the highway is, we focus our attention on the red line we are travelling and look for a thinner black line heading to the black dot labeled by the name of our destination. And, if we're riding shotgun and are the trip's designated navigator, then, once we've completed our mission, we often can derive great pleasure from idly exploring a map's other riches simply by focusing on one of the *overlays* that has been encoded on a sheet of paper. Why is it, for example, that one valley appears to invite a dense population while others are more sparsely inhabited?

Parallelism and Small Multiples

The kind of layering just described actually can stimulate rather complex forms of comparison. While we usually are interested in how one type of information illuminates something else, the very presence and propinquity of a second type of information on the display encourages us to wonder about the significance of the relationship between the two variables. For example, the fact that major roads seldom cross mountain ranges and often follow river valleys is not coincidence and can initiate the reader of the display into further reflection about the relationship between the human-built environment and the rest of nature.

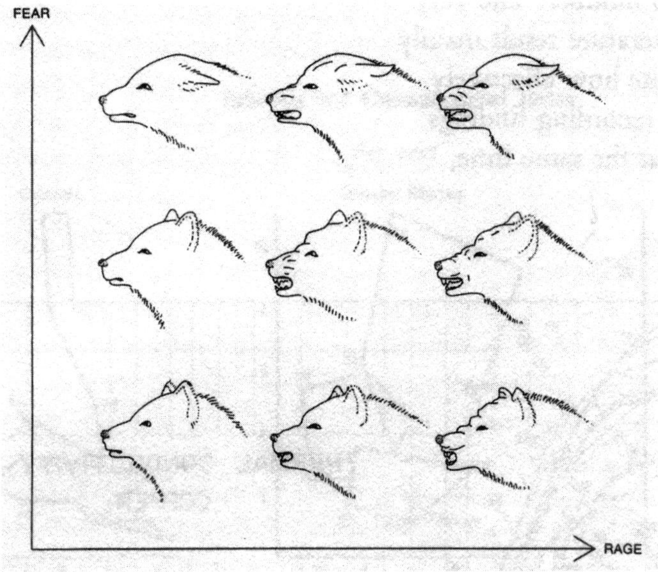

Figure 9.6. Variations of fear and anger. Reproduced from *Scientific American* in Tufte, *VDQI*, 50.

It is only a short step for Tufte to consider the various ways that displays can be designed with the explicit purpose of facilitating comparisons. His discussion about parallelism contains some of the most imaginative and fascinating graphic designs ever published, while his recommendation to use small multiples whenever possible should constitute a major building block for a truly visual social science.

Parallelism refers to a display of information where a variable or some part of an image is represented in close proximity to another representation with a related value, variable, another part of the image, or a different image entirely. However composed, the purpose of the display is to make a point that, if not necessary to an argument, at least enriches it appreciably. Parallelisms can be displayed side by side or can involve overlays or paste-ons. Tufte illustrates *VE* with some marvelous displays of before and after shots that require—or is it entice?—a reader to lift a flap to see a *before* that is otherwise masked (Tufte 1997a, 80–81).

Small multiples are displays that consist of two or more representations built upon the same visual template. A remarkably imaginative example that Tufte cites in *VDQI* comes from *Scientific American* and represents how the simultaneous expression of varying degrees of fear and anger are displayed in the facial display of what appears to be a dog (figure 9.6).

One of the most useful applications of the principle of small multiples is in displaying maps. Because space is a dimension of all social action, identifying the spatial correlates of social information is a methodological necessity for any thoroughgoing analysis. But too much information loaded on a map makes it unreadable. Reproducing just the salient cartographic features (e.g., political units, topography, or the like), and then displaying various variables as correlates on several small maps—with each variable having its own map—not only makes reading easier but pleasurable as well.[2]

Visual Narratives

Tufte's passion is most exercised by customized displays. These one-of-a-kind representations of a phenomena require, as a condition of their creation, a consciousness that has integrated a deep and penetrating knowledge of a particular subject matter. This includes an appreciation of its physical properties; an ability to communicate the essentials of a situation; and an aesthetically informed delight in visual perception and display. This is an apt description of Tufte himself and it probably explains his attachment to kindred souls from the nineteenth century like Minard and Marey.

What is most striking about displays like Minard's various *Carte figuratives* is that they either document or, most importantly, invite explanation (see figure 9.1). They explain how things work, are related to other things, or came to pass. What makes them narratives for Tufte is that they locate important variables in space and in time and so tell us how something moved from one when and where to another.

Figure 9.7, which graphs the life cycle of the Japanese beetle, uses space as Minard's chart does to document change over time, except that here the use of terrain is metaphoric, whereas in Minard's it is strictly historic and geographic. Customized designs are exemplars of creativity and can, like great art, tell stories so compellingly that they evoke a cathartic response from those who view them. Minard's graph of Napoleon's 1812 retreat from Moscow was published in 1869, fifty-seven years after the fact. Yet, as Wainer tells us, "The story of the tragedy is clear. We can see the bodies frozen into the snow. Marey told how this graph 'brought tears to the eyes of all France.' No wonder, there were few families unaffected" (Wainer 1997, 65).

One of the things that Tufte has made clear in the evolution of his work is that while customized designs can provide us with appropriate and

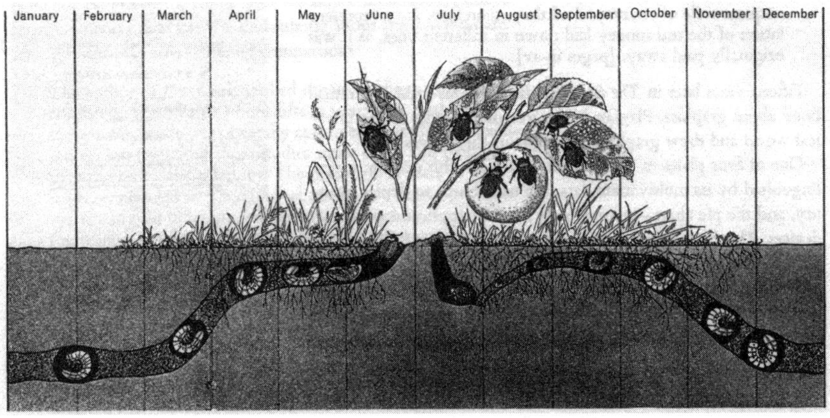

Figure 9.7. Life cycle of the Japanese beetle. Reproduced in Tufte, *VDQI*, 43.

effective ways of understanding complex relationships of cause and effect, sustained arguments most often require the use of a number of varied visual displays. Some of these additional representations also may be customized, but most arguments would require only a series of simple displays of information in a conventional format—like scatterplots, maps, bar and line charts, and so on. In any event, the combined effect of using many different displays helps an investigator develop, modify, and communicate a complex argument.

Nowhere is this clearer than in Tufte's retelling of the story of Dr. John Snow and the Broad Street Pump. At its simplest, John Snow proved in 1854 that a contaminated water pump on Broad Street caused a lethal cholera epidemic that occurred during late August and early September in London. Key to the discovery of this correlation was his use of a map (see figure 9.8). Snow used dots to represent the addresses of everyone who had died in the few days after the outbreak began. The vast majority lived in close proximity to the pump. Furthermore, investigation of anomalies in the map revealed exceptional circumstances that strengthened the correlation. Dr. Snow then convinced the local authorities to remove the pump handle and almost immediately the deaths due to cholera were reduced to a trickle. The story has been enshrined as a miracle tale that a secular society tells itself about the redemptive role of science in modern life.

In his first use of this story in *VDQI*, Tufte presents it much as I have just told it. His major purpose, of course, is to demonstrate the role that visual thinking played in this triumph of modern science. By *VE*, however, the story has expanded into a more complex and powerful tale. While

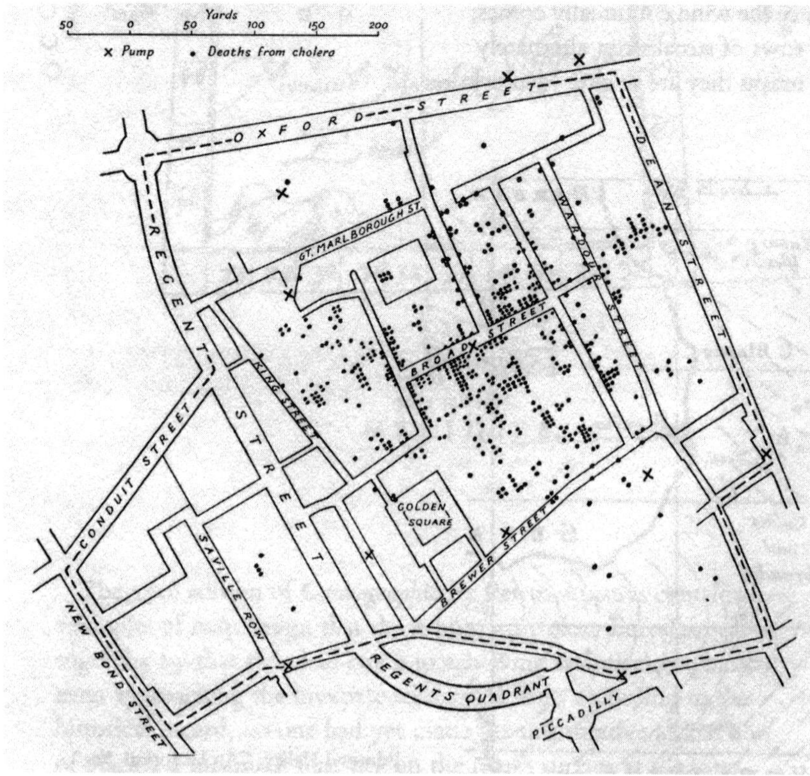

Figure 9.8. Dr. John Snow's map of cholera epidemic. Reproduced in Tufte, *VDQI*, 24.

Snow's science has been confirmed by history, the reasoning in the original version of the story was flawed. A glance at figure 9.9 shows that while the deaths from cholera *did* decline after the pump handle was removed on the morning of September 8, the death rate actually had begun to decline *five days earlier*, and taking off the handle did not affect the rate of decline.

Tufte has become a great storyteller and he lets us see figure 9.9 a page before the map, and fully four pages before he draws our attention to the flaws in the original version of the tale. One of the effects of having seen figure 9.9 first is that a perceptive reader will be agitated by contradictory evidence that Tufte appears to have overlooked. But, when Tufte acknowledges the significance of the data displayed in figure 9.9, and revises the hagiography of Dr. John Snow, he only proves that an insufficient representation does not make a case against the visual display of information, but instead that even more visual displays not only enrich the content of a tale but also shape its telling.[3]

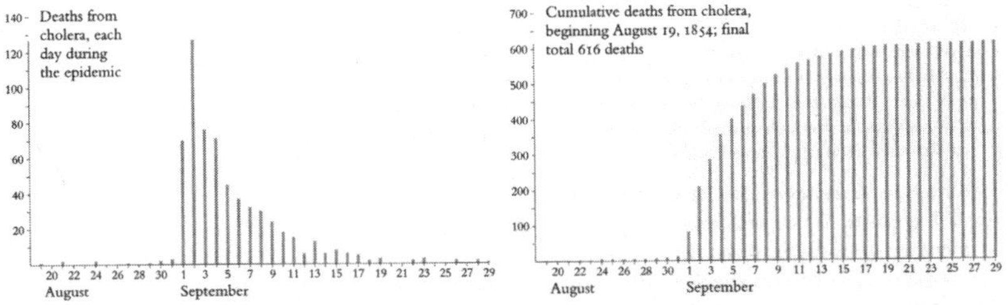

Figure 9.9. Deaths from cholera each day during epidemic. Reproduced in Tufte, *VE*, 29.

3. Visual Explanations

One of the major purposes of *Visual Explanations* is to demonstrate how telling a story with clear visual display can be of great value to policy makers.[4] Tufte especially highlights Chapter 2, which he now distributes as an inexpensive pamphlet under separate cover (Tufte 1997b) and is subtitled "displays of evidence for making decisions." The chapter addresses two events with life or death consequences where good graphic design made—or could have made—all the difference in the world. The first, John Snow's use of a map to combat cholera and save lives, has been discussed above. The second event that Tufte covers—the disastrous explosion of the space shuttle *Challenger* in 1986—is a cautionary tale about how, for the lack of good graphic, lives were lost.

Tufte takes us through this sad story of bad decision making that left seven astronauts dead and horrified a nation of school children who had been assembled before classroom televisions to witness the first school teacher to enter space. In this account, Tufte shows us one bad diagram after another that apprehensive engineers constructed and disseminated to delay the launch. Put simply, the engineers were working on instinct and constructed a case that did little more than graph anecdotal evidence. Nowhere did they clearly chart the relationship between daily temperature and negative incidents that occurred when rockets were launched. Tufte demonstrates that a simple scatterplot would have made it clear that a successful launch had never taken place below 65 degrees Fahrenheit and that planning one at below-freezing temperatures was courting disaster, as is evident by comparing figure 9.10 with figure 9.4.

The most powerful example of Tufte's commitment to use good design to bring about positive policy outcomes, however, is a proposed design for revolutionizing the display of medical records that he created in conjunction with a physician, Seth Powledge (Powledge and Tufte 1994). This

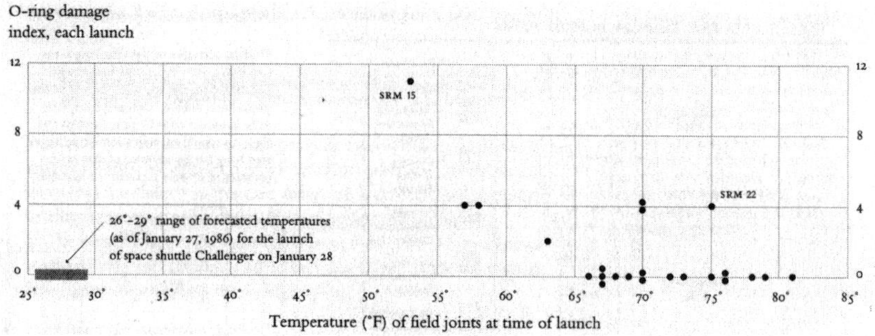

Figure 9.10. Temperature and Space Shuttle launch incidents. Published in Tufte, *VE*, 45.

display (figure 9.11) applies all the principles elucidated in his books. It is composed of small multiples that have maximized their ratio of data to ink. But above all what is important about this design is that it is an intellectual intervention, grounded in acute observation of social practice, which could make a significant difference both for the work of health care providers like doctors and nurses as well as the lives of patients. Because this is the only instance in Tufte's work where he has advanced a display recommendation that would change an institutionalized process, the significance of the innovation cannot be fully appreciated without a more adequate sociological account of the function of the medical record in contemporary health care.

Anselm Strauss and his colleagues (1985; 1988; and Corbin and Strauss 1988) have documented that the central problem of the structure of modern health care delivery is how medical work is organized. One source of this problem is quite clearly the fact that the greatest beneficiaries of the system, and its most powerful vested interests (which include doctors), control the division of labor and most of its rewards. Nevertheless, even if the system's political economy were reformed radically, how people go about their jobs, communicate with each other, and use the technology that medicine increasingly depends upon would still remain issues that are difficult to manage.

The delivery of health care, whether in the hospital or without, depends upon the careful monitoring of a flow of information that is becoming more and more unwieldy. First, there are simply rivers of it. Stanley Reiser has shown in *Medicine and the Reign of Technology* (1978) that physicians are now expected to monitor countless tests as a necessary adjunct to observation, palpitation, and the continuing technologies of a simpler age, like the thermometer and the stethoscope. As if this wasn't enough, medical records are kept in a binder form, which tends to be organized chronologically. In fact, it is probably more accurate to say that medical records are not so much kept as accumulate. Additionally, episodes of acute care usually are managed in hospitals where the medical record functions as the primary means of communication between physicians (who make decisions) and floor nurses (who both carry them out and monitor them). This means that they provide very important evidence in malpractice suits, which in turn encourages practitioners to attend to the legal consequences of medical record management rather than fine tuning it as an instrument for diagnosis and treatment.

The principal effect of the medical record binder, as it is currently constructed, therefore, is that it is one of the major bottlenecks to the delivery of *safe* health care. Thousands and thousands of fatal cases of medical malpractice each year probably are best categorized as communication breakdowns. Doctors often have only the sketchiest memory of the long-term history of a patient's illness trajectory, and when they are reminded that it might not be what they assume it is, they have a devil of a time finding it in the binder. This is especially true with patients who are sicker and more at risk, because their binders tend to be bigger and more unwieldy than most. Accordingly, a widely accepted rule of thumb for patients dealing with serious long-term illnesses that require recurring medical intervention is to have an informed advocate for their interests—often a family member or close friend—who can help them manage what is being communicated and how. But surely more could be done to protect patients?

Tufte and Powledge's display is just such an innovation:

> [Figure 9.11] . . . shows a *graphical* view of patient status, an overlay summary of the traditional medical record designed primarily to assist in the clinical care of the patient. Illustrated is a long and complex history, involving two medical and two psychiatric problems. Some 24 small images depict laboratory readings, medicines, and X-rays (showing an improvement in a cloudy lung). [Figure 9.12] . . . describes the basic graphical elements in the multiple. A common horizontal time-scale, strongly emphasizing recent events, orders the flow of data. On the vertical scale, all measurements are referenced to normal limits (within the whiskered band); the most recent reading is shown both numerically and graphically (in red). Each column of multiples generally represents each medical or psychiatric problem; the left-hand column deals with pneumonia, for example. Thus, data are seen in short-run and long-run contexts, and in relation to other measures (allowing, say, comparisons up and down a column between dosages and responses).

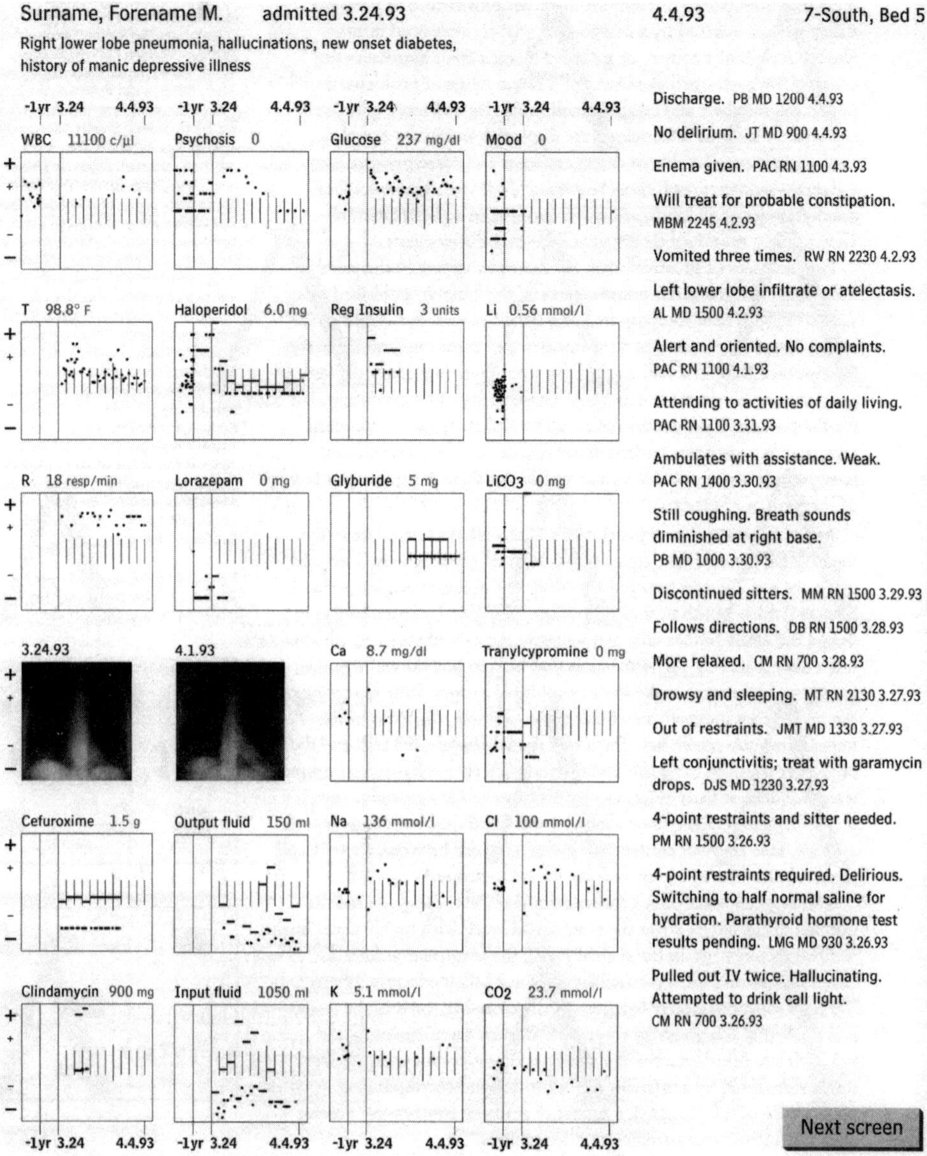

Figure 9.11. Complete medical record display. Published in Tufte, *VE*, 111.

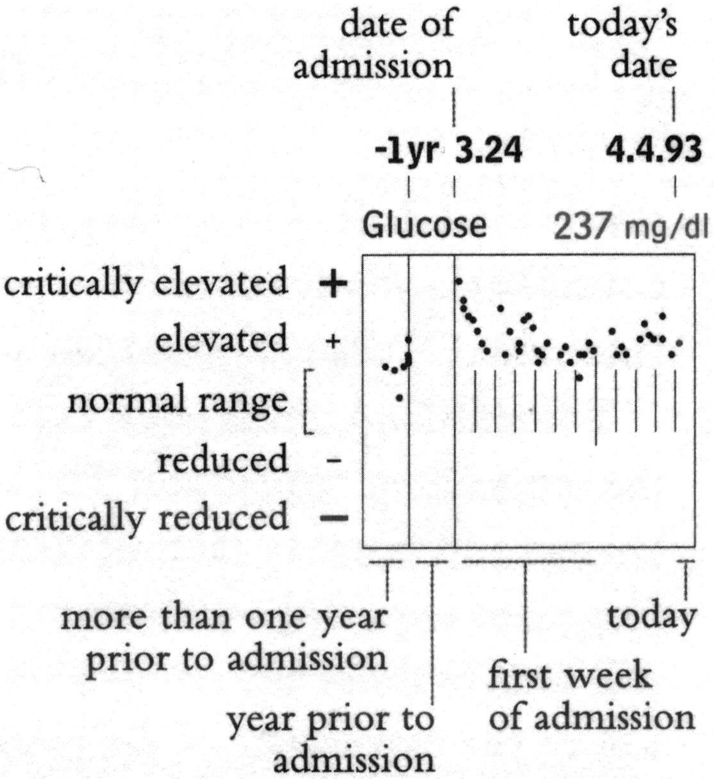

Figure 9.12. Small multiple legend for medical record display. Published in Tufte, *VE*, 110.

> Combining overview with detail, this one-page, high-resolution set of multiples makes sense of thousands of items (the spreadsheet for this display contains 11,616 cells and 1,786 values) now scattered throughout standard medical records. Such graphical summaries will be especially helpful during case conferences or teaching exercises; all of the participants, each with a copy, can quickly assess the history and treatment—with an improvement in the rate of information transfer of perhaps 10 to 20-fold compared with the standard talk talk talk presentation. (Tufte 1997a, 110)

For social scientists, this is a truly amazing display. Not only is it capable of encoding an extraordinary amount of information in a form that is readily legible, but it represents an intervention in the system of medical care that is significant for many reasons.

In the first place, the display not only is consistent with what the sociology of medicine has established about problems in the organization of health care, but also it is based on observations and assessment that are well documented in the sociological literature. Second, Tufte's design is a work of applied knowledge. It is not intended to provide information about a system, but rather for use within a system. Medical sociology, however, is a subfield of the larger discipline where much work of an applied nature is carried on and valued. What this implies is that an important reason that medical sociologists did not come up with this idea long before Tufte is that they lacked the visual imagination and skill to do it.[5]

Tufte closes his account of the medical record display by telling us that: "This architecture—blending quantitative multiples, narrative text, and images—may prove useful for monitoring other data-rich processes" (1997a, 110). He is, I am sure, quite correct, although defining the process and the exact form of the design would depend upon expert knowledge of the situation to be monitored.

Nevertheless, the medical record display might have wider applicability as a template because the information needs it addresses is shared by many non-medical institutions: facilitating decision-making by service providers and providing a means of oversight for their supervisors. Social scientists working in public health, education, and the criminal justice system, for example, could readily design displays based on the medical record model. One aspect of the model that lends it flexibility is that it does not privilege numbers, and more frames could be assigned for the display of photographs of clients, their living environments, diagrams of kinship charts, extended clinical notes, and the like. In this sense, this architecture could be modified into something like a portfolio, which might be more relevant for various social services and educational institutions.

4. Beyond Minimalism

There is an almost palpable tension in Tufte's latest work. On the one hand, the most important message in *Visual Explanations* is, predictably, the need for intellectual discipline. Visual display of any causal narrative must be shaped by the same kind of rigor that the argument itself logically requires. Or as Tufte himself puts it:

> Many of our examples suggest that clarity and excellence in thinking is very much like clarity and excellence in the display of data. When principles of design replicate principles of thought, the act of arranging information becomes an act of insight. (Tufte 1997a, 9)

The displays that Tufte recommends to our attention are, like the medical records above, truly *elegant*. By this I mean that they are not only beautiful, but also, to borrow from the lexicon of the philosophy of science, parsimonious. In this sense, then, Tufte's aesthetic is Minimalist and celebrates the subordination of the analyst (or artist) to the task, and resources, at hand.

On the other hand, *Visual Explanations* is an entrance into "a cognitive paradise of explanation, a sparkling and exuberant world" (Tufte 1997a, 9). It vividly reminds the reader that creativity has a playful mode. Thus, there seems to be another voice in Tufte that is mischievous and can jostle high purpose aside for what appears to be a more or less post-modern exploration of the pleasures derived from performance, embellishment, and the possibilities of low novelty art. How else to explain an entire chapter on magic reveling in the construction of fictive worlds? How else to make sense of the chapter on "Multiples in Space and Time," which, after a sober page or two, veers off onto a romp that leads us through a guide book for identifying tropical fish, an advertisement for gag-shop fluorescent eye glasses, and Ad Reinhardt's tongue-in-cheek history of modern art that purports to represent how well-known modern artists might have depicted a wine glass? (see figure 9.13). Finally, what is to be made of the matching design (see figure 9.14) on the page that faces Reinhardt's display that Tufte produced in collaboration with Bonnie Scranton?

In part, figure 9.14 is vintage Tufte as he strips away the decorative frames that Reinhardt has used to surround each picture of a wine glass. It is the persona we have become accustomed to that admonishes us to note that:

> Each intriguing and smartly annotated image is surrounded by a noisy border, administrative bloat that consumes an astonishing 42% (*!!!*) of space in each framed rectangle. It resembles all too many computer displays, where a cramped window showing the user's work is framed by bureaucratic debris of small bars, buttons, titles, icons, and over-produced drawings . . . [M]y redesign strips away all the frames (for the edge of each picture defines itself well enough). (Tufte 1997, 118–119; italics mine)

Tufte knows what he calls "chartjunk" and he knows what to do about it when he sees it! But, then, there is something quite new in the redesign. Tufte has introduced color into the equation and rendered each picture with the artist's own distinctive palette. It's quite amazing; but why is Tufte spending so much time expanding the interpretative possibilities of a fiction?

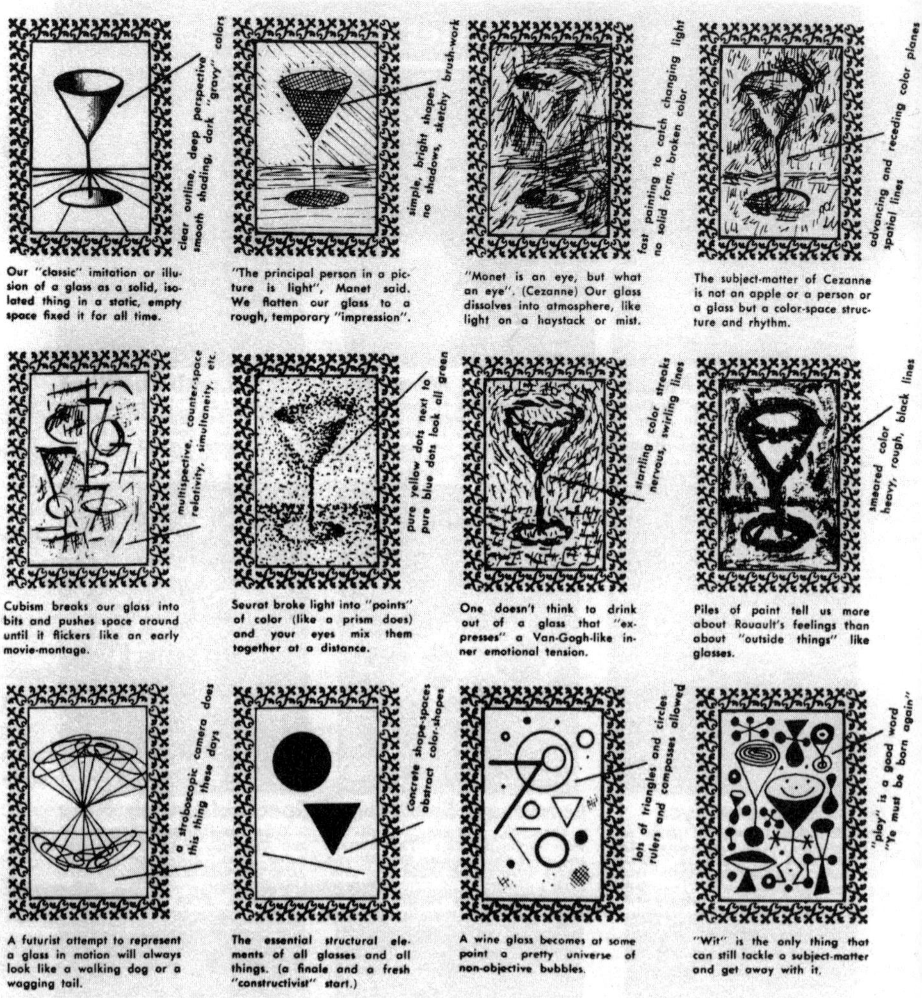

Figure 9.13. Ad Reinhardt's wineglasses. Reproduced in Tufte, © 2004 Estate of Ad Reinhardt Artists Rights Society (ARS). New York. *VE*, 118.

I think the answer is that Tufte is playing. He's pushing an envelope of some sort by tinkering with what one can make out of frivolity and whimsy. Reinhardt's display was intended as a spoof, but Tufte's playing with it reinvests it with more serious intent so that a clever joke becomes, in fact, a useful study guide to how some of the most important figures in modern art varied in technique and sensibility. Tufte, then, may be playing but it has been in the service of a higher purpose: information not distraction. Perhaps the tension in *Visual Explanations* is something else entirely, the electricity, perhaps, of Tufte's creativity?

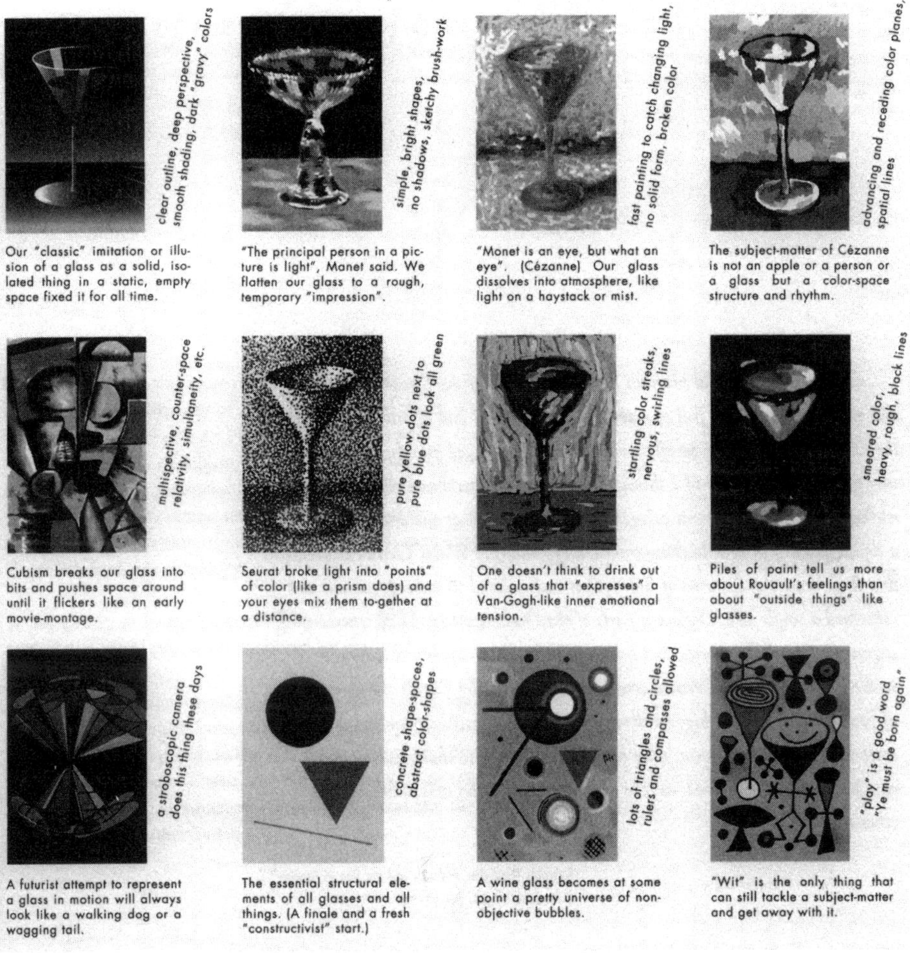

Figure 9.14. Tufte's version of Reinhardt's wineglasses. Tufte, *VE*, 119.

The reader might be justified in coming to this conclusion were it not for the final chapter of the book, entitled "Visual Confections: Juxtapositions from the Ocean of the Stream of Story." This chapter builds on Tufte's abiding interest in visual narrative—the telling of complex tales in just a few images—to address new opportunities, and new challenges, posed by the visual display of information.

This chapter begins with a long quote from Salman Rushdie's *Haroun and the Sea of Stories,* which concludes:

> [T]he Ocean of the Streams of Story was in fact the biggest library in the universe. And because the stories were held here in fluid form, they retained the ability to change, to become new versions of themselves, to join up with other stories and to become yet other stories. (Tufte 1997a, 120)

Tufte uses the image of the "Ocean of the Streams of Story" as a figure of speech to cover all of the information in the world and how it is organized.

> Here [figure 9.15], as time flows across the page from left to right, is a tiny part of the Ocean of the Streams of Story, a handful of the thousand thousand thousand and one tales. As I have drawn it, an event is the intersection of noun and verb, of subject and action—*something happens*. A plane of events, a slice at a particular time, consists of all possible combinations of nouns and verbs at that time, with as many planes as there are times. A story is a progression of noun-verb incidents; each long strand in the diagram represents one story.
>
> A short story or image sequence (like Marey or Muybridge) is a brief period in a single time-strand, shown by the red line [which is barely visible to the left of the plane of events, superimposed on the orange line]. And a *confection* is an assembly of many visual events, selected (at the red dots for example) from various Streams of Story, then brought together and juxtaposed on the still flatland of paper. By means of a multiplicity of image-events, confections illustrate an argument, present and enforce visual compositions, combine the real and the imagined, and still tell us yet another story. (Tufte 1997a, 121)

The chapter consists mostly of illustrations and explicitly symbolic paintings and designs. There is no data waiting patiently to be adequately displayed, nor is there much by way of visual narrative. What this chapter showcases are visualizations of ideas or, in some of the most striking instances, the concrete illustration of quite complex ideas. The illustrations—and that is what most of them are—have not been chosen because they are a way of telling a story, but rather because they encode theories about the nature of the universe and the social order.

Tufte has a tendency to use words in unfamiliar ways and then to endow them with special conceptual importance.[6] Change some of Tufte's idiosyncratic vocabulary in this chapter—for example, "confections," "compartments," "imagined scenes," and so on—however, and the terrain becomes quite familiar.

For Tufte, *confections* appear to refer to little more than:

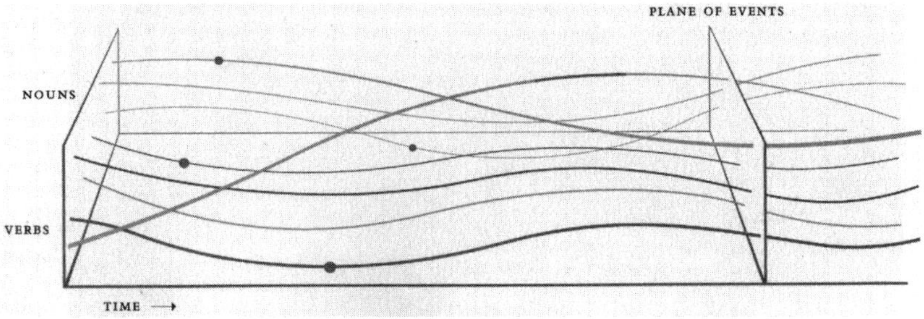

Figure 9.15. Part of Ocean of the Streams of Story. Tufte, *VE*, 121.

- qualitative stories;
- that are depicted with multiple frames, or complex events in a single frame;
- that may involve imagined moments, landscapes, and figures; and
- that convey an argument about the world.

In other words, Tufte has invited the reader to enter a world where stories are constructed and told that rely on metaphor and layered meaning, and which are driven by dramatic tension and an author's voice. At this point in the text, the tension that we encounter is genuine—and unresolved—because it is not clear if Tufte is fully aware what the implications of this turn to story telling might be for the trajectory of his work and its aesthetics.

Figure 9.16 diagrams the difference between what Tufte has told us about the creation of displays so far in his work (left panel) and what he is considering now (right panel). An *explanation* that relies on clear concepts to define graphic integrity is quite a different kettle of fish than a *story* that uses metaphors to define how various graphic elements should be combined in a narrative. In one, the state-of-the-art communicative tools are derived from logic, while in the other the rhetoric of exposition is shaped inescapably by the conventions of fictional story telling.

Tufte concludes *Visual Explanations* by telling us that "by virtue of their arguments, confections make reading and seeing and thinking identical" (Tufte 1997a, 151). This program may not be either possible or desirable! True, carrying out analytic work so that these competencies, and what they afford, are consistent, complementary, and mutually translatable, is a worthy goal and one that Tufte has shown us how to reach. But,

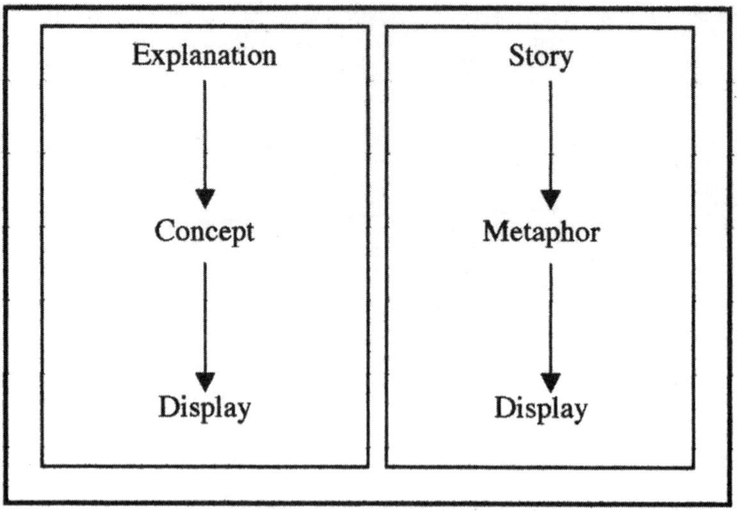

Figure 9.16. Explanations and stories.

this is a goal that analysts can achieve only by writing, displaying, and conceptualizing, which is a process that proceeds by trial and error because very different ways of exercising the imagination are involved. In addition, if the enterprise is to tell stories about and/or with people as characters, it is not at all clear whether the conventions for displaying variables (e.g., numbers, things, and events) constitute an adequate aesthetic vocabulary. In other words, while Tufte's Minimalist aesthetic has served him well for what he has done so far, it may be too constraining in the future, which should require a more figurative approach to the display of information.

Tufte's choices of illustrations for the final chapter of *Visual Explanations,* which are all figurative, compel us to revisit the definition of "chartjunk." Put simply, Tufte has defined chartjunk as unnecessary data ink that adds nothing to the information in the chart (Tufte 1983, 107). If this is true, there is a sense, then, that many of the illustrations in the final chapter of *Visual Explanations* are a form of chartjunk. Each element requires a narrative account to explain not only what information might be there, but also why it is there.

Tufte is right to exhort us when he says:

> In a paragraph accompanying his illustrated history of 20th-century art, Ad Reinhardt wrote what is probably the single best sentence ever written

about the point of images for information design: "As for a picture, if it isn't worth a thousand words, the hell with it" (Tufte 1997a, 119).

But the various frontispieces and illustrations by Scheiner, Pugin, Buno, Durer, and others as well as the posters and paintings by Klutsis, Lissitsky, and Taney require reams of words just to make their authors' intent intelligible. Anscome's whole idea, with which this paper began, was that an argument carried out with words, numbers, and graphic displays was better because it created a fruitful intellectual division of labor. Graphs and charts did things that numbers alone couldn't, and both freed the writer to spend more time and space analyzing and less describing. As beautiful as these displays in *Visual Explanations* may be, they are still like a joke that you only get when someone tells you the punch line and why it is supposed to be funny. What most characterizes these displays is that they tend to be didactic rather than explanatory, are static and tableaux-like in design, and organize their content as allegories. A close look at what Tufte entitles "an eternal confection" should clarify the limits of making these kinds of choices.

Figure 9.17 was engraved on a 6- by 9-inch plaque and placed aboard each of the two Pioneer spacecraft that were launched by NASA to explore the reaches of deep space in the early 1970s. This design includes a number of diagrams that are intended to identify who, what, and where we are.

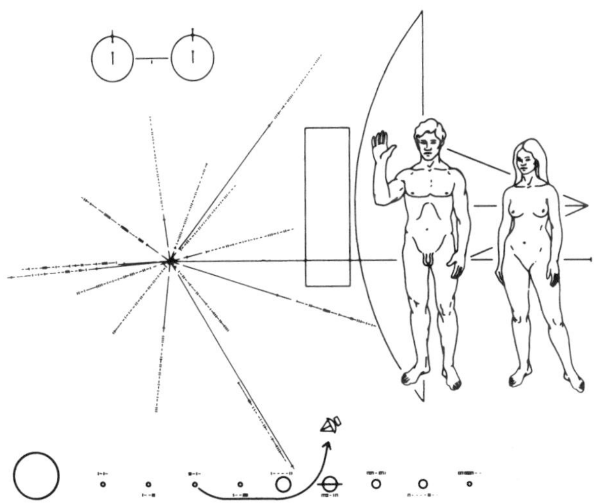

Figure 9.17. Plaque on the Pioneer Spacecraft. Reproduced in Tufte, *VE,* 150.

The assumption that the designer made—and that Tufte shares—is that this display contains information that would be meaningful to some form of intelligent life other than us. Eminent art historian E. H. Gombrich, however, has so many reservations about the design that he does not believe that it was intended seriously:

> We have seen that reading an image, like the reception of any other message, is dependent on prior knowledge of possibilities; we can only recognize what we know. Even the sight of the awkward naked figures in the illustration cannot be separated in our mind from our knowledge. We know that feet are for standing and eyes are for looking and we project this knowledge onto these configurations. It is this information alone that enables us to separate the code from the message; we see which of the lines are intended as contours and which are intended as conventional modeling. Our "scientifically educated" fellow creatures in space might be forgiven if they saw the figures as wire constructs with loose bits and pieces hovering weightlessly in between. Even if they deciphered this aspect of the code, what would they make of the woman's right arm that tapers off like a flamingo's neck and beak?. . . As for the fact that "the man has his right hand raised in greeting". . . not even an earthly Chinese or Indian would be able to interpret correctly this gesture from his own repertory.
>
> The representation of humans is accompanied by a chart: a pattern of lines beside the figures standing for the fourteen pulsars of the Milky way, the whole being designed to locate the sun of our universe. A second drawing (how are they to know it is not part of the same chart?) "shows the earth and the other planets in relation to the sun and the path of Pioneer from earth and swinging past Jupiter." The trajectory . . . is endowed with a directional arrowhead; it seems to have escaped the designers that this is a conventional symbol unknown to a race that never had the equivalent of bows and arrows. (Gombrich 1996, 54–55)

What might the designers of the Pioneer plaque have done instead? Well, they might have taken a page from any of Tufte's books and designed something with a narrative of space and time that was realized with small multiples. This might be something like the one "confectionary diagram" that can be understood easily without a library because it was constructed to communicate a clear message.

Figure 9.18 accompanied an article that was written in the Washington Post. Tufte considers this illustration to be "brilliant," marvels at the fact that it was churned out by a journeyman illustrator on the job, and asks us to:

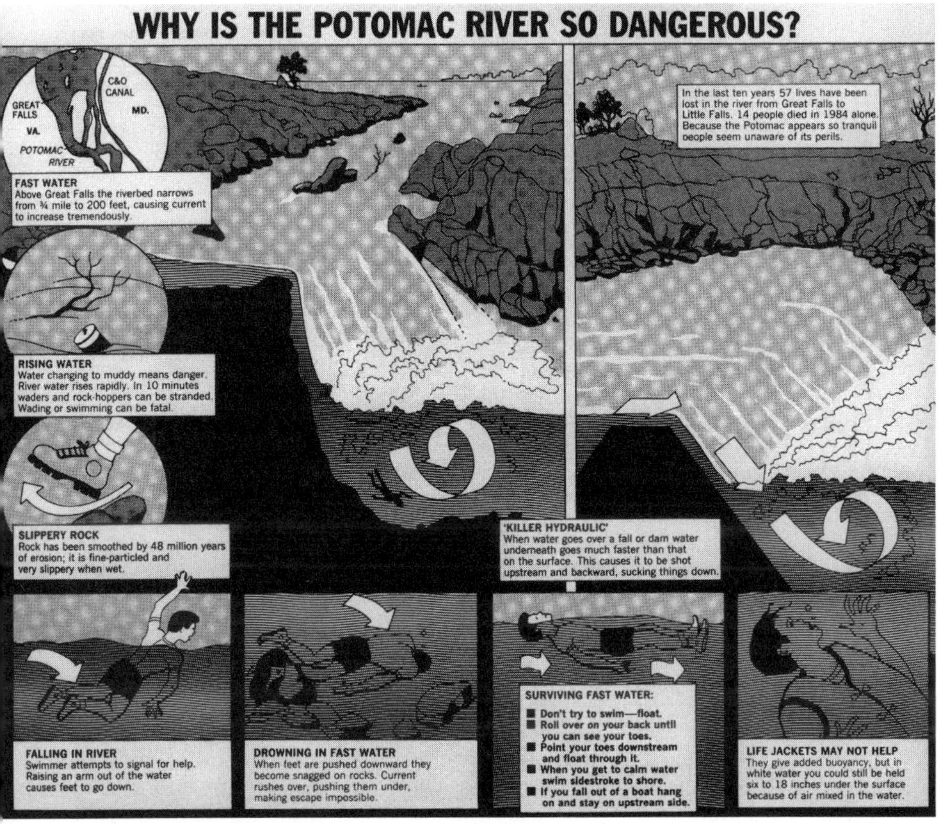

Figure 9.18. Why is the Potomac River so dangerous? Reproduced in Tufte, *VE*, 144.

> Note how the illustration explains why the river, with its dangers lurking beneath the surface, is so deceiving. Drawings can present multiple and cut-away views, sequences of actions, focused annotation; in this case, it is better than any other method for explaining what is going on; it is certainly better than being there . . . Whenever possible, give your audience words and images written down on paper, even if only to supplement spoken words. (Tufte 1997a, 145)

What makes this story of the Potomac River so compelling is that the story is spoken in a clear voice (in this case we imagine an experienced park ranger); it features a protagonist with whom we identify and about whom we worry, a place where terrible events occur, and a story with an uncertain outcome. In short, the display is an informative and complex story about a real situation.

Tufte might want to call it a "confectionary diagram," but perhaps it is best defined as "juxtaposed pictorial and other images in deliberate sequence." Scott McCloud tells us that this definition of this kind of display might be shortened to "Sequential Art" or, to be even more succinct, the comics (McCloud 1993, 9). Note that the art of comics draws on many of the same lessons of good design that Tufte has promulgated, but also on filmic conventions that shape narrative pacing and establishing point of view (POV). In addition, the fact that comics are texts that are meant to be read means that the designer has an advantage that filmmakers don't. They can count on readers being able to pace themselves and to look with greater care at complex data rich graphics.

It's a long shot that other life forms on another planet galaxies away might make sense of any of our texts. After all, look at the trouble we have had with the artefacts of our own prehistory. But might it not have made sense for NASA to show a more realistic picture of the globe and more information about us? Wouldn't a family group that included a suckling child have been more appropriate? Shouldn't there have been some indication of the life course and that "we" includes a world shared with other family groupings, and so on? In many ways, a well-designed and edited graphic novel might have been far more useful for those with whom we are trying to communicate but with whom we share virtually no frames of reference. Minds learn by comparing and contrasting, as Tufte never ceases to remind us, and the narrative sequences of a realistic comic would provide much that could be gleaned using the most elementary forms of analysis.

Comics have progressed beyond the few frames available in the daily newspaper to thirty- or forty-page serials, and now the *graphic novel*. In addition, genres have expanded to include confessional autobiographies, documentaries, narrative histories, instructional manuals, and so on. If what Tufte has said about the power of charts to encode information and of visual narratives to explain complexity is true, it suggests that the next step in his intellectual trajectory should be an exploration not just of how visual metaphors are used to build a narrative, but also of how the authorial voice shapes and transforms those metaphors for the sake of a story. In this sense, the telling of a true story requires becoming comfortable with the role of the author, a fact for which Tufte's minimalism ill prepares him. However, it may be that when a final accounting is done, that Tufte's minimalist rules for displaying data are just a special case of a more general rule about truth telling. Minimalism is a good guide for most of the issues that Tufte has addressed, but a more flexible orientation is needed for complex narratives. Scott McCloud tells us as much in figure 9.19.

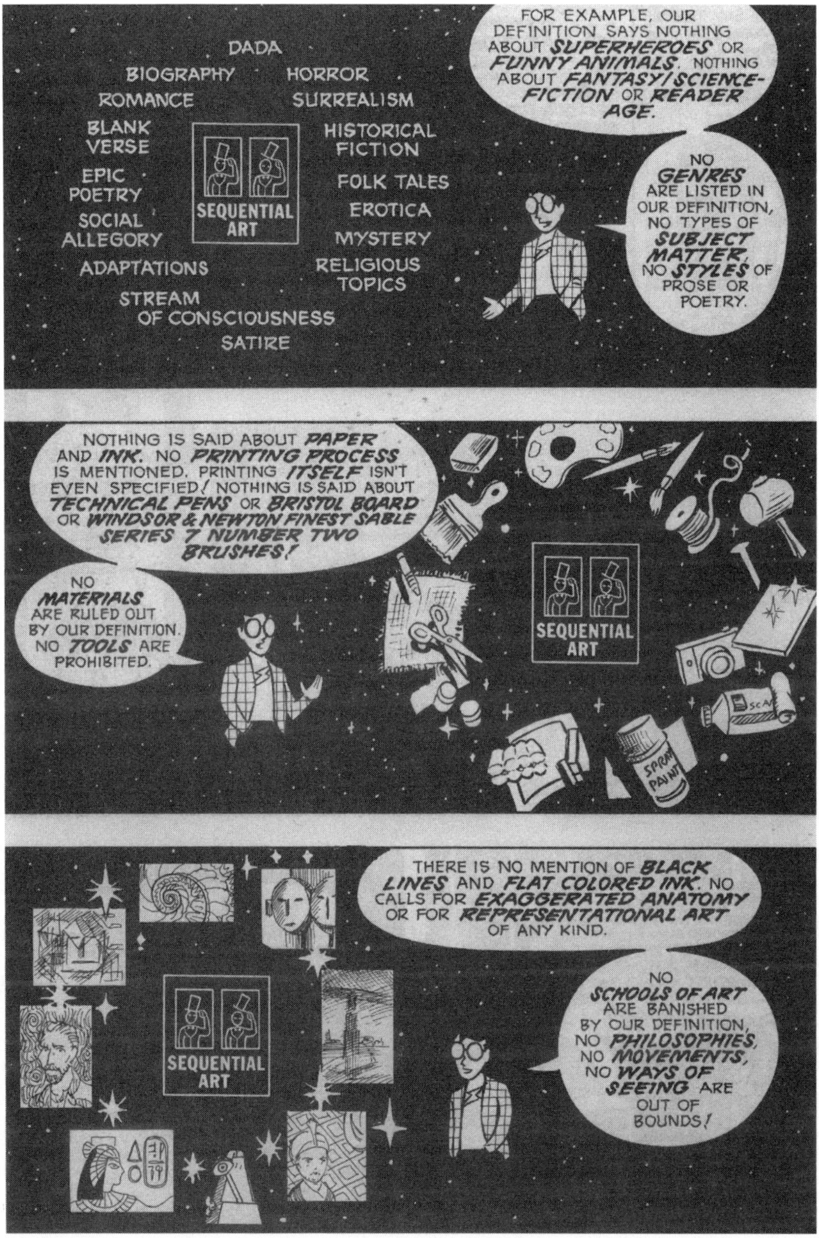

Figure 9.19. Understanding comics. Scott McCloud, *Understanding Comics* (New York: HarperCollins, 1993), 22.

5. Visual Social Science After Tufte

The last several years have witnessed the emergence of a growing and diverse literature concerned with describing or defining the relationship between the social sciences and the visual.[7] Almost everyone writing on this topic finds appropriating the visual to be as unsettling as it is exciting. Empiricists construct visual displays to produce data about patterned behavior that are in some ways richer sources of information about some topics than surveys. But defining in a general way what those topics may be and what makes visual information richer has not been articulated in a form that has convinced the quantitative social science mainstream to incorporate visual research methods into their respective disciplines. Then there are those students of society who believe that the primary function of a discipline should be interpreting the meanings embodied in human action and its products. They have found that the inevitable muteness of the image, however, introduces layers of complexity into the process of interpretation, which not only reduce the credibility of any given analysis but also make it hard to codify methods of interpretation with any coherence. Finally, some social scientists have been impressed by the expressive potential of the visual narrative—whether in photography or film—to explore directly the experience of the social. But they have found that the conventions and opportunities that visual narrative affords persistently challenge accepted protocols for guaranteeing the integrity of social science research findings.

The hard truth of the matter is that the image—and beyond that, the visual dimension of existence in general—as a product of knowing and as a source of information poses a complex and varied set of challenges to the major ways in which social analysts conduct inquiry. It would, perhaps, be comforting for visual social scientists to imagine being on the verge of some kind of revolutionary overthrow of everything the social sciences traditionally do. At least that way they wouldn't have to worry about how to integrate their new passion with the various legacies they have inherited. But reform is always harder than revolution and the question that visual social scientists cannot avoid is how to integrate the visual into what has stood the test of time in social analysis.

Edward Tufte is perhaps the most important figure on the contemporary landscape to address this challenge and has laid the groundwork for addressing many of the problems that visual social scientists face. His approach differs from most others who have confronted these issues, and who focus on resolving the philosophic issues that both visual information and the social sciences pose to the inquiring mind. In contrast, Tufte begins pragmatically with a simple question: What should someone do with an image in order for it to make an invaluable difference in an inquiry? This simple question—and others that follow from where it leads—has enabled Tufte to produce a number of enormously rich warranted assertions about how and why information should be displayed. Put simply, Tufte's three books constitute the only masterpieces that the visual social sciences have yet produced.

Tufte's mode of exposition departs from the communicative conventions of contemporary social science. However, this has not prevented several gifted social scientists from producing work, influenced by Tufte, which can be consulted as manuals. These books should be in the possession of every social scientist and include Stephen Kosslyn's *The Elements of Graph Design;* Anders Wallgren et al.'s *Graphing Statistics and Data: Creating Better Charts;* Gary Henry's *Graphing Data;* and Howard Wainer's *Visual Revelations: Graphical Tales of Fate and Deception from Napoleon Bonaparte and Ross Perot.* Familiarity with these works prepare the analyst to address—and resolve—most issues that anyone constructing and using displays in their professional work would face.

Of greater importance for the development of a more visual social science is the fact that Tufte's work is part of an intellectual trajectory that is far from spent and that is beginning to encompass ever more complex issues and challenges. One of the most interesting things about how Tufte's books are organized is that each book's concluding meditations appear to pose the central topic for the following book. Thus, a chapter on "Aesthetics and Technique" that concludes the *Visual Display of Quantitative Information* sets the stage for *Envisioning Information,* while that book's final chapter, "Narratives of Space and Time," anticipates *Visual Explanations.* Tufte's trajectory is an iterative one, so the reader can't predict exactly what will be in a subsequent book from a close reading of the earlier one's final chapter. Thus, Tufte does not outline a new book in his final chapters. Rather, he identifies the issues that appear to be a next step in a process of discovery.

If this pattern holds true—and, by the way, each book follows the previous one by a seven-year interval—then the next one should appear in 2004 and address the issues broached in "Visual Confections: Juxtapositions from the Ocean of the Streams of Story." We already know from Tufte's Web site (www.edwardtufte.com) that this book will be entitled *Beautiful Evidence.* We should not be surprised if this book addresses the conceptual and display problems of complex arguments that must be made visually.

Whether Tufte will still be repeating Ad Reinhardt's mantra of *No exploitation! No Mixing Things Up!* remains to be seen. Minimalist or not, Tufte's expansion of the classic commitment of the social sciences to produce reliable and usable knowledge for, and as a way of informing, action, should be a source of pride, as well as a challenge, for visual social scientists. Tufte's ideas in *Visual and Statistical Thinking: Displays of Evidence for Making Decisions* (1997b) as well as the medical records display model the kind of work that should characterize visual social science as a whole and that visual social scientists should be doing. We have much to offer our respective disciplines and they have much to offer the world.

Notes

1. Tufte does not refer to himself as a Minimalist, but most commentators on his work do. From one standpoint this is puzzling. After all, "Minimal art strove to focus attention on the object as an object, reducing its historical and expressive content to the bare minimum" ("minimalism," *The Columbia Encyclopedia,* 5th Edition, 1975, 1783), while for Tufte the communicative content of the design is of central importance (Tufte 1983, 177). This apparent inconsistency evaporates, however, when the reader realizes that the key element in both Minimalism and Tufte is what Robert Hughes refers to as "a pull toward . . . self-effacement" (Hughes 1991, 386). Minimal art and design, thus, is not about self-expression. For the Minimal artists there was only a blank canvas, while Tufte enlists the designer in the service of data. The case for Tufte as a Minimalist is strengthened by how closely his aesthetic seems to resonate with work like Frank Stella's *Die Fahne Hoch* and Agnes Martin's *Praise*. Recently, Tufte has begun to exhibit in galleries and to construct and market large (6'x15'x30') steel sculptures, which are very reminiscent of Barnett Newman's *Zum Zum* and work by both Donald Judd and Don Flavin. And, of course, as we shall see later in this paper, Tufte's kinship with the Minimalist painter Ad Reinhardt is very strong. Given Tufte's current influence, perhaps an argument could be made that Tufte should be seen as the greatest of the American Minimalists and the one who discovered its true metier.
2. Currently, some textbooks in sociology have begun to selectively apply the principle of small multiples by using various kinds of dot and hatch maps. Macionis' *Sociology* is a case in point. He uses the Peters' projection map of the globe as the basis for displaying a wide array of social indicators. Dot maps are useful to display exact locations of frequencies, while hatch maps are "suited to showing ratios such as proportions, intensities and averages for areas" (Wallgren et al. 1996, 58).
3. Tufte's is concerned not only with visual display, but also with rhetoric, or how an argument should be presented. *Visual Explanations* includes six rules for giving a talk or paper:
   1. Near the beginning of your presentation, tell the audience:
      What the problem is
      Why the problem is important
      What the solution to the problem is . . .

2. To explain complex ideas or data, use the method of PGP:
   Particular General Particular . . .
3. No matter what, give everybody in the audience one or more pieces of paper, packed with material related to your presentation. . .
4. Analyze the details of your presentation; then master those details by practice, practice, practice.
5. Show up early. Something good is bound to happen.
6. Finish early. (Tufte 1997a, 68–70)

Not only does Tufte model these rules in his one-day workshops, but also he talks about them and provides everyone with a handout listing the rules.

4. Tufte is not completely consistent in how he uses the terms "narrative," and "story." Throughout his books, he seems to use the word "story" to include accounts that are either descriptive or causal or both. In his earlier books "narrative" refers exclusively to stories that take place in time and in space, while in *VE* it is often used as a synonym for "story." Thus, Tufte may discuss a story derived from one image or an assembled set of images (which may or may not refer to data in time and space) and call them all narratives. He may also refer to his account as a "story" or a "narrative." Nevertheless, his over-riding purpose is invariably to display information so clearly that it can be used as evidence in explaining something of significance to someone.
5. There may be more than merely a lack of visual imagination and skills involved. Virtually no attention is paid to health, illness, and medical institutions in classic sociological theory. Sociological interest developed, however, when the medical profession enlisted them to study social problems of concern to physicians and especially the issue of gaining patient compliance. What is today called the sociology of health thus began as the deployment of *sociology in medicine*. For many reasons—growing interest in professional work, the organization of medical care, and concern about the excessive power of physicians in their relations with patients—sociologists took a more critical approach to medical institutions and values. Those sociologists who took this approach redefined their field as the *sociology of medicine,* in which doctors went from being those who established research priorities to becoming just another subject of study. It must be acknowledged that an important motivation for this change was that it redefined the relationship between doctors and sociologists in a way that improved sociologists self-estimate of their own status. The major drawback for sociology, unfortunately, was a trained aversion to applied research.
6. The most striking example of Tufte's predilection for inventing new connotations for terms is found in *Visual Explanations* where he redefines the subject matter of his three books. *VDQI* is about "pictures of numbers," *EI* is about "pictures of nouns," and *VE* deals with "pictures of verbs" (Tufte 1997a, 10). It's unclear if *catchiness* is worth the confusion produced by an inappropriate linguistic metaphor.
7. Some of this new work includes: Banks 2001; Emmison and Smith 2000; Grady 1991 and Grady 1996; Pauwels 2000; Pink 2001; and Van Leeuwen and Jewett 2001.

References

Anscombe, F. J. 1973. Graphs in statistical analysis. *American Statistician* 27 (February): 17–21.

Banks, Marcus. 2001. *Visual methods in social research*. Thousand Oaks, Calif.: Sage.

Becker, Howard. 1998. *Tricks of the trade*. Chicago: University of Chicago Press.

Cleveland, William S. 1985. *The elements of graphing data*. Monterey, Calif.: Wadsworth.

———. 1993. *Visualizing data*. Murray Hill, N.J.: AT&T Bell Laboratories.

Corbin, Judith, and Anselm Strauss. 1988. *Unending work and care*. San Francisco: Jossey-Bass.

Davis, James. 1987. *Social difference in contemporary America*. San Diego: Harcourt, Brace and Jovanovich.

Denzin, Norman, and Yvonne Lincoln. 1994. *Handbook of qualitative sociology*. Thousand Oaks, Calif.: Sage.

Dewey, John. 1934. *Art as experience*. New York: Milton Balch and Company.

Emmison, Michael, and Philip Smith. 2000. *Researching the visual*. Thousand Oaks, Calif.: Sage.

Erikson, Kai. 1976. *Everything in its path*. New York: Simon and Schuster.

Gombrich, E. H. 1996. The visual image: Its place in communication. In *The essential Gombrich*, edited by R. Woodfield. London: Phaidon Press.

Grady, John. 1991. The visual essay and sociology. *Visual Sociology* 6 (2): 23–38.

———. 1996. The scope of visual sociology. *Visual Sociology* 11 (2): 10–24.

Henry, Gary. 1995. *Graphing data: Techniques for display and analysis*. Thousand Oaks, Calif.: Sage.

Hughes, Robert. 1991. *The shock of the new*. New York: McGraw-Hill.

Kosslyn, Stephen. 1994. *The elements of graph design*. New York: W. H. Freeman and Company.

Lieberson, Stanley. 1985. *Making it count: The improvement of social research and theory*. Berkeley: University of California Press.

McCloud, Scott. 1993. *Understanding comics*. New York: HarperCollins.

Norman, Donald. 1998. *The invisible computer*. Cambridge: MIT Press.

Pauwels, Luc. 2000. Taking the visual turn in research and scholarly communication. *Visual Sociology* 15: 7–14.

Pink, Sarah. 2001. *Doing visual ethnography*. Thousand Oaks, Calif.: Sage.

Playfair, William. 1786. *The commercial and political atlas*. London: Corey.

———. 1805. *An inquiry into the permanent causes of the decline and fall of powerful and wealthy nations*. London: Greenland and Norris.

———. 1821. *A letter on our agricultural distresses: Their causes and remedies*. London: William Sams.

Powledge, Seth, and Edward Tufte. 1994. Graphical summary of patient status. *The Lancet* 344 (August 6): 386–389.

Prosser, Jon. 1998. The status of image-based research. In *Image-based research*, edited by J. Prosser. London: Falmer Press.

Ragin, Charles. 1994. *Constructing social research*. Thousand Oaks, Calif.: Pine Forge Press.

Ragin, Charles, and Howard Becker, eds. 1992. *What is a case? Exploring the foundations of social inquiry*. Cambridge: Cambridge University Press.

Reiser, Stanley. 1978. *Medicine and the reign of technology*. New York: Cambridge University Press.

Schmid, Calvin S. 1954. *Handbook of graphic presentation*. New York: The Ronald Press.

Strauss, Anselm, et al. 1985. *The social organization of medical work*. Chicago: University of Chicago Press.

Strauss, Anselm, and Judith Corbin. 1988. *Shaping a new health care system.* San Francisco: Jossey-Bass.

Tufte, Edward. 1974. *Data analysis for politics and policy.* Englewood Cliffs, N.J.: Prentice-Hall.

———. 1983. *The visual display of quantitative information.* Cheshire, Conn.: Graphics Press.

———. 1990. *Envisioning information.* Cheshire, Conn.: Graphics Press.

———. 1997a. *Visual explanations.* Cheshire, Conn.: Graphics Press.

———. 1997b. *Visual and statistical thinking: Displays of evidence for making decisions: John Snow and the cholera epidemic; the decision to launch the space shuttle Challenger* (Separate imprint of Chapter 2 of *Visual Explanations*). Cheshire, Conn.: Graphics Press.

Tukey, John. 1977. *Exploratory data analysis.* Reading, Mass.: Addison-Wesley.

———. 1988. *The collected papers of John Tukey,* edited by W. S. Cleveland. Pacific Grove, Calif.: Wadsworth and Brooks.

Van Leeuwen, Theo, and Carey Jewitt. 2001. *Handbook of visual analysis.* Thousand Oaks, Calif: Sage.

Wainer, Howard. 1977. *Visual revelations: Graphical tales of fate and deception from Napoleon Bonaparte and Ross Perot.* New York: Copernicus.

Wallgren, Anders, Britt Wallgren, Rolf Persson, Ulf Jorner, and Jan–Aage Haaland. 1996. *Graphing statistics and data: Creating better charts.* Thousand Oaks, Calif.: Sage.

10

Making Science Visible
Visual Literacy in Science Communication

Jean Trumbo

> Imagination or visualization, and in particular the use of diagrams, has a crucial part to play in scientific investigation.—Rene Descartes

The media are saturated with scientific images—visualizations of discoveries made through scientific inquiry as well as complex representations created with technologies to describe science not yet discovered but theoretically possible. Comets and asteroids are depicted believably in movies with special effects that transcend the visual ability of many modern space telescopes. Advertisers use images of X-rays or CAT scans to invoke the authority of science to sell products. Network television series give us glimpses into forensic science. And, segments on the nightly news feature animated images of cells (or mineral deposits or tumors) blasted by lasers as well as weather graphics with Doppler radar–inspired fly-ins of storm systems. These visual images of science range in form from the purely symbolic to the highly representational. The question is whether any of these images are accurate representations of science and, more importantly, what special literacy skills are required to understand them.

Visible science varies in purpose from data sharing among scientists to entertainment for the public. In many cases, scientific information is theoretical or conceptual with no obvious physical form, so technology, symbolic representation, and creative expression are employed to craft a visible interpretation (Gershon, Eick, and Card 1998). This is true in both mediated science fiction as well as scientific inquiry conducted in laboratories. In any case, an attentive audience interprets these images and the science behind them with the assumption that scientific principles are represented accurately. In fact, this chapter will suggest that effective scientific visualization can only be shared, regardless of the level of audience expertise, if there is a sufficient level of visual literacy among those interpreting the image.

Visual literacy in science communication is defined broadly to include that which occurs among scientists as well as the mediated communication that occurs from scientists to the public. In this context, visual literacy is a holistic construct that includes visual thinking, visual learning, and visual communication (Randhawa 1978; Seels 1994). While it is possible to distinguish the literacy needs of scientists, science communicators, and the public in specific ways, this article will define a basic framework from which visual literacy can be understood. Behind this framework is the conviction that the visualization of science information places increased literacy demands on all participants in the communication process.

Visual literacy in science has been examined indirectly by communication and journalism scholars stressing the importance of clarity in the representation of scientific data (Friedman, Dunwoody, and Rogers 1986; Nelkin 1995). Visual communication educators have addressed the perceptual, cognitive, and design requirements of images in the media (Lester 1995; Messaris 1994; Moriarty and Kenney 1995) but with few exceptions have not addressed science visualization specifically (Trumbo 1999; 2000). Information designers such as Tufte (1983; 1990; 1997) and Wurman (1989) provide both historical and contemporary examples of information graphics to emphasize the design principles that help or hinder their effective presentation. Statisticians such as Wainer (1997), mathematicians such as Devlin (1998), as well as others interested in numerical literacy have written about the communication potential of information graphics such as charts and diagrams (Meyer 1997). The relationship between visual interpretation and scientific inquiry is explored by Massironi (2002), as a psychological process of creating correspondence between scientific hypothesis and graphic marks. Science historians (Ford 1992; Robin 1992) offer glimpses of the rich heritage of scientific illustration and instrumentation. And, within the diverse disciplines of life, physical, and social sciences, the role of scientific visualization and new technology is explored by biologists, geographers, astronomers, chemists, engineers, and physicists (Pickover and Tewksbury 1994; Sprott and Pickover 1997).

This overview suggests the range of interest in the visual representation of science as well as the diversity of viewpoints that make up this multidisciplinary enterprise. It also presents a challenge for those concerned about the effective communication of science. There is currently no common theory or agreement on how to represent scientific data, processes, or discoveries. There is also an evolving assortment of complex technologies that facilitate the creation of captivating images. Nonetheless, the resulting images may not express the appropriate scientific principle or may be incomprehensible to a non-expert audience. In essence, there is a lack of a cohesive foundation in visual literacy among those relying on the communication potential of science images.

Chemist Roald Hoffmann laments that, in spite of the need for visual literacy, the techniques of visual representation are not part of the scientist's training (Hoffmann 1995). Hoffmann describes representation in chemistry as visual shorthand and the ability to use three-dimensional models and graphic language as essential for the scientist. He explains that the choice of how molecules are depicted influences how scientists think about them. A two-dimensional diagram presents a different orientation from that provided in a three-dimensional illustration depicting the overall shape of the molecule. This is important to the chemist, since the potential for synthesizing a derivative may be more or less clear depending upon the illustration. Hoffmann says,

> It is fascinating to see the chemical structures on the pages of every journal and to realize that from such minimal information people can actually see molecules in their mind's eye. . . . The molecule is certainly seen, but it may not be seen as the chemist thinks (in a dogmatic moment) that it is seen. It is represented as he chooses to see it, nicely superimposing a human illogic on top of an equally human logic. (Hoffmann 1995, 76, 78)

Methane

verbal label

Figure 10.1. A verbal label for methane identifies the colorless gas but tells us nothing about its chemical make up.

CH_4

formula

Figure 10.2. The chemical formula for methane identifies the molecules that make up the gas but does not describe its overall structure.

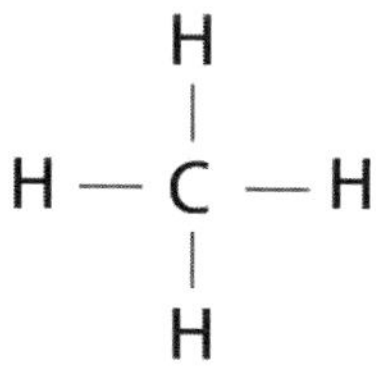

Figure 10.3. A structural diagram tells the scientist something about the relationship of molecules but does not describe the three-dimensional shape.

The scientific process is a formal method of investigation with careful strictures built in and designed to minimize the publication of "wrong" results. Goodfield (1981) notes that there is reluctance among scientists to popularize scientific data through the media. Yet, the visualization of science is *everywhere*—on the World Wide Web, in science-oriented television shows, in magazines in the popular press, and in newspaper information graphics. Priest (1998) asserts that maintaining the integrity of science representation involves training the public and the professional community in "information literacy." This article supports that view with a specific focus on *visual literacy*. The visual representation of scientific information relies on the credibility of the source, the degree to which the image is comprehensible, and the degree to which the image is an accurate reflection of the scientific principle or data. Visual literacy is a critical part of "information literacy."

The conceptual structure for visual literacy in this context includes three operational constructs: visual thinking, visual learning, and visual communication. *Visual thinking* involves the incorporation of visual images as part of conscious and preconscious thought. Visual thinking also involves how we organize mental images using shape, line, color, and composition to make them meaningful (Wileman 1980).

Visual learning is a process of developing visual images for instructional purposes and it is also the process by which we use visual information to

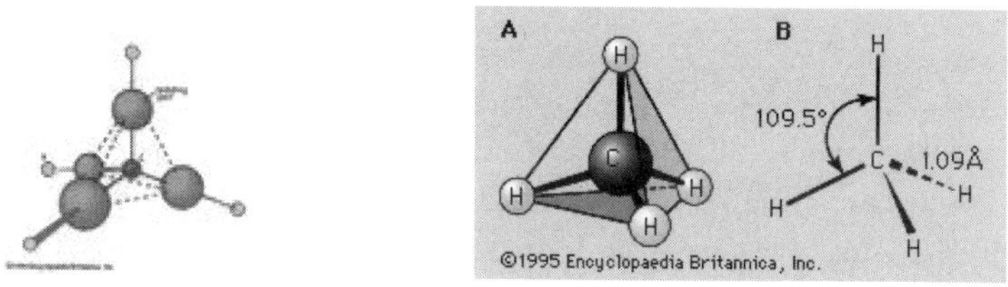

Figure 10.4. Molecular models attempt to show the three-dimensional shape of the molecule. There are a number of styles of three-dimensional representation used in science. In addition, animation programs allow the scientist to visualize potential movement. (Retrieved 11 October 2004 from Encyclopædia Britannica Premium Service, http://www.britannica.com/eb/article?tocId=9052328)

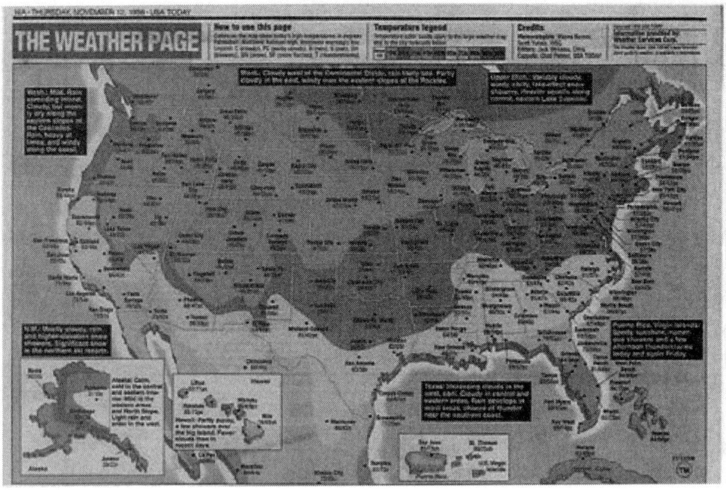

Courtesy of USA Today

Figure 10.5. The weather map that appears in *USA Today* is a simple graphic illustration that depicts temperature changes across the United States. Color is used intuitively to give the reader a quick glimpse of the weather. Warm colors—yellow to red—represent warm weather, and cool colors—blue to green—represent cooler temperatures.

learn (Dwyer 1972, 1978; Randhawa, Bach, and Myers 1977). *Visual communication* involves the use of visual symbols to express ideas and convey meaning to others (Wileman 1980). Each of these aspects will be described in more detail with commentary on how these concepts can inform research and promote the examination of the visual aspects of science communication.

1. Visual Thinking

Leonardo da Vinci considered direct observation and experience to be the gateway to discovery. He called his visualization process *saper vedere,* "knowing how to see" (Boorstin 1992). Science that is in a conceptual stage of development often requires active visual thinking on the part of the scientist. This formative visualization is a creative, intuitive process that incorporates idea generation and question formation. The scientist works in a speculative way to examine possibilities from among the foundation of known scientific principles while synthesizing these possibilities through creative thought. Images may be manipulated in the mind or scratched out on paper, but they are working tools rather than conclusive representations.

Da Vinci's visualization process had two components. The first step was to learn the thing by heart and to create a detailed internalized impression so complete that it need not be reproduced either in the mind or on paper. This process of focused observation secures the subject in the mind. Da Vinci's second step was to allow the *deepest essence* of the thing to be revealed as a universal form that could be combined with other universal forms to create new ideas.

Visual thinking can occur at a number of levels of consciousness. It may be highly conscious, with great, focused mental effort, or it may be barely conscious as in daydreaming. In any case, one of the major uses of graphic language is to communicate the result of visual thinking to other people. Visual thinking is a kind of synthesis or elaboration that allows speculative imagery to be refined again and again as the data become clearer. While the scientist may employ representation that has a high or low level of correspondence to the *real world*, that is not the point. Rather, visual

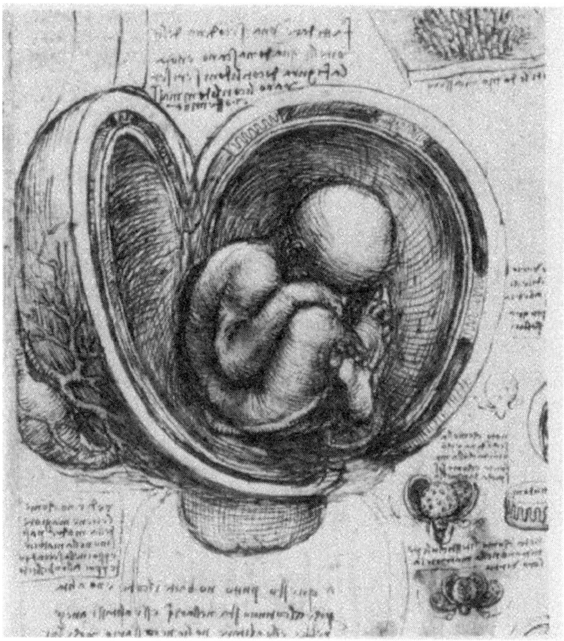

Figure 10.6. From the sketchbooks of Leonardo da Vinci, this drawing of the womb is both an aesthetically beautiful rendering and an example of Leonardo's scientific exploration. The drawing is accompanied by text that Leonardo used to chronicle his observations. Both the text and the sketch are manifestations of his process of visual thinking. (Study of a Womb, c. 1489, http://www.visi.com/~reuteler/leonardo.html)

representation as a product of visual thinking is meant to prompt innovation and exploration. It may result in loose, undeveloped marks in the margins of a lab book or it may be as aesthetically beautiful and intricately developed as the explorations of Leonardo da Vinci.

Design methodologist John Zeisel (1984) describes the process of "imaging" as using available information to form a vague mental picture that is refined in visual representation to become more specific as a problem is resolved. Zeisel suggests that visual representations (images) are deductive constructs or conjectures that parallel the working hypotheses of researchers. As such, these images represent subjective knowledge rather than scientific fact.

Leonardo da Vinci believed that true knowledge comes from synthesizing a variety of points of view into a whole. He referred to his drawings as *dimostrazione* (demonstrations) and believed that they could only be rendered when the subject was understood deeply. His notebooks are filled with detailed scientific exposition drawn from multiple perspectives—below, above, beside. The sheer volume of his work is testament to Leonardo's curiosity about the visible world. Perhaps more importantly, da Vinci had a keen ability to select the appropriate form of visualization. He understood that some explanations were best conveyed through drawings and some were most appropriately recorded using text. Much of his work is a synthesis of written language and graphic diagrams (Boorstin 1992; Mattimore 1994).

A distinction can be made between visual thinking and language thinking. For example, Einstein felt that language had the potential to be a "dangerous source of error and deception." To get beyond the inherent assumptions of language and to keep the concepts that are encoded by words closer to the world of impressions, Einstein used imagery as a way to conduct his *thought experiments* (Mattimore 1994).

There are many notable anecdotes of inspiration prompted through visual thinking in the sciences. One story recounts a dream of chemist Friedrich Kekule, who in 1865 imagined a snake biting its own tail as he struggled to understand how the six carbon and six hydrogen atoms in the chemical benzene were aligned. This insight led him to envision the ring structure of the molecule (Friedhoff 1988).

In searching for an explanation of the atomic structure of elements, Niels Bohr envisioned tiny spheres circling in orbits like miniature planetary systems (Randsepp 1983). James D. Watson describes a complex visual thinking process that allowed him and his colleagues to develop a large three-dimensional model of the DNA molecule revealing its structure—the double helix—for the first time (McKim 1980).

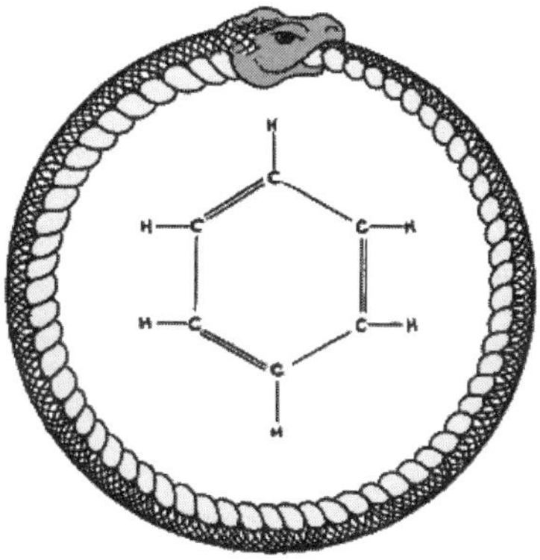

Figure 10.7. Kekule's dream has been visually represented in a number of ways. This simple diagram positions a structural diagram of benzene within a graphic representation of a snake biting its tale. (http://members.ozemail.com.au/~caveman/Creative/Brain/kekule.com)

Loehle (1994) argues that science is not based on a fixed recipe or method, but rather scientific discoveries come from a complex process of pattern-synthesizing and elaboration. These patterns are networks of relationships among facts, assumptions, mathematical relationships, measurement, and observation. Science, says Loehle, is largely about the struggle to convert an intuitively perceived pattern into something definite. This idea is not necessarily at odds with the scientific method. Rather, it acknowledges the importance of visual thinking in the conceptual and interpretive process. It assumes that scientific data, in the conceptual stage, may be rendered in the symbolic language of science or in expressive visual forms of graphic marks on paper that serve as abstract images of association in thought.

Each individual builds knowledge structures or schemata that are personal, internal representations about the nature of the world. As we combine and add schemata, we create new knowledge (Halpern 1996). Thinking is a building process and knowledge is the building. Visual

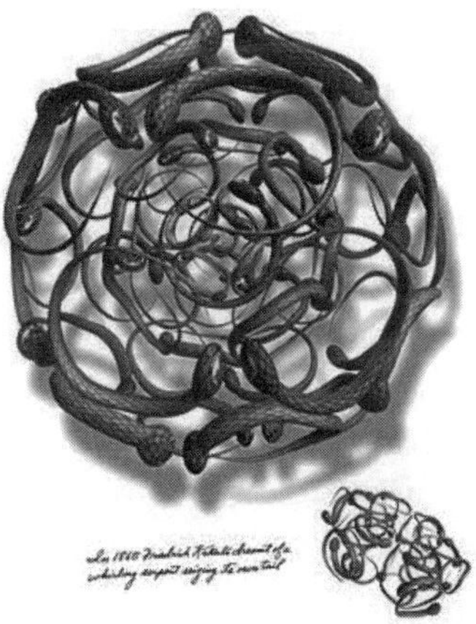

Figure 10.8. The imagery that makes up the story of Kekule's dream has been represented by a number of artists. Kenneth Nelson created this 1996 painting as his own interpretation of Kekule's dream. (http://www.kennethsnelson.net/comp/05.htm)

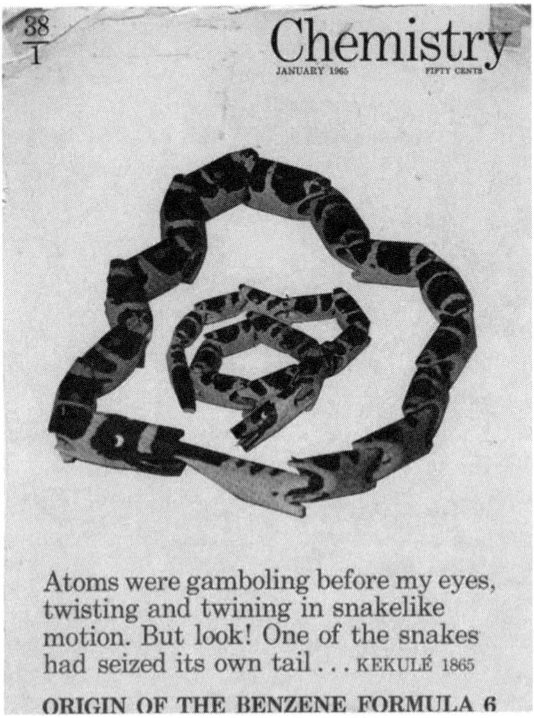

Figure 10.9. In January 1965, the journal *Chemistry* described Friedrich August von Kekule's dream about the structure of benzene. Kekule envisioned a snake biting its own tail and related this to the ring of carbon atoms that make up the organic chemical benzene.

thinking therefore is conceptualized as both a cognitive process and as a set of skills that can be described through a variety of perspectives.

Cognitive scientists and education researchers approach the process of thought in largely objective ways. For example, differences in preferred modes of thinking are described. Education theorist Howard Gardner (1993, 1982) argues that individuals have greater or lesser degrees of "multiple intelligence" and others support the idea that some people prefer spatial over verbal modes of thought (Halpern 1996). Perceptual psychologist Richard Gregory (1974) describes the inner logic of perception in visual problem-solving and believes that it is an ability to see patterns in things that allows a meaningful whole to be created. Gregory and Gombrich's (1973) work in understanding visual illusion has many implications for the creation of effective science visualization. Perhaps one of the most important aspects of illusion is the ease with which intelligent humans can be fooled by complex imagery. The principles of Gestalt perceptual psychology have been applied to visual representation to explain the tendency of individuals to group individual elements into cohesive wholes (Koffka [1935]; 1963; Kohler 1938). Friedhoff suggests that two consistent findings persist in research on visual thinking: the way people think varies among individuals; and the degree to which individuals rely on visual thinking is measurably distributed in the general population. In fact, the bulk of research in this area suggests that some individuals simply think in more visual ways than others (Friedhoff 1988).

2. Visual Learning

There are two components to visual learning: the process of gaining awareness of the meaning of visuals and the process an individual uses to interpret meaning from a visual representation (Couch, Caropreso, and Miller 1994). The dual responsibility of visual representation as a teaching/learning tool and as an integral component in the cognitive process employed by the learner suggests the tremendous importance of visual literacy in the sciences. We use images to learn about science and while we are learning about science we visualize these concepts in images.

Visual learning involves gaining familiarity with the icons and systems of symbols that constitute specialized vocabularies within science disciplines and then interpreting the meaning associated with a particular representation. Latour (1990) takes a controversial approach to the scientific process by advocating a semiotic perspective that regards science as a largely symbolic enterprise. In his studies of laboratory settings, Lynch (1985) observed that lab scientists are obsessed with using representation, diagrams, archives, and abstracts as part of the process of science. He notes that communication with colleagues is only attempted when an effective visual representation is developed. Latour observes that with every scientific dispute, great pains are taken to invent a new instrument or process of visualization that will support the argument of the scientist. In this respect, the tools of visualization shape the scientist's ability to learn and ultimately define how this acquired knowledge is shared with others.

There are a variety of ways that visual representation performs to enhance learning. For example, images illustrate concepts, verify research, solve problems, clarify ideas, assist in theory development, serve as a source for comparison and contrast, correct misconceptions, or summarize a topic (Dwyer 1978). A subtle hierarchy of learning is involved as the scientist, then the science communicator, and ultimately the public, rely on visual images to make information about science accessible. Each link in the chain is a new opportunity for visual representation to hinder or augment learning.

The power of a visual scientific representation in shaping visual learning is evidenced in its ability to dominate the written word when images and text appear together. Some argue that written language must be processed cognitively while the image is processed along the same perceptual pathways as direct experience. We react emotionally to the image before it is cognitively understood (Barry 1997). This factor is important in scientific visualization where the sensory potential of a representation is engaging on a variety of perceptual levels. For example, computer graphic animation makes it possible to create models of substances or processes that have three-dimensional qualities, movement, color, as well as a potentially high level of realism. Yet, their correspondence to the actual science behind the representation may be strictly conceptual. Learning to interpret these images is essential for both the scientist and science communicator. As perceptual psychologist R. L. Gregory (1989) notes,

> Like language, pictures can project into the past or imagined future and create new or even impossible worlds. But they work only for observers with knowledge and intelligence to create meaning from the pictures, as one creates meaning from the words of a language. So the key to understanding the power of pictures, or any symbols, lies in the human brain or the mind.

In its purest sense, visual learning occurs as the scientist attends to the artefacts of exploration: the experimental data or the process itself, the plant or animal specimen, the microscopic image, or the chemical reaction. Visual learning is a process of synthesizing what the scientist *knows* (expert knowledge) with what the scientist *perceives* (conceptualization or perception). The internal image in the *mind's eye* is combined seamlessly with external images—scientific notation, diagrams, photographs, sketches, or computer models. The expert uses visual representation to advance science and to enrich the process of discovery.

Visual learning for the non-expert involves competencies that are somewhat distinct from those of the expert. The non-expert may have only limited access to the specialized notation and vocabulary within a scientific discipline and may not be privy to the constraints of various forms of visual representation. For example, sense of scale or distance may not be comprehensible to an average individual looking through a telescope at the Orion nebula. The subjective nature of an artist's expressive illustration of DNA sequencing may not be understood easily by those unaware of the science. An animated, three-dimensional computer model of a new virus may use color and movement to enhance the structural features of the organism when, in fact, the organism's exact color is not apparent and its movement is only known through speculation. These representations indeed may provide attractive, attention-getting visual illustrations for the audience, but they are also merely interpretations. The correspondence to the science information behind the interpretation may not be clear to the non-expert.

Visual learning in the public arena involves educating observers on the conventions of visual representation, the nature of the medium through which it is presented, as well as the science itself. The science communicator is a mediator who must make careful choices about the form of visual representation in an effort to share the science in accurate, articulate ways. This can be a difficult undertaking.

Visual literacy sometimes is equated with language literacy and the subsequent assumption is that learning about or through visual images is a similar process to learning a language. McKim (1980) notes it is common to move from one graphic language to another while visual thinkers apply the built-in mental operations of each language. In this respect, visual thinkers use graphic languages to expand the range of their thinking. Dondis (1973), on the other hand, concludes that visual literacy could never be a clear-cut, logical system similar to language. Languages are made-up systems, constructed by humans to encode, store, and decode information. According to Dondis, language structure has a logic that visual literacy is unable to parallel. Messaris (1994, 1998) also argues against thinking of visual images as language. He suggests that learning to understand images does not require the lengthy initiation or education required to learn verbal language. Pictures, says Messaris, make sense to inexperienced viewers despite the discrepancies between image and reality. Visual literacy, in Messaris's view, regards images as sources of aesthetic delight, instruments of potential manipulation, and conveyors of some kinds of information. Visual literacy from this point of view is simply a valuable complement to verbal language.

Barry (1997) distinguishes between the linear, logical structure of verbal language and the holistic, amalgamated, synthetic, dynamic, and open structure of visual language. For example, written equations and visual diagrams are created in accordance with the formal vocabulary and grammar of the discipline. Further, the function of visual representation in each of these disciplines is directed toward recording, communicating, and processing information. The communication potential of both forms of expression—visual and verbal—may involve the same subject, but the languages are distinct. Language, visual or verbal, is key to making science communication possible.

3. Visual Communication

> We speak (and hear)—and for 5,000 years have preserved our words. But, we cannot share vision. To this oversight of evolution we owe the retardation of visual communication compared to language. Visualization by shared communication would be much easier if each of us had a CRT in the forehead.
>
> —B. H. McCormick, T. A. DeFanti, and M. D. Brown 1987

Visual representation can express scientific principles, experimental data, or discoveries by augmenting text to convey meaning or to clarify ideas. As argued previously, effective visual representation is an important tool in the communication of science for both expert and public audiences. However, visual representation does provide a challenge because it often functions as one-way communication. Most examples *received* by an audience do not provide for a response. We see an illustration or diagram, we attend to the animated model within a Web site, but there is typically no dialogue with the scientist or producer about the visual decisions that went in to the representation. Without some understanding of how visual representation functions, we are left out of an important level of discourse. We receive messages but have no parity in response; as such we are visually illiterate (Saunders 1994).

The communication of scientific thought historically has been conveyed through the printed and spoken word with visual images performing a strictly illustrative function. That is, they are used to amplify or fill in details of processes that are articulated in the text. Jay Ruby (1975) suggests two possible reasons that visual images have acquired such limited function. First, there may be inherent limitations in images that curtail their communicative value. Another possibility is that the perceived limitations of images are part of culturally derived attitudes. Ruby argues that while humans have examined the nature of spoken/written communication for thousands of years, the technologies to produce images are recent and the scientific examination of their communication potential is still in its infancy.

Another possibility can be argued. Visual literacy is not a priority within the traditional education system. Seels (1994) suggests that visual literacy in education has gained attention only as concern grows about the influence of television and more recently video games on behavior and learning in children. In fact, visual literacy and scientific or mathematical literacy often are neglected. An article in the *Chronicle of Higher Educa-*

tion argued that universities need to do a better job of training scientists to explain their work verbally and graphically while also training science communicators or journalists to become proficient in basic sciences (Chappell and Hartz 1998).

According to the chapter on public understanding in the *Science and Engineering Indicators Report* (National Science Foundation 2000), only 13 percent of Americans can explain what a molecule is. Only one in three Americans can provide a minimally acceptable definition of DNA despite the public discussion of genetics, cloning, and forensic blood evidence in widely publicized court trials and on dramatic television shows, although this is up slightly from 1998. Researchers conclude that only 21 percent of Americans understand the nature of scientific inquiry well enough to make informed judgements about the scientific basis of results reported in the media. While the researchers point out that the level of understanding of basic scientific terms and concepts is associated closely with formal education, they also emphasize that Americans receive most of their information about public policy issues from television news programs and newspapers. In 1997, Americans watched an average of 432 hours of television news programs and read an average of about 200 newspapers. During this same period, Americans watched about 70 hours of televised science programming (National Science Foundation 1998). It seems logical to conclude that visual literacy and scientific literacy are both important prerequisites of effective science communication.

Communication is a process of sharing information with others; it is also a means of sending and receiving messages through a variety of channels. The definition of what constitutes communication varies according to the theoretical perspective that is used. Watson and Hill (1993) suggest that there are five fundamental factors (elements) in the process of communication regardless of the theoretical framework. These elements involve an initiator; a recipient; a mode or vehicle of communication; a message; and an effect. The communication process begins when a message is conceived by a sender, encoded into a sequence of signals, transmitted via a particular medium or channel to a receiver who decodes it (Littlejohn 1996).

Visual communication may involve how a message is encoded and it may also dictate the vehicle within which it functions. Lester (1995) defines visual communication as an optically stimulating message that is understood by the receiver. The goal, says Lester, is to produce powerful images that enable the viewer to understand and remember their content. Berger (1998) notes that much of the information we encounter is of a visual nature and it is important that everyone know something about how images function and how people learn to *read* or interpret images. Studying the effects of visual communication is both a diverse theoretical enterprise and a methodologically difficult issue. Yet, it is of great importance, particularly in the communication of science.

There is a diverse stylistic and technical range of visual representation in science. Photographs, diagrams, symbolic notation, and computer visualization are among the choices from which the scientist and science communicator select in an effort to *show* rather than *tell* the science. Each form of representation carries its own conventions and potential for interpretation or misinterpretation. It is crucial that the appropriate form of representation is selected. Yet, it is often the availability of a technology that dictates how the science is rendered. If animation software is accessible, it is very difficult to prevent it from being used even when a static form of representation is more understandable.

4. Conclusion

Contemporary science communication relies on visual representation to clarify data, to illustrate concepts, and to engage a public informed through an ever-increasing arsenal of computer graphics and new media tools. Examples of visual representation in science communication are abundant, but relatively little attention has been directed toward the challenge of building visual literacy among scientists, communicators, and the public. This article offers an introduction to the complexity of achieving visual literacy in science communication by attending to three dimensions: visual learning, visual thinking, and visual communication.

The challenge of achieving visual literacy among scientists, communicators, and the public has been acknowledged across disciplines as diverse as the arts, computer science, cognitive psychology, communication, engineering, and the life and physical sciences. It is important to recognize that while the tools of data visualization and visual representation are evolving quickly, the potential for visual representation of science information to carry meaning or to be understood in an accurate way by an audience is an important issue for researchers to examine. Communication researchers can contribute to the effectiveness of such efforts by turning a critical eye toward the function, purpose, and effects of visual representation in the sciences.

References

Barry, A. M. S. 1997. *Visual intelligence: Perception, image, and manipulation in visual communication.* Albany: SUNY Press.

Berger, A. A. 1998. *Seeing is believing: An introduction to visual communication.* Mountain View, Calif.: Mayfield Publishing.

Bloom, B. S., M. D. Engelhart, E. J. Furst, W. H. Hill, and D. R Krathwohl, eds. 1956. *Taxonomy of educational objectives: The classification of educational goals. Handbook I: Cognitive domain.* New York: David McKay.

Boorstin, D. J. 1992. *The creators: A history of heroes of the imagination.* New York: Random House.

Chappell, C. R., and J. Hartz. 1998. The challenge of communicating science to the public. *Chronicle of Higher Education,* March 20, B7.

Couch, R. A., E. J., Caropreso, and H. B. Miller. 1994. Making meaning from visuals: Creative thinking and interpretation of visual information. In *Visual literacy: A spectrum of visual learning* by D. M. Moore and F. M. Dwyer. Englewood Cliffs, N.J.: Educational Technology Publications.

Devlin, Keith. 1998. *Life by the numbers.* New York: John Wiley & Sons.

Dondis, D. A. 1973. *A primer of visual literacy.* Cambridge: MIT Press.

Dwyer, F. M. 1972. *A guide for improving visualized instruction.* State College, Penn.: Learning Services.

———. 1978. *Strategies for improving visual learning.* State College, Penn.: Learning Services.

Ford, B. J. 1992. *Images of science: A history of scientific illustration.* London: The British Library.

Friedhoff, R. M., and W. Benzon. 1988. *Visualization: The second computer revolution.* New York: W. H. Freeman and Company.

Friedman, S. M., S. Dunwoody, and C. L. Rogers. 1986. *Scientists and journalists: Reporting science as news.* American Association for the Advancement of Sciences, Washington, D.C.

Gardner, H. 1982. *Art, mind, and brain: A cognitive approach to creativity.* New York: Basic Books.

———. 1993. *Creating minds.* New York: Harper Collins.

Gershon, N., S. G. Eick, and S. Card. 1998. Information visualization. *Interactions* 5 (2) (March/April): 9–15.

Goodfield, J. 1981. *Reflections on science and the media.* Washington, D.C.: American Association for the Advancement of Science.

———. 1989. Foreword. In *Visualization: The second computer revolution,* edited by R. M. Friedhoff and W. Benzon. New York: W. H. Freeman and Company.

Gregory, R. L. 1974. *Concepts and mechanisms of perception.* London: Duckworth.

Gregory, R. L., and E. H. Gombrich. 1973. *Illusion in nature and art.* London: Gerald Duckworth & Co.

Halpern, D. F. 1996. *Thought and knowledge: An introduction to critical thinking.* Mahwah, N.J.: Lawrence Erlbaum Associates.

Hoffmann, R. 1995. *The same and not the same.* New York: Columbia University Press.

Koffka, K. [1935] 1963. *Principles of gestalt psychology.* New York: Harcourt, Brace & World.

Kohler, W. 1938. Some gestalt problems. In *A source book of gestalt psychology,* edited by Willis D. Ellis. London: Routledge & Kegan Paul.

Latour, B. 1990. Drawing things together. In *Representation in scientific practice,* edited by Michael L. Lynch and Steve Woolgar. Cambridge: MIT Press.

Lynch, M. 1985. *Art and artifact in laboratory science.* London: Routledge & Kegan Paul.

Lester, P. M. 1995. *Visual communication: Images with messages.* Belmont, Calif: Wadsworth Publishing Company.

Littlejohn, S.W. 1996. *Theories of human communication.* Belmont, Calif.: Wadsworth Publishing Company.

Loehle, C. 1994. Discovery as a process. *Journal of Creative Behavior* 28 (4, 4th Quarter): 239–250.

Marzano, R. J., R. S. Brandt, C. S. Hughes, B. F. Jones, B. Z. Presseisen, S. C. Rankin, and C. Suhor. 1988. *Dimensions of thinking: A framework for curriculum and instruction.* Alexandria, Va.: Association for Supervision and Curriculum Development.

Massironi, M. 2002. *The psychology of graphic images: Seeing, drawing, communicating.* Mahway, N.J.: Lawrence Erlbaum Associates.

Mattimore, B. 1994. *99% Inspiration.* New York: AMACON.

McCormick, B. H., T. A. DeFanti, and M. D. Brown. 1987. Visualization in scientific computing. *Computer Graphics* 21 (6) (November).

McKim, R. H. 1980. *Thinking visually: A strategy manual for problem solving.* Belmont, Calif.: Wadsworth, Inc.

Messaris, P. 1994. *Visual literacy: Image, mind, and reality.* Boulder, Colo.: Westview Press.

———. 1998. Visual aspects of media literacy. *Journal of Communication* 48 (1) (Winter): 70–80.

Meyer, E. 1997. *Designing infographics: Theory, creative techniques and practical solutions.* Indianapolis: Hayden Press.

Moriarty, S. E. 1997. A conceptual map of visual communication. *Journal of Visual Literacy* 17 (2) (Fall): 9–24.

Moriarty, S. E., and K. Kenney. 1995. Visual communication: A taxonomy and bibliography. *Journal of Visual Literacy* 15 (2) (Autumn): 7–156.

National Science Foundation. 1998. *Science and Engineering Indicators Report,* chapter 7, Science and Technology: Public Attitudes and Public Understanding. Washington, D.C.: National Science Board. http://www.nsf.gov/sbe/srs/seind98/access/toc.htm.

———. 2000. *Science and Engineering Indicators Report,* chapter 8, Science and Technology: Public Attitudes and Public Understanding. Washington, D.C.: National Science Board. http://www.nsf.gov/sbe/srs/seindoo/access/c8/c8s1.htm#c8s114.

Nelkin, D. 1995. *Selling science: How the press covers science and technology.* Revised edition. New York: W. H. Freeman & Company.

Pickover, C. A., and S. K. Tewksbury. 1994. *Frontiers of scientific visualization.* New York: John Wiley & Sons, Inc.

Priest, S. H. 1998. Public opinion, expert opinion, and the illusion of consensus: Gleaning points of view electronically. In *The electronic grapevine: Rumor, reputation, and reporting in the new on-line environment,* edited by D. L. Borden and K. Harvey. Mahwah, N.J.: Lawrence Erlbaum and Associates.

Randhawa, B. S. 1978. Visual trinity: An overview. In *Visual learning, thinking, and communication,* edited by B. S. Randhawa and W. E. Goffman. New York: Academic Press.

Randhawa, B. S., K. T. Bach, and P. J. Myers. 1977. Visual learning. Paper presented at the annual meeting of the Association for Educational Communications and Technology, April, Miami Beach, Florida: (ERIC Document Reproduction Service No. ED 143319).

Randsepp, E. 1983. Profile of the creative individual, part 2. *Creative Computing* (October): 201.

Robin, H. 1992. *The scientific image: From cave to computer*. New York: Harry Abrams, Inc.

Ruby, J. 1975. Is ethnographic film a film ethnography? *Visual Communication* 2 (2) http://www.temple.edu/anthro/ruby/is.html.

Saunders, A. C. 1994. Graphics and how they communicate. In *Visual literacy: A spectrum of visual learning,* edited by D. M. Moore and F. M. Dwyer. Englewood Cliffs, N.J.: Educational Technology Publications.

Schur, A., K. A. Keating, D. A. Payne, T. Valdez, K. R. Yates, and J. D. Myers. 1998. Collaborative suites for experiment-oriented scientific research. *Interactions* 5(3) (May/June): 40–47.

Seels, B. 1994. Visual literacy: The definition problem. In *Visual literacy: A spectrum of visual learning,* edited by D. M. Moore and F. M. Dwyer. Englewood Cliffs, N.J.: Educational Technology Publications.

Sprott, J. C., and C. A. Pickover. 1997. Scientific visualization in mathematics and physics. *Interactions* 4 (1) (January/February): 78–79.

Trumbo, J. 1999. Visual literacy and science communication. *Science Communication* 20 (4) (June): 409–425.

———. 2000. Seeing science: Research opportunities in the visual communication of science. *Science Communication* 21 (4) (June): 379—391.

Tufte, E. 1983. *The visual display of quantitative information.* Cheshire, Conn.: Graphics Press.

———. 1997. *Visual explanations*. Cheshire, Conn.: Graphics Press.

———. 1990. *Envisioning information.* Cheshire, Conn.: Graphics Press.

Wainer, H. 1997. *Visual revelations*. New York: Copernicus Press.

Watson, J., and A. Hill. 1993. *A dictionary of communication and media studies*. London: Edward Arnold.

Wileman, R. E. 1980. *Exercises in visual thinking*. New York: Hastings House.

Wurman, R. S. 1989. *Information anxiety*. New York: Doubleday.

Zeisel, J. 1984. *Inquiry by design: Tools for environment-behavior research*. Cambridge: Cambridge University Press.

CONTRIBUTORS

Massimiano Bucchi (Ph.D. Social and Political Science, European University Institute, Florence 1997), teaches Sociology of Science at the University of Trento, Italy; he has carried out research and given seminars at several universities, such as Sussex University, Universität Bielefeld, ETH Zurich, University of Wisconsin, University of California Berkeley, Royal Swedish Academy of Sciences, University of Edinburgh, University of Tokyo, and Museu da Vida Rio de Janeiro. He has received several awards and recognitions, including the RAI television prize for research in mass communications (1996), the Mullins award for the best Sociology of Science graduate paper (1997), and the Lelli prize for the best Sociology dissertation discussed in Italy (1998). His main fields of interest include public communication of science, representation of science and technology in the mass media, and the role of citizens in the governance of innovation. He has published articles in journals such as *Public Understanding of Science, Nature, History and Philosophy of the Life Sciences,* and *The British Journal of History of Science,* and six books, including *Science and the Media* (London and New York: Routledge, 1998) and *Science in Society: An Introduction to Social Studies of Science* (London and New York: Routledge, 2004).

Alberto Cambrosio is an Associate Professor in the Department of Social Studies of Medicine at McGill University. He holds a Ph.D. degree in the History and Sociology of Science from the University of Montreal. His work focuses on the sociology of biomedical practices and innovation, with a special interest in the material culture of biomedicine. He co-authored (with Peter Keating) *Exquisite Specificity: The Monoclonal Antibody Revolution* (New York: Oxford University Press, 1995) and co-edited (with Margaret Lock and Allan Young) *Living and Working with New Biomedical Technologies: Intersections of Inquiry* (Cambridge: Cambridge University Press, 2000). He has published numerous articles examining the role of tools, instruments, and devices in the development of laboratory and clinical practices, the role of visual imagery in immunological research, and the production of facts during public controversies in health-related fields.

John Grady is the Hannah Goldberg Professor of Sociology at Wheaton College in Norton, Massachusetts, as well as a Visiting Scholar at the Salt Institute for Documentary Studies in Portland, Maine. He received his Ph.D. from Brandeis University, where he studied under Egon Bittner, Everett Hughes, and Charles Fisher. His interests include urban sociology; technology and the built environment; and comparative social organization and the institutional order. He offers several courses on visual sociology, including a production course on sociological moviemaking. He has produced numerous documentary films, including *Mission Hill and the Miracle of Boston* (1979), *Water and the Dream of the Engineers* (1983), and *Love Stories: Women, Men and Romance* (1987). He has written numerous articles on visual social science, including "The Visual Essay and Sociology" and "The Scope of Visual Sociology," which have appeared in the journal *Visual Sociology*. He is a Past President of the International Visual Sociology Association.

Daniel Jacobi is Professor in the Department of Information and Communication Sciences at University of Avignon (France). He graduated in Life Sciences and obtained a State Doctorate in Sociolinguistics. His work focuses on the popularization of life and medicine sciences. He has published many papers examining the linguistic forms of scientific terms formulation and also their figurability. Author of *La Communication Scientifique; Discours, Figures, Modèles* (Grenoble: Presses Universitaires de Grenoble, 1999).

Peter Keating is an Associate Professor in the History Department of the University of Quebec at Montreal. He has a Ph.D. in the History and Sociopolitics of Science from the University of Montreal and specializes in the history of modern biomedicine. He has maintained a special interest in the history of immunology and immunological disorders as well as in the role of techniques and instruments in the development of biomedical research and its modern institutions. Author of *La Science du Mal: L'institution de*

La Psychiatrie au Québec, 1800–1914 (Montréal: Boréal, 1993), he has also co-authored (with Alberto Cambrosio) *Exquisite Specificity: The Monoclonal Antibody Revolution* (New York: Oxford University Press, 1995) and (with Yves Gingras and Camille Limoges) *Du Scribe au Savant: Les Porteurs du Savoir de l'Antiquité à la Révolution Industrielle* (Paris: Presses Universitaires de France, 2000).

Michael Lynch is a Professor in the Department of Science and Technology Studies at Cornell University. He is a sociologist by training, with a specialized interest in ethnomethodology, the study of the social organization of situated discourse and embodied practice. His research examines day-to-day practices in scientific laboratories, courts of law, and other organizational settings. He focuses closely on the relationship between social interaction and the composition of material (and visual) evidence. His publications include *Art and Artifact in Laboratory Science* (Routledge & Kegan Paul, 1985), *Scientific Practice and Ordinary Action* (Cambridge University Press, 1993), and *The Spectacle of History: Speech, Text and Memory at the Iran Contra Hearings* (with David Bogen) (Durham, N.C.: Duke University Press, 1996). His current research investigates the intersection between science and law in controversies about forensic uses of DNA profiling.

Francesco Panese is Assistant Professor in the University of Lausanne (Switzerland), where he teaches Sociology of Science and Technology, and Director of the Claude Verdan Foundation (Lausanne), a museum for scientific culture, where he organizes exhibitions. He has been visiting scholar at the universities of Bologna and Berkeley. His research interests include the uses of visualizations in the production of scientific knowledge and the social construction of the body as a scientific and medical object. His current research with Alain Kaufmann investigates the uses of the human genome maps. His publications include: "Reading the Will of the Gods: Inquiries on Graphical Divination" (*Ecumene,* special issue); *Spaces of Truth* (forthcoming); "Rationalisation Scientifique et Images du Merveilleux: Brève Enquête sur la Photographie Expérimentale des "Esprits" au Tournant du Siècle" (*Traverse. Revue d'histoire,* 1999); "Disciplines Scientifiques et Disciplines du Regard au Xviie Siècle" (Réseaux, CNET, Paris, 94, 1999); "Entre Connaissance et Émotion: Un Épisode de L'histoire du 'Sentiment De La Nature' Des Lumières Au Romantisme"

(*Traverse. Revue d'histoire*, 1997/2); "La Part Maudite de L'iconographie Scientifique" (*EspacesTemps*, nos. 62–63, 1996); "Transferts de Modèles Entre les Sciences de la Vie et L'informatique: Le Cas des Algorithmes Génétiques. Une Approche Sociologique," *Computer Science, Communications and Society: A Technical and Cultural Challenge* (Lausanne, 1993) (with Alain Kaufmann); "L'effet Paysager," *Paysage et crise de la lisibilité* (Lausanne, 1992) (with Lorenza Mondada and Ola Söderström).

Bernike Pasveer is Assistant Professor at the Faculty of Arts and Culture of the University of Maastricht, the Netherlands. She is a sociologist by training. Her research focuses on technologies of/for the body: X-ray images, technologies of obstetrics, reproductive technologies. Her current interest is in elite sport and technology. Her research has been conducted at the University of Amsterdam, the Centre de Sociologie de l'Innovation (CSI) in Paris, and the University of Maastricht. She has published in national and international journals (*Sociology of Health and Illness, Theoretical Medicine* (with M. Akrich), *Kennis & Methode/Krisis, Traverses, Techniques et Culture*), in edited books (*Birth By Design*, London: Routledge, 2001, with M. Akrich (edited by R. DeVries, C. Benoit, E. R. van Teijlingen, and S. Wrede), and the books *Comment la Naissance Vient Aux Femmes* (with M. Akrich, Paris: Les Empêcheurs du Penser en Rond, 1996) and *De Voortplanting Verdeeld* [Reproduction Divided] (with S. Heesterbeek, Den Haag: Rathenau Instituut, 2001).

Luc Pauwels is Professor and Chair of the Department of Communication Sciences at the University of Antwerp, Belgium, as well as Program Director of the "Master in Film Studies and Image Culture." He studied sociology, communication sciences, and philosophy at the universities of Antwerp and Leuven, holds a Ph.D. in Social and Cultural Sciences from the Free University of Amsterdam, and currently teaches visual communication, image theory, and aesthetics of film, scientific and documentary film theory, and visual culture at the undergraduate, graduate, and postgraduate level. His research focuses on methodologies of visual research, nonfiction visual cultures, and the social impact of new media technologies. Previous publications include *De Verbeelde Samenleving* (Imaging Society) (1996), *Visual Cultures and Visual Literacies*, a double issue of *Visual Sociology* (now a Routledge journal called *Visual Studies*). He also has published his work in several readers and in international journals (including *Visual Sociology/Visual Studies, New Cinemas, Visual Anthropology,* and the *Journal of Visual Literacy*). In 1999, he organized the international conference "Visual Cultures and Visual Literacies: Changing Ways of Imaging Science and Society," on behalf of the International Visual Sociology Association (IVSA) (www.visualsociology.org).

Jean Trumbo is currently an Associate Professor in the School of Journalism at the University of Missouri, Columbia. She previously served as an assistant professor at the University of Wisconsin, Madison, visiting professor at Cornell University, and as an assistant professor at the University of Nevada, Reno. Trumbo recently served as Product Development Director for an Internet publishing company, Illustrated Ideas, Inc., where she directed a team of individuals in the creation of science- and health-related multimedia. She has fifteen years of professional experience as an art director and designer. Trumbo currently teaches visual communication and new media with an emphasis on the role of technology in the communication of science information. Her research is directed toward visual literacy in computer-mediated communication, the role of new media technologies in communication, and science visualization. She earned a B.A. in Visual Design and an M.S. in journalism from Iowa State University and a Master of Fine Arts in Visual Design from the University of Massachusetts at Dartmouth.

INDEX

Page numbers in **bold** face indicate pages of particular significance to the discussion of a term.

abstraction, 35
accompanying tables, by Dally, 51
algorithmic processes, 8, 10
algorithmic representations, 18
algorithmic visual images, 14
anatomical body, 46, 47, 49, 59n5
anatomical drawings, 11, 81
anatomization, 45
anatomy, 8, 10, 44, 45, 46, 49, 52, 54, 55, 56, 60n6, 87n14, 209, 215; anatomy atlas, 45; graphic anatomy, 210; human anatomy, 11; pathological anatomy, 44, 46; visible anatomy, 201; X-ray anatomy, 49, 54
Anschauung, 107
Anschauungsmaterial, 107
Anschauungsmittel, 102, 117n11
Anschauungsunterricht, 99, 107
anthropological fieldwork, 121
anthropological research, 120
anthropology, 123, 128, 131, 135, 137; of signifier and signified, 60. *See also* visual anthropology
antibodies, 153–190
antibody formation, theory of (template or instructionist), xv, 153–54, 158, 168, 173, 174, 180
antibody theory, 158
antigen, 153–90
artefact (artifact), 9, 32, 33, 38, 258, 276; cultural, 143; material, 28; pictorial, 28; staining, 207; technical, 11; textual, 30; visual, 2
artefact (artifact) of instrumentation, 9, 34
auscultation, xiii, 51, 54, 57
automated visual image, 14

bar codes, 36, 37
biochemistry, 186; protein, 156; structural, 155
biological shapes, Ehrlich's, 189
biologist(s), 94, 96, 170, 173, 267
biology, ix, xiv, xv, 103, 107, 154, 170; cell biology, 216n2; cellular biology, 155; molecular biology, 28, 154, 155, 182; teaching of, 94, 97, 107, 115; textbook, 17
biomedicine, 155
botanic systematization, project of, 86n10
botanical: atlas, 108; knowledge, 108; scholarship, 90; teaching, 108; wall charts, 97, 106
botany, 64, 90, 94, 109
boundary object: definition, 21; map as boundary object, 21

Cartesian separation/division, 207
CAT (Computed Axial Tomography) scans, 266
chain(s) of custody, 36, 38
chart(s), 2, 18, 139, 200, 201, 214, 217n12, 225, 236, 237, 238, 240, 254, 255, 256, 258, 267; bar chart, 232, 241; kinship chart, 248; line chart, 241; pie chart, 232, 237; recordings, 28; surface chart, 232
chartjunk, 232, 249, 254
chemistry, 79, 87n14, 168, 170,

chemistry *(continued)*
192n22, 268; laboratory experiments in, 94; Nobel prize for, 153; protein, 155; structural, 155, 188–89
chemists, 267, 268, 272
cinema: nonfiction, 120; observational, 127
clonal selection theory, 155–56
coagulation, 157
cognitive psychology, 280
cognitive scientists, 274
collaboration, xiv, xv, xviii, 6, 157, 173, 249
comic strip, 166, 171, 172
comics, 258, 259
communication, 280; researchers, 280; scholars, 267
complementariness, 153–55, 167, 171, 172, 180, 186, 190, 191n1. *See also* molecular complementariness
complementarity, 191n1. *See also* complementariness
computer science, 280
computer simulations, 139
computer visualization, 280
conceptual constructions, 7, 138
conceptualization, x, 60n11, 276; verbal, 208
confection(s), 252, 253, 255
confectionary diagram, 256, 258
constructivist arguments, 32, 34
constructivist literature, 31
constructivists, 34. *See also* social constructivists
contextualization: (image-)external, 138, 141; internal, 138
continuity, 36
conversation analysis, ix
CPK (Corey-Pauling-Koltun) model(s), 178, 181
critical theory, 224
CT (Computed Tomography), 16
Cubism, 12
cultivation, 35
cultural object, 215n1
cultural repertoire, xii
cultural self-portrayal, 133, 147. *See also* native image production
cultural studies, 224
curiosity, culture of, 77, 86n7

data/ink maximization, 237. *See also* smallest effective difference, principle of
data/ink ratio, 237–38, 244. *See also* smallest effective difference, principle of
de-narrativization, 71
denaturation, 157
depiction, xiii, xv, 3, 8, 13, 14, 36
depiction process, xiii
description(s), xii, 27, 29, 30, 33, 64, 67, 71, 72, 73, 75, 81, 86n5, 94, 98, 108, 121, 125, 172, 183; textual, 72; verbal, vii, 75, 218n22; visual, 72
design methodologist, 272
design strategies, 237
diagram(s), ix, 16, 26, 27, 29, 90, 139, 142, 158, 171, 172, 173, 174, 175, 180, 190, 195, 197, 199, 201, 208, 214, 217n8, 233, 236, 243, 248, 252, 255, 266, 267, 275, 276, 277, 278, 280; circle diagram, 232; conceptual, 30; graphic, 272; mechanical, 81; structural, 269, 273; visual, 277
didactic communication , 91
digital base formats, 139
digital image, 28
digital media, 149; technology, 139
digital photo restoration, 36
digitized images, 139
directly observable phenomena, 7, 8, 9
DNA (Deoxyribonucleic Acid), 36, 37, 276, 279; molecule, 272; profiling techniques, 31
docile bodies, 217n12
docile object, 6, 201, 217n12, 218n18
docile record, 217n12
documentaries, 122, 258; drama documentaries, 123

domesticated, 72, 81
domesticating: natural objects, xvi; project of, 215
domestication, 85; of perspective, 79, 85
Doppler radar, 266

education researchers, 274
educational theory, 91
electrochemical events, 208
electron micrographs, 202, 210, 215
electron microscopy, xvi, 33, 196, 207
embodied actions, 196
embodied "modalities" of perception, 216
empirical science, 107
empiricists, 260
encoded by nature, 43, 58
encoding, 16; anatomical, 51; practice of, 58; strategies of, 44; the new images, 55; work of, 59
enframing, 34, 35
engineering, 280; students of, 11
engineers, ix, 233, 243, 268
ethnocinematography, 142
ethnocommunication, 142
ethnography, ix, 135, 197
ethnomethodological perspective, 6
ethnomethodological studies, 217n12, 218n14
ethnomethodology, ix
experiment(s), ix, 26, 33, 34, 63, 85, 172, 197, 199, 202; chemical, 106; hydrodynamic (ultracentrifuge), 176; of physics and chemistry, 79; philosophical, 80; ultracentrifugation and electrophoresis, 157
experimental approach, 79, 80
experimental method, Boyle's notion of, 86n4
experimental point of view, 80
experimental practices, 27
experimentalism (English tradition of), 67, 77, 80, 87n18
experimentalists' perspective, 80
experimentation, 121, 197
expression, ix, x, xiv, 3, 19, 60n10, 75, **120–22**, 126, 187, 267, 277; chemical, 106
expressiveness, **124**, 145

familial images, 133
feminism, 224
figuration, 156, 187–88, 189, 190
figurativization, 192n23. *See also* figuration
film elicitation, 131
film(making): anthropological, xiv, 6, 120–21, 122, 128, 139, 140, 141, 144–46; educational, 122, 123; ethnographic, 127, 128, 144; postmodern, 127, 146; scientific, 7, 121, 122, 123, 124, 125, 127, 139, 140, 146; social scientific, 121; sociological, xiv, 120–21, 127, 140
forensic science, 266

Galenic body, 45
geographers, 267
geography, 51, 92, 186
geology, ix, 29, 217n11
geometry, 157, 186, 209, 217n9; molecular, 156
Gestalt perceptual psychology, 274
Gestalt psychologists, 13
graphic: competence, 224; conventions, 225; design, 222, 228, 232, 236, 239, 243; diagrams, 272; display(s), 197, 213, 214, 219n24, 224, 232, 255; language, 268, 271, 277; novel, 258; representation, xii, 12, 167, 197, 223, 273
graph(s), vii, ix, x, 26, 27, 28, 30, 31, 35, 37, 195, 196, 199, 200, 201, 202, 204, 205, 209, 210, 211, 212, 213, 214, 217n11, 225, 230, 235, 255; conceptual, 18; historical origins of, 217n11; mathematical, vii; Minard's, 240

Hamburger Theses, 107
histological artefact (artifact), 216n2
histology, 214
historian of education, 106

historians of radiology, 43
historians of science, 30
history of science, ix, 26, 28
human science disciplines, 196
human sciences, 216n4
hybrid: nature of the sciences, 79; product, 139; representational space, 170; semiotic code, 180
hybridity, x
hyperrealism, 12

iconoclasm, 85
iconographic elements, 71, 189
iconography, xiii, 63, 67, 68, 77, 171
ideal-type images, 80
ideal-type representations, 80
ideology of representation, 33
illustrating, 27, 28, 122
illustration(s): scientific 11, 190, 196, 267; zoological, 81
image-based research, 224
imagery, 135, 176, 179, 271, 273; handmade, 8; machine-generated, 8; Pauling's, 176, 190; scientific, 156; structural-chemical, 182; technological, 28; textual, 187; transfers of, 208; visual, 28, 154, 187
images: of science, xiv, 90, 266; scientific, xii, 26, 29, 43, 156, 266; visual, 30, 37, 275, 277
imagined scenes, 252
imaging, 27, 28, 272; technology, 44, 58, 59. *See also* medical imaging
immuno-chemical visual tradition, 173
immunochemistry, 156
immunological imagery, 156, 176, 182. *See also* Pauling's imagery
immunology, xv, 155–56, 170, 188; cellular, 156; textbooks, 171; visual tradition of, 180
indexing, 27, 28, 205, 207
information design, ix, xvi, 237, 255
information designers, 267
inscription, 4, 12, 24, 28, 33, 196, 216n7. *See also* literary inscription
instructionist theory. *See* antibody formation, theory of
intentional positioning, 17
intentional process(es), 10. *See also* non-algorithmic processes
intentional visual images, 14
interaction studies, ix
intersubjective agreement, 67
intersubjective site(s), of agreement/disagreement, 29
intersubjective visible image(s), 7, 37

journalism scholars, 267

laboratory: activities, 213; animals, 33; conditions, 81; conversation, studies of, 216n5; equipment, 178; ethnographies of, 29, 31; experiments, 94; laboratory-based phenomena, observations, 172; practice, 29, 31, 116, 196, 215; practitioner, 196; procedures, 32; readings, 245; research, 27, 29; researcher, xvi, 33, 196; science(s), 32, 34, 35, 36; scientific, 29, 31; scientists, ix, 28, 195, 202; specimens, 34; studies 216n3; work, 197, 216n1; workers, 215n1
lattice hypothesis, 175, 176, 177, 178, 180, 188
layout, 165–66, 171, 210
life sciences, 188, 267, 280
literacy: information, xvii, 269; media, xvii. *See also* visual literacy
literary inscription, 29, 196, 216n1; theory of, 216n7. *See also* inscription
lithography, 105

Manichaeism, 79, 171
map(s), 21, 27, 29, 35, 139, 198, 199, 200, 202, 217n12, 218n19, 225, 234, 238, 240, 241, 242, 243; dot maps, 262n2; hatch maps, 262n2; Peter's projection map of the globe, 262n2; weather, 270
materializing, 27, 28
mathematics, ix, 87n14
mechanical diagrams, 81

mediation: processes of, 43; technological, 59; work of, 43
medical images, 44, 59
medical imaging, ix, 44
medical-imaging techniques, 16
medical sciences, 216n4
medical sociology, 248
medical visualizing technologies, 43
medicine, xii, 41, 44, 46, 52, 55, 60n11, 73, 170, 244; clinical, 46
mentalistic perception, 214
metadiscourse, 136–38
metaphors, 3, 186, 187, 253; textual, 155; visual, 111, 189, 258
methodological principles and praxis, 137
methodology, 1, 121, 124, 138, 142, 214
micrographs, 202, 210, 214
microscope(s), 2, 30, 55, 106, 159, 208, 210
microscopic image, 276
mimesis, xiv, 3, 121–22, 143
minimalism, 248, 258, 262n1
minimalist aesthetic, 254
molecular complementariness, 153–54, 189. *See also* complementariness
morphological-systematic approach, 94
MRI (Magnetic Resonance Imaging), 16
multiformity,121, 123
multivocal narrative, 130
multivocality, xiv, 120, 128

naïve realism, 127, 144
naïve realistic tradition, 124
narrative sequences, 165–66
native image production, 133, 135. *See also* cultural self-portrayal
natural history, 73, 91, 94
natural science disciplines, 196
natural sciences, viii, xi, xviii, 197
naturalist's perspective, 73
naturalists, 67, 68, 71, 72, 73
neo-Platonists, 72
neuroscience(s), 207; journal, 201; research project, 197
new media, x, 22, 140, 280
new photography, 42, 43, 46, 49, 50, 51, 55, 58
non-algorithmic processes, 10. *See also* intentional process(es)
non-art images, 156

objectivity, 7, 42, 55, 128; mechanical, xiii, 42, 54, 59n2; of the camera, 124
object-semaphores, 75
observation(s): covert, 138; direct, 7, 8, 13, 270; electron-microscopic, 216n2; experimental, 173; systematic, 128, 197; visual, 127
open representation, 144
ornithologists, 10

palaeontology, 109
palpation, xiii, 50, 54
palpitation, 244
parallelism, 238–39
participatory democracy, 235
participatory idea, 130
particle physics, 28
pathology, 46, 60n6
paving (technique), 167, 168, 176
perception(s): sense perception, 28; sensory perception, 215; visual perception, vii, 38, 240
perceptual psychologist, 274, 276
percussion, xiii, 45, 50, 51, 52, 54, 57
performance, 249
performativity, 171
PET (Positron Emission Tomography), 16
phenomenology, ix
philosophers, 30, 34, 67, 85; of science, 26
philosophy of science, ix, 26, 28, 249
photo elicitation, 131, 135
photograph, ix, 8, 12, 14, 17, 27, 37, 42, 195, 201, 207, 208, 210, 211, 212, 225, 248, 276, 280; advertising, 5; aerial, 202; color, 10;

photograph *(continued)*
digital, 8; higher-power, 211, 215; lower-power, 215; Polaroid, 135. *See also* X-ray photograph
photographic: camera, 5, 7; image, 212; (imaging) technique, xiii, 42, 120, 123; particularity, 10; plate, 28, 50, 54; styles, 5, 6

photography, 8, 17, 42, 208, 260; black and white, 16; cine-photography, 122; high-speed, 2; news, 6
physical examination, 45, 46, 51, 53, 54, 56; instruments of, 52; methods of, 45, 50, 51, 54
physical sciences, 267, 280
physicists, 185, 191n1, 268
physics, ix, 79, 87n14, 94
physiognomy, 79
physiology, 45, 52, 60n10, 79
pictorial: data, 29; documents, 27; essay, 29; form, 28; games, 31; images, 27–28, 258; markings, 156; means, 30; phenomena, 29; presentation, 197; renderings, 37, 180; strategies, 156
point of view (POV), 148, 149, 258
policy research, 225
positioned subject, 136
positivistic tradition, 124
postmodern (post-modern) approach, 128, 144
postmodern (post-modern) conceptions of science, 32
postmodern (post-modern) critiques of scientific realism, 32
postmodern (post-modern) exploration, 249
postmodern (post-modern) literary theory, 32
postmodernism (post-moderismn), 224
poststructuralist (post-structuralist) approach, 155
potter's object, 215n1
projective drawing techniques, 136
protestant reformation, 72
psychoanalyst(s), 136
psychology, 30, 214, 223

radiograph, 37, 41
radiography, 52, 56, 57, 60n7
radiological examination, 54
radiological image, 57
radiology, 43, 51, 55, 58, 60n11
radioscopy, 52, 56
ready-mades, 35
realist-constructivist debate, 31
referencing, 28
referent, viii, x, xi, xiii, xv, xviii, **2–4**, 5, 7, 8, 9, 12, 13, 16, 18, 21, 24, 43, 44, 47, 49, 52, 55
reflexive action, 138, 145
reflexive approach, xiv, 138
reflexivity, xiv, 120, **136–38**, 144
regime of credibility, xiii, 85
rendered body, 47
rendering(s), viii, 11, 29, 44, 53, 55, 56, 177, 178, 180, 187, 188, 199, 215, 216n6, 217n12, 218n22, 271; artistic, 13; auditive, 54; computer, 1, 12; diagrammatic, 175; graphic, 37, 201; high-resolution, 167; objective, 42; pictorial, 180; realistic, 9; series/sequence of, 33, 36, 204; visual, 29, 170, 172, 180
rendering practice(s), 55, 56, 196, 216n7
rendering procedures, 201, 213
representation(s), 1, 5, 7, 9, 10, 12, 27, 30, 32, 275; conceptual, 139; graphic, xii, 12, 167, 197, 223; hand-drawn or driven, 8; in science, viii, xi, xvii, 3, 19, 269, 280; of scientific data, x; (visual) scientific, xiii, xvi, xvii, 4, 14, 16, 19, 21, 31–32, 276; verbal or numeric, 126
representational: activity, x; capability, 17; commitment, 16; devices, vii, 7; practice(s), vii, viii, ix, x, xi, 1, 2, 21, 148; production process, 4; systems, x
representativity, 17

restoration, 35
retrospectroscopy, 48
rhetoric, x, xv, 55, 253, 216n4, 262n3
rhetorical analysis, 190
roentgenology, 55
roentgen-ray picture, 58
roentgen rays, 56

scatterplot, 230, 231, 241, 243
scholarly communication, x
Schulwandbildern, 116n1. *See also* wall charts; wall diagrams; Wandtafeln
science communication, xi, xvii, 1, 266, 267, 270, 277, 279, 280
science communicator, 267, 275, 276, 277, 279, 280
science education, 90, 94, 116
science historians, 267
scientific argument(ation), x, 141
scientific communication, 108, 116, 121, 125, 129, 140
scientific discourse (visual) vii, xiii, 1, 18, 19, 20, 21, 67, 121, 127
scientific legitimacy, 111
scientific perspective, 79, 125
scientific practice, 6, 27, 31, 34, 63, 115, 121, 124, 149, 155
scientific praxis, 124, 137
scientific theory, 1, 126
scientific visibility, 195, 205
seismograph, 208
self-representation(s), 135
semiotic: analysis, ix, 190, 216n7; code, 168, 180; elements, 197; hybrid, 1, 167; inquiry, xv; means, 167, 190; perspective, 275; uncertainty, 179
sense cognition, 28
shared production, 129
side-chain theory, 168, 171
similitude, 180
sketch(es), 30, 158, 159, 175, 180, 182, 218n22, 276
skiagraph(s), 47, 50, 57, 58
skiascopic image, 54
small multiples, 238–40, 244, 247, 256, 262n2
smallest effective difference, principle of, 237. *See also* data/ink maximization; data/ink ratio
social constructivists, 32, 34. *See also* constructivists
social historians, 26
social history, 83
social object, 12
social psychology, 214
social science(s), viii, 141, 223, 224, 225, 228, 260, 261, 262, 267; knowledge, 141; methods and approaches, 135. *See also* visual social science
sociology, 135, 218n23, 223, 225, 262n2, 263n5. *See also* visual sociology
sociology in/of medicine, 247, 263n5
sociology of health, 263n5
sociology of science, ix, xii, 26, 28
statisticians, xvi, 222, 232, 267
stethoscope, 45, 51, 244
studies of science: ethnographic, 28, 34; social 12; social-historical, 28; sociological, 28

style of execution, 12
stylistic features, x
subjective selectivity, 124
subjectivity, 54, 55, 124, 137, 138, 144
symbolic notation, 280

table(s), 51, 81, 175, 196, 216n4, 217n13; Anscome's, 231; of classification, 71, 72. *See also* accompanying tables
telescope detectors, 36
telescopes, 2, 30, 266, 276
template theory. *See* antibody formation, theory of
theory of species, 72
transcription, 4, 7, 67

ultrasound, 16; imaging, 35; scans, 8

validity, 17, 18; of self-produced videos, 136

video article, 147–48
video self-productions, 136
virtual witnesses, 63
virtual witnessing, 30, 172
visible assessment, 212
visual aids, 132
visual anthropologists, 135
visual anthropology, ix
visual approach, xiv, 121, 127, 135, 142, 148, 149; mimetic, 144
visual argument(s), 155–56, 183, 190
visual argumentation, 190
visual-argumentative strategies, 187
visual-argumentative structure(s), 190
visual cognition, 28, 223
visual communication, 24, 90, 127, 134, 267, 269, 270, **278–80**, 279, 280; educators of, 267; of science, 105
visual component, 165
visual continuum, 180
visual conventions, 188, 216n4
visual culture(s), xi, 149, 155
visual cultures of science, x, xiii, xvii
visual data collection, 127
visual dimension, x, 154, 260
visual display(s), 195, 196, 211, 216n6, 223, 224, 228, 232, 236, 237, 240, 241, 242, 243, 248, 251, 260, 262n3
visual document, 125, 196
visual domain, 172, 175
visual elicitation, 133
visual enigma, 189
visual experience, 80
visual feedback, 130, 133, 143; methods, 130; techniques, 130, 134, 135, 144
visual fields in science, 216n6
visual grammar, 149
visual illusion, 274
visual imagination, 248, 263n5
visual information, 126, 149, 260, 269
visual interview, 131–32. *See also* film elicitation; photo elicitation
visual language, of scientific representation, 16, 29, 277
visual learning, 267, 269, **275–77**, 280
visual (scientific) literacy, x, xi, xvii, xviii, 21–22, 266–68, 269, 275, 277–80
visual manifestations, 126
visual material, 107, 131–33, 136, 217n8
visual meaning, xviii, 179
visual mechanism, 189
visual media, 14, 17, 18, 116, 135, 139, 140, 149
visual narrative(s), 237, **240**, 251, 252, 258, 260
visual natural phenomena, 3
visual participation, 147
visual pedagogy in science, xiv
visual phenomena, 2, 148
visual portrait, 187
visual practices, 134
visual presentation, xiii, xv
visual product(s), xii, 12, 136, 138, 147
visual productions, 121, 126, 137, 156, 168
visual quotations, 190
visual reconstruction, xvi
visual representational latitude, 14–16
visual research, 6; methods, xvii, 260; participative or interactive forms of, 129
visual science, xiv, xvi
visual shapes, 176
visual signatures, 71
visual skill, 248, 263n5
visual social science, xvi, 222, 224, 225, 228, 239, 260–62
visual sociologist, 135
visual sociology, 224
visual statements, 155, 180
visual stimuli, 12, 132
visual study, 141, 146
visual symbols, 270
visual techniques, x, 14, 29, 136, 148
visual technologies, x
visual thinking, 154, 241, 267, 269, **270–75**, 280
visual transcription, 7

visualization(s): cross- and interdisciplinary nature of, xi; in education, xiv; in/of science, ix, xi, xii, 1, 27, 216n4, 269; of discoveries, 266; of ideas, 252; of natural objects, xv; of science information, 267; process of, viii, 275; products and media of, ix; re-visualization, 68; scientific, ix, xii, xiv, xviii, 6, 19, 30, 267, 274, 276; study of, viii, 26, 27, 29, 31, 32, 34, 37, 38; technologies of, 43
visually literate scholar(s), xvi, 22, 24
visually literate science, 148

wall charts, xiv, 90–116, 116n3, 117n19. *See also* Schulwandbildern; wall diagrams; Wandtafeln
wall diagrams, 90. *See also* Schulwandbildern; wall charts; Wandtafeln
Wandtafeln, 90, 116n1, 117nn21–22. *See also* Schulwandbildern; wall charts; wall diagrams
weather graphics, 266
working object(s), 44, 49, 50, 52, 55

X-ray(s), 8, 28, 41, 42, 48, 51, 52, 54, 55, 56, 59n2, 245, 266; astronomy, 28; body, X-rayed, 46; examination, 55; images, xii–xiii, 36, 42, 43, 44, 47, 48, 49, 51, 52, 53, 54, 55, 56, 58, 59n4; in medicine xii–xiii, **41–62**; machine, 50; methods, 51; organs, 46; photograph(s), 41, 44, 47, 48, 54, 55, 56, 57; photography, xii; picture(s), 46, 47, 51, 52, 54, 57; qualities, 52, 56; technology, 42, 51, 58; workers, 46, 47, 48, 50, 51, 55

zoology, 94, 103, 109